The Oil Crisis

Essays by

Raymond Vernon
Joel Darmstadter
Hans H. Landsberg
Edith Penrose
George Lenczowski
James W. McKie
Romano Prodi
Alberto Clô
Yoshi Tsurumi
Marshall I. Goldman
Norman Girvan
Mira Wilkins
Robert B. Stobaugh
Zuhayr Mikdashi
Ulf Lantzke
Klaus Knorr
Ian Smart

The Oil Crisis

Edited by RAYMOND VERNON

*Written under the auspices of the Center for
International Affairs, Harvard University*

W · W · NORTON & COMPANY · INC · *New York*

ISBN 0 393 05576 0 (cloth edition)
ISBN 0 393 09186 4 (paper edition)

1 2 3 4 5 6 7 8 9 0

Contents

Preface

THE READER who picks up this book a decade from now may find the title a trifle ambiguous. *Which* oil crisis, he may ask—1974, 1977, or 1980?

Crises such as the one that surfaced in the world's oil markets after the Egyptian attack on Israel in October 1973 are usually ephemeral events. They tower in our consciousness as we live through them. Then, within a few years, our memories have performed in their accustomed pattern. They have erased the mood of the era before the crisis, have distorted the sequences of the crisis itself, and have garbled its details.

That fate is exactly what most crises deserve. With the passage of time, most events that appeared earthshaking at their inception tend to lose their initial high profiles. The uninhibited outpourings of commentary, analysis, and projection that such crises precipitate usually prove quite disproportionate to the significance of the crisis.

The events in the oil market that drew the world's attention in the months following October 1973, however, may prove to have a more enduring significance. That was the strong assumption, at least, that drew a group of us together in Cambridge in March 1974 under the joint auspices of the Center for International Affairs of Harvard University, and of *Daedalus*, the journal of the American Academy of Arts and Sciences. The objective of the assembled group was to try to see the crisis in terms of its larger significance. The question was how to achieve that goal. There were obvious risks in the exercise. Although most of us were aware that the crisis had long historical roots, nevertheless the dramatic phase was only a few months old; the impact of near events could obscure the meaning of what had gone before. Besides, it seemed inevitable that the professional background of each of the participants would limit his or her approach to the crisis: that economists would want to describe the crisis as an economic event, political scientists as a political event, and so on. Accordingly, the essence of the crisis might escape between the interstices of a series of such scholarly treatments. Then there was the perennial problem of ethnocentricity: people from different countries would see the crisis each in their own way.

To deal with these risks, the group that was invited to the March 1974 meeting at Cambridge and the group that assembled at a later meeting at Turin in January 1975 were drawn from various countries and from diverse professional backgrounds. There were businessmen, journalists, economists, historians, and political scientists; Frenchmen, Italians, Britons, and Americans, as well as a Lebanese and a West Indian. Most of us shared a long-time interest in international affairs, along with a professional understanding of the international oil industry; a few, in fact, were life-long specialists in these fields.

The task was broken down by parts, on lines that would encourage the individual contributors to avoid a narrow professional focus. Those that had failed to do so in their first drafts faced an uninhibited flood of advice and criticism in the two-day Turin meeting, which ultimately led to extensive revisions. At this meeting, the commentary came not only from other contributors, but also from outside critics selected for their independence of viewpoint and knowledgeability. Although overlapping among the various studies was not encouraged for its own sake, we were prepared to entertain two versions of the same question wherever two versions appeared better than one. The section of the book labeled Synthesis, consisting of the efforts of three different authors, should be read in the same spirit as the Japanese novel *Rashomon—Three Versions of an Incident.*

The contributions of the Center for International Affairs to this collective effort consisted of various things: my services as general editor of the volume; the Center's facilities as a meeting place; and a substantial contribution of funds, in turn provided to the Center by various donors. These included funds from the Ford Foundation earmarked for innovative activities, from the Rockefeller Foundation for research on transnational conflicts, and from the Fritz Thyssen Foundation for research on critical problems of postindustrial society.

Daedalus, our partner in the venture, contributed the services of its extraordinary staff, notably of Stephen R. Graubard, editor, Geno A. Ballotti, managing editor, and Margaret B. Ševčenko, manuscripts editor. In addition, *Daedalus* was responsible for securing a generous grant from the Alfred P. Sloan Foundation, which represented some of the first seed money for the venture.

Finally, all of us participating in this volume owe a debt to the Fondazione Giovanni Agnelli, which provided both funds and conference facilities for our indispensable January 1975 meeting, all ably orchestrated by Mr. Giovanni Granaglia of the Fondazione.

R.V.

The Oil Crisis

SYRIA

BAGHDAD ★

IRAQ

KUWAIT

Major Oil Fields:
The Middle East

SAUDI ARABIA

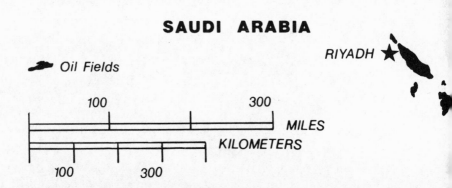

RIYADH ★

🖝 Oil Fields

| 100 | | 300 |

MILES

KILOMETERS

100   300

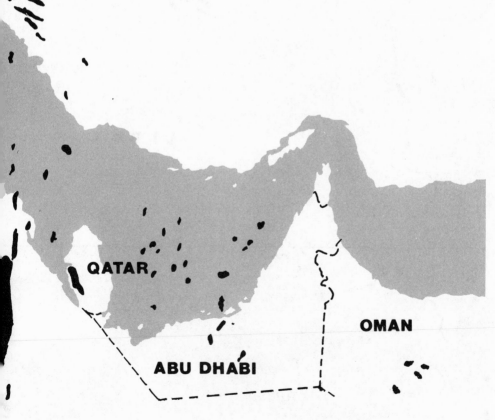

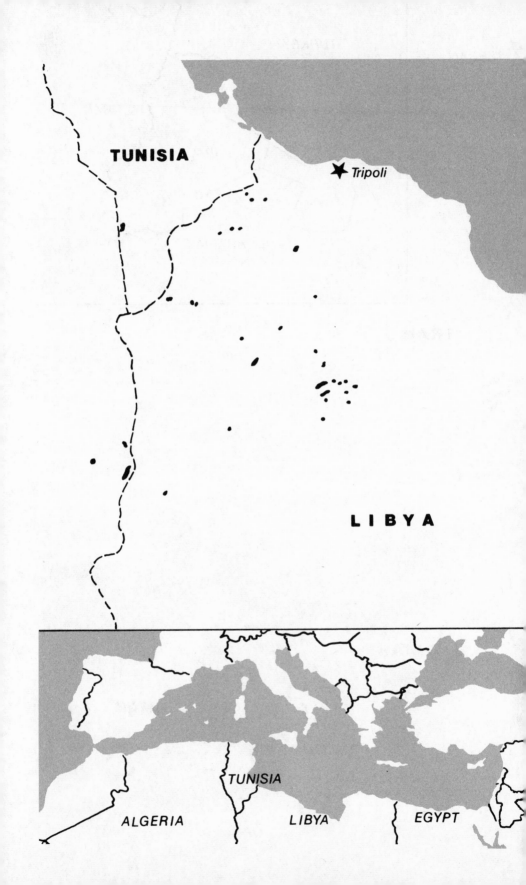

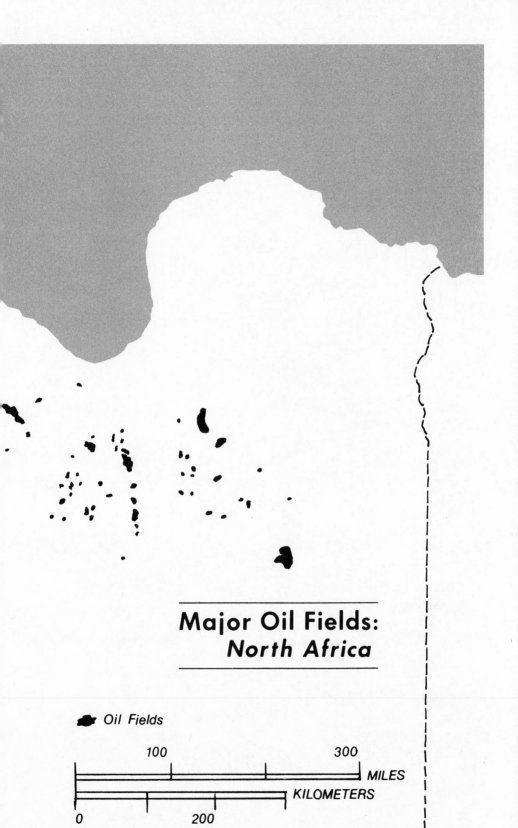

Major Oil Fields:
North Africa

🖤 Oil Fields

100   300

⊢————————————⊣ MILES
KILOMETERS
0   200

Part 1: INTRODUCTION

RAYMOND VERNON

An Interpretation

THE CRISIS IN THE WORLD'S OIL MARKETS that began in October, 1973, it is now generally agreed, was a symptom of something more profound. If the problem had simply been that of the oil trade, it would have been important enough to command world attention. But the meaning of the crisis was bigger than oil—indeed, bigger than the Middle East war that had triggered it.

This collection of essays tries to ferret out the meaning of that crisis. Though most of our analysis is intended to illuminate the meaning of the fifteen-month period that began in October, there is no implication that the crisis is now over; quite the contrary. Serious analysis, however, demands reflection; publication cannot yet be achieved with the speed of light. We have resigned ourselves to placing some terminal date on our several essays.

During those fifteen months of crisis, a series of transformations occurred with regard to how the crisis was perceived. At first, in the two or three months after October, 1973, the crisis was generally seen as a threat to the security of the oil importers' supplies. Then, around the beginning of 1974, after the price of crude oil had moved violently upward, the focus of concern in most countries shifted to the question of price. By midyear of 1974, however, the concern over price had been converted into a concern over the international monetary mechanism as a whole; considerable doubt arose over whether institutions existed or could be created that would be capable of handling the massive shifts in financial resources that were developing.

Though perceptions of the nature of the crisis went through several changes, the period could still be seen as a single traumatic process, stemming from the play of a basic set of forces. This first chapter offers an interpretation of what I think was learned

1

about the crisis period itself. But it is a personal interpretation, the interpretation of an editor who has been involved in the subject far too intensively and far too long to be able to claim Jovian detachment. Besides, happily for me, my colleagues did not always agree on the significance of certain key events of the crisis, thereby providing me with a pretext for indulging my own conclusions. In some cases, I shall not hesitate to do just that; but wherever I am conscious of having slighted some conclusion widely held among my colleagues, I will try to mention that fact. On some points, a reader may find himself in closer agreement with one or several of the essays that offer alternative explanations of these same phenomena.

The crisis that surfaced in 1973, as every reader knows, was signaled by an extraordinary event: the emergence of a group of unindustrialized oil-producing countries with both the aspirations and the seeming power to control the international oil market. As Girvan emphasizes in his paper, this bid for power had been a long time coming. Some would say it began with Mexico's nationalization of the foreign-owned oil companies in 1938; some would date it with the formation in 1960 of OPEC, the Organization of Petroleum Exporting Countries; still others would associate it with the abortive efforts of the world's leading oil companies in 1970 and 1971 to extract some sort of agreement from the oil-exporting countries that might have given them a few years of stability. But the immediate event, the issue that was to raise oil to the level of high politics, was the seeming capability of the oil exporters after the outbreak of war in October, 1973, to control the supply of oil to the United States, Europe, and Japan.

How had this new power developed? Could it be that the reasons were quite narrow, technical, and specialized, relevant only to the particular commodity and the particular circumstances of that moment? The papers of Darmstadter and Landsberg, Penrose, Smart, and Stobaugh indicate that in many ways the crisis was based on quite special circumstances. The conditions of demand and supply as they evolved in the decades after World War II, for instance, did place oil in a somewhat special category.

Demand and Supply

By now, almost every businessman, politician, journalist, and scholar has been exposed to a dozen studies of what happened to the demand for oil and its products in the period after World War II. The bare bones of these developments are straightforward and simple. As the incomes of most people grew over most of the world, the demand for the main sources of energy—coal, hydropower, nuclear power, gas, and oil—grew as well. But the demand grew faster for oil than for any of the others.

The reasons for the especially rapid growth in the demand for oil were various, but one overwhelming fact dominated: the cost of producing oil was lower—much lower—than the cost of producing practically any other source of energy. Here and there in the world, hydroelectric power or gas could be produced more cheaply. But these exceptions were of no great importance. When oil could be produced at ten or twenty cents a barrel, as it could in large parts of the Middle East, no other source of energy could come close to competing.

Of course, the sales price of the world's oil reflected much more than the ten or twenty cents of production costs. For one thing, the oil companies paid taxes to the countries in which the oil was produced—taxes that in 1969 were about 90 cents a barrel in the Middle East, and, in 1974, about $10 a barrel. Tariffs and import restrictions imposed on the crude oil imported into certain countries also raised the price. In most cases, such measures were used to protect local high-cost oil producers or the remnants of the coal-mining industry in the importing country. The United States, for example, imposed a system of import restrictions on oil in the middle nineteen-fifties, and this system lasted until the early nineteen-seventies. In addition, the large international oil companies, responsible for most of the production of the Middle East, Africa, Venezuela, and Indonesia, were not so imprudent as to allow quite all of their production to become concentrated in the Middle East, even if it was the cheapest source of oil. Prices to the ultimate buyers were based on sources of energy considerably more costly than that of Middle Eastern oil. That fact, in the end, provided the lure for a stream of producers to the Middle East; it also represented the point of vulnerability of the producers who were already there.

Before the oil firms became aware of this vulnerability and before the oil-exporting countries understood their strength, basic shifts in the production of coal and oil were allowed to take place. As the costs of producing coal in Europe and of discovering oil in the United States soared during the nineteen-fifties and -sixties, the attraction of cheap Middle East oil managed to generate the obvious response: the high-cost coal industry of Europe and the domestic oil industry of the United States gradually lost their position in world markets. Until the late sixties, this gradual shift did not register greatly in the consciousness of the general public. In the Suez crisis of 1958 and in the Six Day War of 1967, enough underused capacity still existed in the United States and elsewhere to absorb the transitory problems of shortage. By the late sixties, however, the shift to reliance on Middle East oil was very far advanced. It was at about this time that a series of other trends greatly heightened the monopolistic potential of the Middle East countries. There was a stiffening in the demand for energy in general, and for oil in particular. The increase in energy demand was the consequence of a remarkable surge in industrial growth that hit Europe, Japan, and the United States simultaneously in 1972 and 1973, a very rare convergence of cyclical timing. The sharp increase in the demand for oil in particular came about partly because of delays in bringing nuclear power plants into operation, and because of various antipollution controls. In the United States, for example, the controls on automobile emissions reduced somewhat the efficiency of gasoline. Also, as the United States began belatedly to worry about its environment, some of the oil development that might otherwise have gone rapidly ahead on the continental shelf and in Alaska was considerably slowed.

History may one day indicate that these were evanescent factors, unlikely to converge with quite the same force and effect for any other product at any other time; most of my colleagues and I would agree that they helped set the stage in the early nineteen-seventies for the seeming show of monopoly power that was soon to come.

The Independents and the Majors

The next point in our collective analysis needs to be introduced with a certain deliberate emphasis, lest its significance be overlooked. It has become commonplace in current discussions of the oil crisis to refer to the monopoly power of OPEC, suggesting that it was the key element in explaining the behavior of the world's oil markets in the mid-nineteen-seventies. Some of my colleagues, including Penrose and Wilkins, draw attention to quite another change in market structure as important during the fifties and sixties: the growth of the so-called "independent" oil companies. In my view, the rise of the independent companies is central to an explanation of the events of the seventies. Stated differently, the crisis that commenced in October, 1973, would be much harder to picture if the six or seven oil companies that dominated the international oil trade at the end of World War II had continued to maintain the same degree of domination. This is a view shared by others who have written in this volume, though only a few would give it the same weight that I have.

The word "independent" in the context of the oil industry is not always understood. As a rule, the term is applied to any large oil company standing outside the group that was operating in the Middle East at the end of World War II. The original group—the "majors," so-called—includes Exxon, Mobil, California Standard, Texaco, Gulf, British Petroleum, Royal Dutch/Shell, and the Compagnie Française des Pétroles. Their working relationships, though punctuated by outbreaks of rivalry, go back well before World War II. The other companies, the independents, had to develop relationships both with the majors and with each other.

A factor that prevented some sort of working equilibrium in the industry after World War II was the extraordinary cheapness of oil from the Middle East—and, later, from Libya as well. Few independent companies at the end of World War II had the organizational strength or the financial resources to break their way into the more obscure and remote areas of the Middle East. From the viewpoint of any independent company, whether located in the United States, Europe, or Japan, the threat that the majors might use their Middle East oil as a competitive weapon for enlarging their positions at the expense of the independents was always very real. Such companies were compelled to build their business of refining and distributing oil and oil products on relatively expensive crude oil—oil acquired sometimes from high-cost sources in the United States and elsewhere, or oil purchased from the majors themselves. The oil purchased from the majors came on all sorts of terms, depending on time, place, and relationship; what these terms were are well-kept business secrets. But whatever they were, they left certain large independent buyers acutely uneasy about their competitive position in relation to the majors.

The first entry for the independents in the Middle East came with the reopening of the Iranian oil business in 1954, two years after the expulsion of the British-owned operating company and the shutting down of the Iran oil fields. In the constitution of an Iranian oil consortium, the United States government insisted that a number of independent, American-based oil companies be included, along with Exxon and Mobil. This policy gave certain independents the advantage necessary for acquiring the organization and experience essential to successful operation in the Middle East. From that moment on, a dozen or so independent companies leaped the heretofore difficult entry barriers of the international oil market, appearing as crude-oil producers in Libya, Algeria, and various countries of the Middle East.

To say that the independents leaped the barriers, however, is not quite to describe the effects of their entry. At the time of entry, each independent company had less at stake than any major company in safeguarding the system of agreements in the Middle East and Africa that had existed between the companies and the governments. Having no outstanding agreements with the host countries that might be damaged by a new arrangement, an independent company had little to lose in proposing or accepting an agreement that would cede a somewhat larger share of the rewards to the host country, so long as Middle East oil remained cheaper than that available from other sources. An independent could generally assume that the majors would eventually be obliged to accept the same terms. Accordingly, if the majors were prepared to pay 55 per cent of their profits in taxes to the host country, it made good sense for the independents to offer 57 per cent; if the majors were willing to cede 20 per cent of the ownership of their producing facilities to the host country, it was good business for the independents to offer 40 per cent.

Of course, eventually the independents would develop a stake in the existing arrangements similar to that of the majors, but that would come only later when they had established a large supply of cheap crude oil for themselves. Meanwhile, from the viewpoint of the majors, the presence of the independents in crude-oil activities in the Middle East and Africa created a new sort of vulnerability. Once an independent company succeeded in establishing a supply of crude oil in the Middle East or North Africa, that source was likely to constitute a much larger proportion of the aggregate supplies of the enterprise than would be the case for one of the majors. In difficult circumstances, when foreign oil companies might be facing a negotiating showdown with a host country, the independents in the group would be most vulnerable to the threat that their oil might be cut off. The results were predictable. As Penrose, Wilkins, and others suggest, at critical stages in the negotiations between the oil companies and the oil-exporting countries that took place between 1970 and 1974, the vulnerability of one independent company or another to the pressure of a host government was generally the factor that broke the deadlock.

From 1954 on, the share of the profits retained by the oil companies in the production of crude oil had begun gradually to decline, moving to well under 40 per cent by the late nineteen-sixties. As a result, the profits recorded by the majors fell from about 80 cents per barrel of crude oil to about 30 cents. At the same time, the share of the international crude-oil market accounted for by the majors also fell, dropping from about 90 per cent in the early nineteen-fifties to about 70 per cent twenty years later. For some years, the dampening effects of these trends on the majors' crude-oil profits were offset by the rapid increases in the volume of oil sold. By 1970, however, the scissors effect was becoming too much for the major companies; their bid for a five-year agreement at Teheran can be seen as a defensive effort to reestablish some measure of stability in a situation in which their control seemed to be ebbing.

With hindsight, it may appear that the majors could have delayed the shift in the balance of power between themselves and the oil-exporting countries if they had been prepared to bolster the negotiating position of the independent companies in the period of crisis by offering alternative sources of oil in the face of an oil country's threat. Indeed, that is precisely what the major companies did, though for only brief periods and in special situations; to sustain a strategy of that sort would not have been easy. For one thing, certain of the independents were genuinely entitled to their sobriquet, and preferred to hold the majors at arm's length, hoping to snatch a larger share of the market from them. Moreover, the habits of accommodation and forebearance among the majors, though well developed, do not appear to have been strong enough to prevent their having certain qualms about how the others might behave in a crisis. There were other difficulties involved in agreeing jointly about a long-term policy to deal with an assorted bag of independent companies. Besides, especially with respect to the American companies, the antitrust laws continually cast their shadow.

Whatever the reasons, the relatively tight control that the majors had been able to exercise over the international oil market in earlier years was probably bound to erode, crisis or no crisis. Signs of that trend appeared long before October, 1973, and multiplied as the majors began, in the early nineteen-seventies, to adjust to the growing shortage of crude oil. It is difficult to say how strongly they favored their own captive customers in the adjustment process; Stobaugh, for example, believes that, where outstanding commitments existed, the majors tried to fulfill them. Still, some large-scale users of crude oil, such as public utilities and industrial processors, were obviously growing anxious.

What happened as a result of that anxiety in the early nineteen-seventies would have been unthinkable in the fifties; as supplies dwindled, a number of customers went to the Middle East and Africa to see if they could satisfy their crude-oil needs directly. That such a possibility was entertained was the consequence of many developments, including certain changes in the capabilities of the oil-exporting countries themselves. The trend, however, can be ascribed principally to the revolutionary shrinkage in international space, and also to the striking improvements in international communications that have become common in recent decades.

In the immediate aftermath of the October, 1973, war, the needs of certain of these bulk buyers of crude oil grew so powerful that they were prepared to pay any price. By that time, also, state-owned companies in some of the oil-exporting countries were directly engaged in the sale of a small part of the oil sold in international markets. For a brief time, free oil offered by the exporting countries fetched unheard of figures; sales of $17 a barrel and more were reported. Stobaugh guesses, and I am inclined to agree, that the extraordinary size of the price increases imposed by the Shah of Iran at the time was largely triggered by the obvious disposition of certain buyers to pay them. In any case, neither OPEC nor OAPEC—the all-Arab group of oil exporters—was the prime mover.

Some observers have suggested that the reason the major oil companies failed to take steps to reduce the panic of the free-wheeling bulk buyers after the outbreak of hostilities in the Middle East was that they welcomed the higher prices and higher profit margins that this panic-buying induced. Though this assumption may seem reasonable to some, the published record, as reported by Stobaugh, suggests a very different set of facts. At that early stage in the crisis, the leading international companies were deeply concerned by the breathtaking price increases that were being imposed.

Despite the erosion in the position of the majors that had been going on during the fifties and sixties, the habits of accommodation were quite strong enough for them to manage the extraordinary pressures during the first stage of the crisis, when security of supply was thought to be the critical problem. Such accommodation, because it was largely passive in form, consisted mainly of mutual acquiescence in modifying old supply relationships throughout the world.

Indeed, the first and most important level of adjustment of the major companies took place inside their respective international networks rather than between them. The adjustments within each system required the rerouting of crude oil and tankers, the altering of refinery schedules, and the allocation of supplies to competing buyers. The second level at which the international oil companies responded to the crisis involved them in a complex set of mutual interactions. Every major company was linked to practically every other one through a series of joint producing and refining ventures, long-term bulk purchase agreements, and long-term reciprocal supply arrangements. Where joint facilities existed, they normally required joint operating decisions; the output of Aramco in Saudi Arabia, for instance, as well as that of the Iraq Petroleum Company, was partly determined by an intricate bargaining process among the major companies. When output was restricted by governments, as it was during the crisis, there was still room for the reapportioning of the output.

Throughout these processes of adjustment, according to Stobaugh, the major oil companies used a standard of "equal suffering" among importing countries. A standard of that sort cannot be precise; it left considerable leeway in its application. The United States-based majors exhibited no signs of greatly favoring the American market; British Petroleum (half-owned by the British government) showed little sympathy for its government's claims for preferred treatment; France's Compagnie Française des Pétroles insisted on a consideration of the interests of its non-French connections; and so it went.

In general, the international oil companies responded to the emergency situation with the conditioned reflexes of entrepreneurs minimizing their risks at the margin. The patterns of oil distribution in the crisis, dictated by the principles of greatest prudence and least pain, were curiously non-national. Anyone looking for confirmation of the view that it paid a country to have an oil company based within its own jurisdiction would have found scant support for such a hypothesis in this brief episode in the oil industry's history. Whether the outcome would have been the same had the shortage persisted for very long was never tested.

The Oil-Exporting Countries

The perceptions and calculations of the oil-exporting countries themselves were, of course, critical in shaping the oil crisis. These are well described in the essays by Lenczowski, Penrose, Mikdashi, and Girvan. These studies suggest that the oil-exporting countries took many years to prepare themselves for the novel role that they assumed in the crisis. Though the oil-producing activities of foreign-owned oil companies are sometimes represented as an enclave industry that rarely touches the local economy, this description is more hyperbole than fact. Few countries have harbored such activities without experiencing profound consequences in their political and social structures, especially in the effects these have for a key group of elite politicians, technicians, and intellectuals. As the oil-exporting countries have been drawn into the modern world, so have their capabilities grown. Step by step, the leaders of these countries have begun to understand how the world's oil markets were structured, where alternative sources of technical help might be obtained, and how their oil revenues could be converted into productive capital.

This apparent shift in the negotiating power of the OPEC countries can be thought of as an outcome in which OPEC itself at first played only an incidental role. Two basic elements account for the apparent growth in the bargaining strength of OPEC member countries. One, already discussed, was the increase in the number of companies bidding for oil. The other was the increase in the ability of the oil-exporting countries both to collect and interpret the information that affected their negotiating position, a function of their increasing involvement in the world economy. All countries discovered that a technical elite, even if very small, could play a powerful role in negotiations. Libya, for example, learned very early how useful an independent oil company could be as a means of putting pressure on the majors. Later on, the changes in the oil-exporting countries would be registered by the fact that some of them had elevated their annual per capita income from bare subsistence levels to $8,000, $10,000, or even $40,000 per head. The character of these countries was irrevocably altered when they began to acquire health services, schools, factories, and consumer goods at a prodigious rate. Along with their internal growth, these countries developed a network of associations and a store of assets in the developed areas of the world that they could call upon. They would, it was clear, never be the same again.

Though all these countries were affected by strong modernizing tendencies in the nineteen-sixties, each approached the crisis period of 1970 with distinctive perceptions of position and purpose. At the core were the governments of the Arab countries. These were eager to get all they could from their oil resources; they were prepared to act in concert if such action would indeed help. They were bound together by a common desire to defeat Israel. The need to be unequivocal in their commitment was made stronger because the various pan-Arab and Palestinian organizations, straddling national boundaries, could threaten existing state regimes that failed to show the appropriate degree of enthusiasm for the cause.

Certain Arab countries were in a position to cut their oil production if that would serve their purpose; others were not. Algeria and Iraq, for example, were among the more inhibited, precisely because of their need for a sustained flow of foreign exchange. Before they acted, they would have to feel reasonably sure that a decline in shipments would not reduce their gross revenue unbearably.

Beyond the Arab group lay the other oil producers, whose common interest was a desire to capture the largest possible rent. They understood that joint action might on occasion serve that end. In this group, the capacities of the various countries to pursue particular strategies varied enormously. Indonesia, Iran, and Nigeria were virtually committed to producing flat out, no matter what. Venezuela was in a more equivocal position, torn by a short-term need for government revenue and by a longer-term imperative to diversify its economy. Canada and the Soviet Union were in more ambiguous positions still. The Canadians were partly immobilized by their intimate ties to the United States economy, yet were determined to demonstrate that they had some measure of independence and choice, and that they could in fact move toward an economy less dependent on the American market. The Soviet Union, as Goldman's study shows, was busy with various arcane and involuted maneuvers, aimed among other things at holding the East European economies in its economic ambit, while maintaining a favored position in the Arab world, and availing itself of the opportunity to accumulate foreign exchange.

The oil crisis did not test conclusively whether the community of interest among the oil-producing countries was sufficient to provide the cement for concerted action on a sustained basis. There was, to be sure, a certain amount of parallel action in the crisis, action that could scarcely be distinguished from collaboration. Price increases that originated in one country reverberated in the others; changes in concession agreements and national legislation instituted in one place seemed quickly to lead to comparable action elsewhere. Continuous consultation among heads of state and ministers suggested the possibility of hard bargaining and joint positions by the exporters. The activities of the all-Arab OAPEC association and of the more widely based OPEC group indicated collaboration and coordination.

But a critical test was lacking. All these coordinating activities took place in a setting in which none of the principal actors needed to compromise its national interests in any significant way in order to achieve consensus. It was reported in the public press from time to time that Saudi Arabia would have preferred not to raise its oil prices quite so much as it did in the last months of 1973, fearing lest the increases alienate the United States and stimulate the search for alternative sources of energy. If Saudi Arabia in fact demurred, the demur must have been bittersweet, since it did not involve any financial loss; Saudi Arabia received substantially more money for its oil.

One may ask what difference it made whether the action on supplies and prices during the crisis was taken by collective decision or by parallel action; in the end, the central fact was that between October, 1973, and January, 1974, prices rose by four hundred per cent. The question is relevant mainly for its projections about the future. Concerted action among the oil-producing countries may never again be necessary. If Saudi Arabia is prepared to curtail its production, it alone can create an umbrella under which the oil suppliers will be able to exploit their collective position. The experience of the crisis, however, offers little or no guidance for responding to a question that could prove critical in the future: Will the oil-producing countries be able to stand together if that requires significant compromises in the national positions of several of them?

Whatever the outcome of such speculation on the likely course of future events, it is fairly evident that by the time of the 1973 crisis, OPEC had not developed great organizational strength, certainly not sufficient strength to make it an initiating force, distinct from the force of its member states. As Mikdashi's paper points out, OPEC's record of initiatives up to the time of the crisis did not encourage a high assessment of its potential strength as an instrument for action. Its secretariat consisted of national officials seconded to their OPEC jobs by national governments; these officials owed their first loyalties to those governments. In its early stages, its council of national representatives consisted of ministers of finance, usually little interested in much more than the anticipated oil revenues. In its later stages, certain of the national representatives came to be drawn from the top officials of the fledgling state-owned oil companies, who were mainly interested in the prospects of dealing in the international oil business. What was conspicuously lacking was an apparatus that could speak to questions involving the highest political and developmental strategies of the countries concerned. It remained to be seen whether leaders like the Shah of Iran, Qadhafi of Libya, and Boumedienne of Algeria would be prepared to submit to the discipline of any such organization if that discipline required them to tailor their choices of high strategy to the collective wishes of other leaders.

The Oil-Importing Countries

From the political viewpoint of most oil-importing countries, the oil crisis of 1973 could not have come at a worse moment. The rich importing countries, including the United States, Japan, and those in Europe, were in a state of political discord, within and without. The poorer importing countries, notably those of South Asia and East Africa, were confronting balance-of-payments problems that seemed unmanageable.

The state of discord among the rich countries in the early nineteen-seventies followed a period of unusual achievement in international relations—achievements in the military, security, and economic spheres. After World War II, an elaborate set of international institutions was put into place; there was a mandate to reshape the system of world trade and international payments; with it came support for the growth of the developing world. Arrayed against historical precedent and measured by the usual economic yardsticks—growth rates in world product, per capita income, world trade, and world capital movements—the period that followed World War II seemed outstanding. That kind of measurement played only a minor role, however, in shaping the mood of the rich countries in the early nineteen-seventies. The papers of Smart and Knorr offer a wealth of thoughtful speculation about the reasons for the disarray that had seized the rich countries. Let me highlight here what I believe are the central elements.

Though the postwar economic agreements of the rich countries seemed successful, given what they were originally designed to do, their very success was a factor helping to undermine them. So long as the world's markets were fragmented by high tariff walls, import licenses, prohibitions on capital flows, and restraints on the movement of labor, which was the situation at the end of World War II, the rich countries could reasonably place a high priority on the removal of such barriers. Of course, their efforts might not have succeeded quite so well if powerful technological forces had not been pushing in the same direction. This was an era in which international space was rapidly being shrunk. The introduction of reliable civilian aircraft, radio-telephone transmission, and sophisticated information-processing systems was creating some of the conditions for global scanning and global operations by many enterprises that could once have been satisfied to remain national in scope. The results were observable in the spectacular rise of international trade and international capital movements, as well as in the increase of enterprises whose operations extended far beyond their home countries. The shrinkage in international space contributed also to the readiness with which cheap Middle Eastern oil displaced coal in Europe, supplemented the more costly oil production in the United States, and helped fuel the growth of Japan. The ease and efficiency with which this development was achieved, as both Prodi and McKie point out, lowered the guard of the importing countries. Only the United States sought to curb this shift, and its motive was primarily to protect the profit margins of its domestic producers.

The period following World War II, however, was distinguished by more than the rising capacity of man to overcome international space. It was also the period when the concept of the social-service state became firmly established in the countries of Western Europe and North America. In one country after another, the national political process developed elaborate programs for the delivery of public services, the redistribution of income, the development of lagging regions, the avoidance of unemployment, the suppression of inflation, and the achieving of other social goals.

These national programs grew year by year; but they grew in a global economic setting that seemed at times to pose certain threats to their continued vitality. With national boundaries open to goods, money, and enterprise, the rich countries were also progressively exposed to crop failures, inflation, depression, and all the competitive drives that any one of them might generate. The international environment had become one giant tuning fork.

To add to the difficulty, one country within this system—the United States—had two staggering advantages: an early start and an outsized economy. The combination gave it special strengths and immunities, to which other countries reacted with varying degrees of unhappiness. American businessmen were—at least in the beginning—the best placed to respond to the new opportunities of the international environment. The United States dollar seemed—again, in the beginning—immune from international pressures. Even the country's ample supplies of domestic raw materials, as the oil crisis itself would emphasize, seemed to set the country apart from both Europe and Japan. As a result, the American economy appeared to enjoy many of the advantages and few of the vulnerabilities of the open international society.

Over the years, Europeans tried a number of policies that might ease their difficult position; by the early nineteen-seventies, none was offering much promise as an adequate response. One set of policies, represented by the various efforts at European economic unity, was epitomized by the Treaty of Rome. A tightly knit Europe, it was thought, might be able to match the United States with an economy of roughly equal size and invulnerability, and this could pave the way for a system of cooperation among equals. It soon became clear, however, that the national political systems of the various European states were unprepared and unwilling to vest the new European institutions with responsibility for delivering the package of social services and social guarantees that had become the hallmark of the modern state. Accordingly, the European idea proved no answer to Europe's problems, at least not by the early seventies when the oil crisis broke.

Another approach emerged in the late nineteen-sixties, an approach that embraced not only the European states but the United States and Japan as well. This was the idea that nations could have both their open boundaries and their national independence of action if only they would give up one key ingredient of the post-war economic arrangements, namely, a system of fixed exchange rates. If exchange rates could be allowed to vary, so went the theory, the latitude of nations to follow their separate national policies would grow. But the consequences of that approach were already being questioned in the early seventies; and by the mid-seventies, after several disturbing years of exchange-rate seesawing, the premises of that prescription were under attack. At that point, the economies of the rich countries were still exposed to one another to an extraordinary degree—more perhaps than at any time in history. But it was not clear for how long that condition would apply. Both horns of the dilemma seemed sharp. Any nation that closed its borders would seem to undermine its opportunities for expansion and choice; any nation that kept them open would seem to threaten its internal stability and balance.

As Knorr's paper reminds us, it was not only the *economic* assumptions of the post-war period that appeared in peril: the basic *security* assumptions of the rich countries also seemed in doubt. Europe's capacity for military action had declined to such a point that it seemed incapable of contributing much either to its own defense or to the protection of its residual overseas interests. Japan's military capability was never restored after the country's defeat in World War II. The implicit assumption of these countries had been that in a military crunch they could lean on the United States; a succession of American administrations had been careful not to discourage that view. Nevertheless, by the time the oil crisis became tangible, that set of assumptions was no longer altogether credible. The trauma of Vietnam destroyed the illusion of American invincibility; the policy of détente weakened the expectation both in Europe and Japan that the United States would automatically resist any pressures by the Soviet Union. France's defection from NATO and Europe's determination not to follow the American policy with respect to Israel were quite enough to render the old set of security assumptions anachronistic. Yet, by the early nineteen-seventies, no new set of security assumptions had emerged to replace the old. Europe and Japan remained vulnerable, and they were uncomfortably aware of their vulnerability.

This was part of the setting in which the rich importing countries set about formulating their responses to the emerging oil crisis in the early seventies—a setting of uncertainty and disarray, of transition from a too successful past to a too uncertain future. Small wonder that these countries did not emphasize the cooperative approach as if it were in the nature of things. The papers of McKie, Tsurumi, Stobaugh, Prodi, and Lantzke provide a detailed account of the reactions of the rich countries as they confronted the crisis. Though each country seemed to react in its own way, there were certain common elements worth noting.

For some years before the outbreak of the crisis in 1973, there were frequent warnings about the likely shape of future events. Saudi Arabia's representatives, for example, warned privately and publicly that if the impasse between Israel and Egypt went on much longer, the Arab states would have to use the oil weapon. Venezuela and Kuwait were considering production cutbacks for the long-run conservation of their oil resources. The American-based oil companies were very explicit in expressing their apprehensions about the future behavior of oil-exporting countries; so, also, were a few key officials in the United States government.

But certain critical elements for action were missing. For one thing, few military or political specialists anticipated an early resumption of war in the Middle East; the cement that would be necessary for joint Arab action seemed to be lacking. Besides, most of those outside the industry who were knowledgeable about the world's oil trade were acutely conscious—in retrospect, perhaps overly conscious—of the fact that the oil companies' interests and the larger national interests were by no means identical. Finally, it was especially difficult at the time to capture the full attention of the President of the United States, the one official capable of taking the costly and unpopular measures that would be required to blunt the threat. His mind, as we now know, was filled much of the time with the problems of reelection and Watergate.

As a result, the international oil companies were left to define the nature of the problem in terms of their own perceptions and interests, and to grope for their own preferred solutions. From their point of view, understandably enough, the paramount objective was to buy some measure of predictability and stability for their world-wide operations. Governments sometimes helped them in their efforts; sometimes not. The United States government, for example, appears to have given lukewarm support to the all-embracing global agreement that its oil companies were seeking to promote at Teheran in 1970; when the Shah expressed displeasure with such an agreement, the United States government appears to have backed away. The French government was rather more positive in its policies than certain others, alert always to any possibility for asserting its independence from the United States on an important political issue. On the whole, however, governmental roles at this preliminary stage were of secondary importance.

When the crisis finally broke in October, 1973, governments could no longer look away. Nevertheless, most of them discovered very quickly that any relevant response would be politically disconcerting and technically complicated; so long as the threatened shortages did not materialize in any really painful form, the propensity for fobbing off such problems prevailed.

There is a theory widely held in Europe and elsewhere that the seeming inaction of the United States government during these first critical phases of the crisis, when supplies were threatened and when prices leaped upward, stemmed from a conscious decision that the scarcity and costliness of oil would, on the whole, be a good thing for American interests. According to the theory, various American groups hoped to gain: the international oil companies, through increased profit margins; the anti-environmentalists, through a more permissive strip-mining bill and a green light for the Alaska pipeline; and the United States government itself, through an increase in its capacity to dominate the needier European countries.

But there is a much simpler explanation of United States behavior, one that is more in keeping with the scraps of evidence. A full realization of the implications of the crisis did not appear at the highest levels of government until early 1974. To be sure, the international oil companies continued to express their fears over unprecedented price rises, as they had done before the crisis erupted. But their very ability to distribute supplies more or less adequately during the early months of the crisis, from October, 1973, to the end of the year, blunted the effects of their expressions of concern. American public opinion tended to see the oil crisis primarily as a maneuver of the companies themselves, not so much as a threat by the Arabs. A preoccupied President continued to concern himself with Watergate. An involved Secretary of State worked around the clock on peacemaking efforts. Anti-environmentalists saw the crisis as a blessing in their war against the Sierra Club and its allies; environmentalists and conservationists saw high prices as a boon in discouraging oil consumption; independent oil companies saw the crisis as an opportunity to counterattack the majors; tax reformers saw it as a chance to end certain provisions that benefited the major oil companies abroad.

The inability to shift course rapidly was not confined to the United States alone. In many countries, the crisis was seen by various groups as an opportunity to promote any cause that the situation seemed to serve, whether that cause was in fact related to the crisis or not. Tsurumi points out that, in Japan, MITI (the Ministry for International Trade and Industry) used the crisis as a peg for bolstering its waning power to control the Japanese economy. In France, the crisis provided a platform for independent political action in the Arab world, action that had only a partial relation to the oil issue itself. In Britain and Norway, governments cast an eye to their future in the North Sea, cautiously asking themselves what precedents they might be setting that would embarrass them when they themselves became major oil exporters.

So it was business as usual on the political front, at least in the short run. If any effective reaction to the crisis was to be expected from the rich oil-importing countries, a good deal more time would be needed. Indeed, it would be late 1974 or even early 1975 before the full gravity of the crisis would begin to be reflected in governmental policies.

Part 2: THE CRISIS

JOEL DARMSTADTER AND HANS H. LANDSBERG

The Economic Background

Introduction

THIS PAPER ADDRESSES THE QUESTION: What major developments in the energy field, and particularly in oil, predisposed the world economy to the shocks to which it has been subjected since the end of 1973? By "shocks," we have in mind, first of all, the drastic rise in world oil prices and, second, the intimation (based partly on what happened during the October, 1973, war and partly on subsequent OPEC actions and pronouncements) that supply would be manipulated to maintain the new level of prices and for other ends.

The answer has both economic and non-economic components. In essentially economic terms, it comes down to asking what factors on the energy-demand and energy-supply sides provided a receptive setting for the actions of the producer cartel. In its non-economic—and largely political—aspect, it comes down to asking what policies in the Arab-Israeli conflict (and perhaps with respect to other issues as well) made for the cohesiveness that was necessary not only for the implementation of the embargo but, looking beyond that short-term disruption, also for the degree of bargaining effectiveness that, on purely economic grounds, could not have been predicted on the basis of previous efforts at cartelization.

This paper, however, deals exclusively with the economic component—that is, it surveys those world-wide developments that seemed to render the major oil-consuming countries susceptible to the supply and price pressures that began late in 1973. We do not point to such developments as sufficient by themselves to account for the crisis, for we make no pretense either of exhaustiveness or of knowing how the behavior of the producers might have shifted had a different set of economic circumstances prevailed. Our exercise does not even permit a quantitative "feel" for how much of the supply and price crisis has its roots in the developments we will recount. All we suggest is that the phenomena we will review contributed significantly in setting the stage for the events of 1973 and 1974, and that, to the extent that these phenomena are recurrent, they represent enduring issues. Their consideration and discussion are therefore necessary for an appraisal of the shape of things to come. Above all, we find it important to stress that an "energy crisis" was developing, that both decision-makers and the public were aware that it was long before the October, 1973, war, and that many of the factors singled out below were combining toward a new constellation of forces in world energy. The embargo and the ensuing price revolution brought to an early and violent boil what probably was inevitable in the longer run, but what might have been more manageable at lower temperatures and with greatly attenuated disturbances.

Several fairly distinct trends on the world energy scene helped create the climate in which the producer-country actions of 1973-74 could bear fruit. These trends revolve, in their physical manifestation, around the growth in overall energy consumption, the rising share of petroleum within the total, and, in turn, the steadily rising share—already high in the early nineteen-sixties—of Eastern Hemisphere (particularly Persian Gulf and North African) oil in accommodating that rise in oil consumption. In general, a number of these trends accelerated more rapidly than had been anticipated. Concurrently with the growing dependence on energy from this concentrated geographic region, major changes benefiting the producer countries were occurring in oil-pricing and in the financial relationships between the companies and their host governments. The resulting increases and, especially, the prospective accumulation of revenues made it possible at least for the principal exporters to contemplate slowdowns or even cutbacks in production without adversely affecting—and perhaps even benefiting—their economies. Finally, a comfortable reserve-production situation *globally* tended to obscure the political, economic, and distributional aspects of converting these reserves into supplies where and when needed, while at the same time removing urgency from the development of alternative sources of energy.

Table 1
WORLD ENERGY CONSUMPTION AND POPULATION, SELECTED YEARS, 1925-72

	Total energy consumption [a]		*Population* (million)	*Energy consumption* [a] *per capita* (10^6 Btu)
	10^{12} Btu	10^6 barrels/day oil equivalent		
1925	44,249	21.6	1,890	23.4
1950	76,823	37.5	2,504	30.7
1960	124,046	60.5	2,990	41.5
1970	214,496	104.7	3,609	59.4
1972	237,166	115.7	3,747	63.3
	Average annual percentage rate of change			
1925-50	2.2		1.1	1.1
1950-60	4.9		1.8	3.1
1960-72	5.5		1.9	3.5

[a]One barrel of crude oil has a heat content of about 5.6 million Btu. Therefore, 1 million barrels a day equals approximately 2,044 trillion ($=10^{12}$) Btu per year. Oil consumption is sometimes also expressed in metric tons per year and energy consumption in metric tons oil-equivalent per year. Since 1 ton of crude oil is equal to about 7.3 barrels, 1 million barrels a day equals approximately 50 million tons a year. From the foregoing approximate calorific equivalents, we can then derive an additional one: 1 million tons of oil equals approximately 41 trillion Btu.

Source From data shown in Darmstadter and Schurr, "The World Energy Outlook to the Mid-1980's: The Effect of an Alternative Supply Path in the United States," *Philosophical Transactions,* 276 (Royal Society, London, 1974).

Trends in Aggregate Energy Consumption[1]

By 1972, world-wide energy consumption had reached an estimated level of 237 quadrillion Btus—representing, in more familiar terms, the equivalent of 116 million barrels a day of petroleum (see Table 1). The rate of energy-consumption growth appears to have been accelerating throughout much of the twentieth century: from 2.2 per cent between 1925 and 1950 to a postwar rate of nearly 5 per cent annually, rising during the most recent decade to about 5½ per cent. While the more recent change in rate may not seem dramatic, one must realize that, at the aggregate consumption levels prevailing in 1970, each tenth of a percentage point of consumption was equivalent to over one million barrels a day, compared to slightly over half a million a decade earlier—that is, acceleration by even a fraction of a percentage point represented a substantial amount.

At 2 per cent per annum population growth, the aggregate energy growth rate of 5½ per cent during 1960-72 meant a yearly growth in per capita energy consumption of approximately 3½ per cent. The worldwide average spanned a range of per capita energy-growth rates, varying from around 3 per cent in the United States to over 10 per cent in Japan, with a number of regions clustered in the 4 to 5 per cent growth-rate bracket. But regional differences in the *level* of per capita energy use, while narrowing, have remained dramatically wide. In 1970, per capita energy consumption in the United States and in Canada—which was only slightly lower—was more than two and one-half times the level of the next ranking regions—Western Europe, the Soviet Union, and Oceania. An even more extreme disparity is reflected in the fact that North American per capita consumption was between thirty and forty times the levels prevailing in Africa and the developing portions of Asia.[2]

To be sure, trends and levels in per capita energy consumption are not proxies for per capita income or gross national product; nor are the latter measures, in turn, truly reflective of living standards, however defined. Nonetheless, while recognizing clearly the potential for considerable energy savings of a sort that would not inhibit economic aspirations, the existence of such disparities, coupled with a close—if not fixed—correspondence between levels of per capita energy consumption and general economic development, has an important implication for the future: the drive toward a substantial improvement in living standards in the years ahead will exert added claims and growing burdens on world energy supplies.

In the early seventies, an almost unprecedented concurrence of boom conditions prevailed in North America, the European Community, and Japan. Growth rates of real GNP were at peak levels in the two years preceding the oil crisis. For the developed countries as a whole, real GNP rose as follows between 1970 and 1973:[3]

1969–70	3.2
1970–71	3.6
1971–72	5.5
1972–73	6.3

Especially notable was the world-wide harmony of growth rates in the year immediately preceding the crisis. For example, while the change from 1969 to 1970 was

+10.3 per cent in Japan, but –0.4 per cent in the United States, between 1972 and 1973, and especially during the first three quarters of 1973, rapid expansion in the industrial countries coincided. For the year 1973 as a whole, these were the individual GNP growth rates of the major areas:

United States	5.9%
Western Europe	5.4%
Japan	10.4%

Many industries worked at increasingly high rates of capacity utilization; some energy-intensive primary industries in the United States operated at rates approaching 100 per cent in 1972 and into 1973. The near-capacity utilization of industrial facilities can be illustrated by data for the United States compiled by the board of governors of the Federal Reserve System. The index of capacity utilization in materials industries (ranging from steel and copper to paper and fibers) reached 96 per cent in the third calendar quarter of 1973 and averaged 95 per cent for the year as a whole. "For much of 1973," says the *Federal Reserve Bulletin* of January, 1974, "basic materials-producing industries were operating at rates higher than at any other time in the postwar period." Similar conditions seem to have prevailed abroad. Such high utilization rates necessarily involve the operation of less efficient equipment and a strain on the supply capacity of both materials industries and energy suppliers. In Japan, for example, fuel consumption by manufacturing industries in October, 1973, was nearly 30 per cent above the monthly level for calendar 1972; while, for the same month, the rate of capacity utilization in manufacturing stood at a remarkable 102 percent.[4]

Shifts Among Energy Sources to Meet Demand

Table 2 provides a picture of postwar shifts in the role of the different energy sources to supply the rapidly rising demand documented above. A sharp relative decline of coal and marked relative increases in both oil and gas characterize the period. The latter two sources accounted for 38 per cent of world-wide energy consumption in 1950, but had risen to 64 per cent in 1972; concurrently, coal experienced a relative decline, from 56 to 29 per cent. The primary electricity share—hydroelectricity along with a thus-far modest but rising nuclear component—remained essentially unchanged.

With variations, these shifts occurred throughout the world, as can be seen in Table 3. In no area did the share of coal fail to decline. In most areas, the share of oil and natural gas was higher in the early nineteen-seventies than in 1950 and, more often than not, substantially or radically so. Only in two regions—Eastern Europe and China—did coal continue to contribute more than half of total energy consumption in recent years. In Western Europe, there was a sharp decline in coal's relative share from over three-fourths to a little over one-fourth in the short span of less than two decades, and its absolute use declined as well. Concurrently, the share of oil and gas went from

Table 2
WORLD ENERGY CONSUMPTION: DISTRIBUTION BY SOURCE, 1950-72

	1950	1960	1965	1968	1970	1971	1972
AGGREGATE CONSUMPTION							
(10^{15} Btu)	76.8	124.0	160.7	189.7	214.5	223.5	237.2
			Per cent shares				
Coal	55.7	44.2	39.0	33.8	31.2	29.9	28.7
Oil	28.9	35.8	39.4	42.9	44.5	45.2	46.0
Natural gas	8.9	13.5	15.5	16.8	17.8	18.3	18.4
Primary electricity[a]	6.5	6.4	6.2	6.5	6.5	6.6	6.9
	100.0	100.0	100.0	100.0	100.0	100.0	100.0

Average annual percentage rates of increase

	1950–60	1960–72
Coal	2.5%	1.8%
Oil	7.1	7.8
Natural gas	9.4	8.3
Primary electricity	4.8	6.2
TOTAL	4.9	5.5

[a]Comprised of geothermal, nuclear, hydro. For 1972, the 6.9% figure in the table broke down approximately as follows: geothermal, 0.03; nuclear, 0.7; hydro, 6.2
Source: Same as for Table 1.

15 to 62 per cent. Japan's energy patterns underwent similarly dramatic shifts. The Soviet picture is highlighted by a big postwar rise in the share of natural gas.

Changes in the American pattern were different, but eventually aggravated those noted above. The early emergence of oil and, subsequently, of gas had resulted in important shares for these fuels in the country's total energy consumption far ahead of most other regions. The proportion of oil in United States energy consumption rose rather modestly between 1950 and 1970, with the sharply declining relative position of coal being principally compensated for by increases in the use of natural gas. However, significantly for the 1973 events (and discussed more fully below), in the past several years United States oil consumption has advanced at a disproportionately fast rate, as gas output leveled off, power-station coal utilization encountered tight environmental constraints, and oil demand in the transport sector accelerated. The significant factor in the American situation is thus not the sharp shift to oil and gas in the past two decades but rather the recent acceleration in oil demand and the meeting of this increased demand not from indigenous but from foreign sources.

Table 3

DISTRIBUTION OF ENERGY SOURCES FOR MAJOR REGIONS, 1950 AND 1970

(percentage except for Btu columns and row)

	1950						1970					
	coal	oil	natural gas	primary electricity	total %	total 10^{15} Btu	coal	oil	natural gas	primary electricity	total %	total 10^{15} Btu
North America	38.0	38.7	16.7	6.4	100.0	36.86	18.3	43.7	32.1	6.0	100.0	74.48
Canada	40.6	28.6	2.8	27.9	100.0	2.71	9.6	41.3	26.9	22.2	100.0	7.04
United States	37.8	39.5	18.0	4.7	100.0	34.15	19.1	43.9	32.7	4.3	100.0	67.44
Western Europe	77.4	14.3	0.3	8.0	100.0	17.48	27.4	55.6	6.1	10.8	100.0	47.87
Oceania	65.3	27.3	—	7.4	100.0	0.89	39.8	48.6	2.2	9.3	100.0	2.45
Latin America	9.8	72.9	8.3	9.0	100.0	2.40	4.9	67.8	18.4	8.8	100.0	9.13
Asia (excl. Comm.)	53.3	28.5	1.4	16.9	100.0	3.80	23.3	64.1	6.2	6.4	100.0	20.82
Japan	61.9	5.0	0.2	32.9	100.0	1.74	22.4	68.8	1.3	7.5	100.0	11.26
Other Asia	46.0	48.3	2.4	3.3	100.0	2.06	24.3	58.7	11.8	5.2	100.0	9.56
Africa	61.4	36.9	—	1.7	100.0	1.30	43.5	48.7	1.5	6.2	100.0	3.68
U.S.S.R. and Comm. Eastern Europe	81.4	14.6	2.3	1.7	100.0	12.84	49.6	28.7	18.5	3.1	100.0	44.76
U.S.S.R.	75.6	19.7	2.5	2.3	100.0	8.43	40.4	33.3	22.5	3.8	100.0	31.99
Eastern Europe	92.5	4.8	2.0	0.7	100.0	4.41	72.7	17.4	8.4	1.5	100.0	12.76
Communist Asia	92.8	0.9	—	6.3	100.0	1.25	89.4	8.2	NA[a]	2.4	100.0	11.30
WORLD: per cent	55.7	28.9	8.9	6.5	100.0		31.2	44.5	17.8	6.5	100.0	
WORLD: 10^{15} Btu	42.80	22.20	6.82	5.00		76.82	66.90	95.48	38.21	13.91		214.50

[a] The U.N. shows no figures on China's natural-gas production and consumption. A recent analysis puts China's natural-gas production at 0.4×10^{15} Btu in 1965 and 1.2×10^{15} Btu in 1971 (C.S. Chen and K.N. Au, "The Petroleum Industry of China," *Die Erde*, 3-4 (Berlin, 1972), p. 319. Such a quantity would clearly raise the estimate of China's energy consumption—shown above—quite markedly. Another estimate (*Oil and Gas Journal*, August 20, 1973) credits China with only about 20% of this amount of gas output.

Source: Same as for Table 1.

The Growth of Oil Imports

The increasing dependence on oil as the "balancing" element in meeting United States energy demand is reflected in substantially expanded American oil imports dur-

Table 4
ENERGY CONSUMPTION, OIL CONSUMPTION, AND OIL IMPORTS:
UNITED STATES, WESTERN EUROPE, AND JAPAN, 1962 AND 1972

	1962			1972		
	United States	Western Europe	Japan	United States[c]	Western Europe	Japan
	10^6 barrels per day					
Energy consumption (oil equivalent)	23.27	13.96	2.25	35.05	23.84	6.58
Oil consumption	10.23	5.24	0.96	15.98	14.20	4.80
Oil imports [a]	2.12	5.19	0.98	4.74	14.06	4.78
From Middle East/ North Africa [b]	0.34	3.80	0.72	0.70	11.30	3.78
From elsewhere	1.78	1.39	0.26	4.04	2.76	1.00
	Percentage of energy consumption					
Oil consumption	44.0	37.5	42.7	45.6	59.6	73.0
Oil imports [a]	9.1	37.2	43.6	13.5	59.0	72.6
From Middle East/ North Africa [b]	1.5	27.2	32.0	2.0	47.4	57.4
From elsewhere	7.6	10.0	11.6	11.5	11.6	15.2
	Percentage of oil consumption					
Oil imports [a]	20.7	99.0	102.1	29.7	99.0	99.6
From Middle East/ North Africa [b]	3.3	72.5	75.0	4.4	79.5	78.6
From elsewhere	17.4	26.5	27.1	25.3	19.4	20.9
	Percentage of oil imports					
From Middle East/ North Africa [b]	16.0	73.2	73.5	14.9	80.4	78.9
From elsewhere	84.0	26.8	26.5	85.1	19.6	21.1

[a]Imports are "gross" rather than "net"; that is, exports are not deducted. Thus, they exclude product exports from West European refineries. For Japan, excess of imports over consumption arises because of small quantities of product exports, refinery losses, and (presumably) independent construction of the two series. By showing gross rather than net imports, we overstate slightly the degree of foreign dependence. The overstatement matters, if at all, only in the case of Western Europe.

[b]Includes negligible quantities from West Africa in 1962.

[c]For changes in the U.S. import situation after 1972, see accompanying text remarks.

Sources: Data for 1962 based on Joel Darmstadter *et al.*, *Energy in the World Economy* (Baltimore, 1971), and British Petroleum Co., *Statistical Review of the World Oil Industry, 1962;* data for 1972 based on British Petroleum Co., *Statistical Review of the World Oil Industry, 1972.*

ing the past ten years (Table 4). Rising from 2 million barrels a day in 1962 to nearly 5 million barrels a day in 1973—or at a growth rate of 8½ per cent per year—the import share of United States oil consumption has gone from 20 to 30 per cent, and just before the October, 1973, war was running at close to 35 per cent. Nonetheless, as a share in aggregate national energy consumption, American oil imports stand far below comparable levels elsewhere. In Western Europe, oil imports went from 37 per cent of total energy consumption in 1962 to nearly 60 per cent in 1972; in Japan, from 44 to 73 per cent; while, for the United States, the increase was from 9 to 13 per cent. Even so, the high absolute level of United States energy consumption and the pervasive role of oil in transportation make even this relatively small increase a substantial one in the face of supply difficulties.

For Western Europe and Japan, the dominance of the Middle East and North Africa as suppliers of oil, coupled with the importance of oil in West European and Japanese energy consumption, has meant that these two producing areas have come to occupy a crucial role in the *total energy* position of the consuming areas: 47 per cent for Western Europe and 57 per cent for Japan in 1972.

Traditionally, American oil imports have come in the main from Caribbean and Canadian sources, the Persian Gulf-North African region having maintained an approximately constant share of around 15 per cent of American oil imports (and 3 to 4 per cent of American oil consumption) during the period from 1962 to 1972. However—and this is not brought out in the 1962 and 1972 comparison on Table 4—it has now become clear that, because Canadian and Venezuelan expansion possibilities are at best severely limited, further additions to United States imports must bank heavily on Eastern Hemisphere supplies, particularly from Saudi Arabia, but from other Arab and non-Arab suppliers (Iran, Nigeria),as well. Indeed, during the period June–October, 1973, American crude-oil and oil-product imports coming directly from the *Arab countries only* already amounted to an estimated 1.3 million barrels a day, representing 8 per cent of oil consumption and 19 per cent of oil imports for the United States; if one adds imports indirectly traceable to Arab countries,[5] the respective figures for this mid-1973 period become 10 per cent and 25 per cent.[6] With the inclusion of Iran, the entire Middle East-North African region would figure more prominently still —a minimum of 12 per cent of consumption and 30 per cent of imports.

Contemplation of the continued world-wide shift to imports from the Middle East-North African oil fields during the past decade and the more recent American changes just noted may shed a different light on what has been judged by at least some as a panicky and submissive response to producer-country moves during 1973–74. It may instead reflect this prominent characteristic of acute dependency in world oil flows.

The Past Misreading of Future Trends

None of the three principal lines of development sketched above—the rapid growth of world energy consumption as a whole, the continued shift toward oil everywhere, and the rapidly rising volume of American oil imports—was adequately anticipated in the succession of energy projections that have appeared since around 1960. The review of forecasts summarized in Table 5 is designed merely to convey an impression of the

Table 5
REVIEW OF SELECTED PAST ENERGY CONSUMPTION PROJECTIONS

| Region | Actual Data [a] | | Projections | | | |
	Period	Average annual growth rate (%)	Source	Year published	Period	Average annual growth rate (%)
Total energy consumption						
World	1960-70	5.6	(A)	1966	1960-70	4.6
					1970-80	4.8
			(B)	1966	1960-80	5.0
Western Europe	1960-70	6.3	(C)	1960	1955-75	2.8
			(A)	1966	1960-70	4.4
					1970-80	4.0
			(B)	1966	1964-70	4.2
					1970-80	4.1
United States	1960-70	4.2	(D)	1963	1960-70	2.9
					1970-80	2.8
			(A)	1966	1960-70	3.6
					1970-80	3.3
Japan	1960-70	11.9	(A)	1966	1960-70	9.1
					1970-80	7.0
			(B)	1966	1964-70	10.0
					1970-80	6.9
oil consumption						
Western Europe	1962-72	10.5	(B)	1966	1964-80	4.1[b]
United States	1962-72	4.6	(A)	1966	1960-70	3.5
					1970-80	2.7
			(E)	1971	1965-75	3.4
					1975-80	2.9
			(F)	1968	1965-80	3.1
Japan	1962-72	17.5	(B)	1966	1964-70	14.3
					1970-80	8.3
oil imports						
United States	1962-72	8.4	(F)	1968	1965-80	3.2
			(E)	1971	1965-75	3.5
					1975-80	2.6
			(A)	1966	1960-70	4.4
					1970-80	4.2[c]

Note: Differences in definitional practices among the various projection studies impairs exact comparison, even of growth rates.

[a]From Table 8, Darmstadter and Schurr, "The World Energy Outlook to the Mid-1980's," cited in Table 1.

[b]Highest of range shown in source (B).
[c]Midpoint of range shown in source (A).
Sources: (A) European Coal and Steel Community, Review of the Long-Term Energy Outlook of the European Community; (B) O.E.C.D., Energy Policy; (C) O.E.E.C., Towards a New Energy Pattern in Europe; (D) Hans H. Landsberg, Leonard L. Fischman, and Joseph L. Fisher, Resources in America's Future (Baltimore); (E) Sam H. Schurr, Paul T. Homan, and associates, Middle Eastern Oil and the Western World: Prospects and Problems (New York); (F) U.S. Department of the Interior, United States Petroleum Through 1980.

kinds of efforts undertaken. It makes no claim to being comprehensive; a detailed analysis comparing the different projection studies as well as a comparison between the projections and actual performance cannot be adequately undertaken here.

This said, the degree of underestimation disclosed by the table is still worth pondering. As we might expect, there is less error in total energy projections than in the projections for oil only. But even for the total, the projections have been markedly conservative. For example, even though Western Europe's economic growth rate in the sixties had been forecast quite accurately, its energy growth noticeably outstripped expectations. For Japan, the continuation of the postwar economic boom, carrying energy growth with it, surprised people by its unbroken momentum. In the United States as well (its experience is dissected in some detail elsewhere[7]), the progressively declining relationship, observed for at least four decades, between energy growth and GNP growth did not, contrary to expectations, endure into the latter part of the sixties. Among the things not foreseen were the halt to fuel-efficiency improvements in the American electric-power sector and the acceleration in demand for motor-vehicle fuels, based not only on rapid growth of the vehicle population, but also on less efficient fuel use, which in turn was the result of heavier cars, added "extras" such as air conditioning, higher speeds, and urban congestion.[8]

Fairly gross misjudgments, also evident in Table 5, were made with respect to the role of oil in Western Europe's energy balance and of American oil consumption and imports. In the case of Western Europe's oil consumption, one source of error seems to have been that projections—for example, the 1966 OECD study, Energy Policy—allowed for a far greater, though not expanding, role for coal between 1964 and 1970 than actually emerged. The OECD envisaged West European solid-fuel production at between 410 and 440 million coal-equivalent tons in 1970. In fact, 378 million tons were produced—a "slippage" equal to (on the basis of the midpoint projection) 700 thousand barrels a day of oil, or 6 per cent of oil imports for that year. With respect to coal, the 1966 OECD study was only the latest in a succession of projections that failed to anticipate the rapidly shrinking role of the West European coal industry. Since imported-oil prices were decisively competitive with European coal throughout most of the nineteen-fifties and -sixties, this failure was not so much one of economic miscalculation as of misreading the governments' policy intentions regarding the scale of support for the coal industry. But even apart from the extent to which oil (in power generation and heating) filled the incremental demand that coal had been anticipated to meet, the demand for oil in nonsubstitutable uses also progressed at a buoyant clip. In the six-nation Common Market area, for example, the demand for automotive fuels advanced at 10 per cent annually throughout the sixties.

The misjudging of the future role of oil imports in relation to domestic supplies applies only to the United States; neither Europe nor Japan was expected to be a significant producer for the period under consideration. As for the United States, the conventional assumptions had been two. The first was that government import controls could hold the import share to a constant percentage. Instead, these provisions were frequently revised, and the import-control program became so eroded that it collapsed entirely shortly before the events of 1973. It should, however, be noted that even those who were calling for an end to the import-control program did not anticipate a staggering rise in imported oil. Consider the relaxed views expressed by the staff of the United States Cabinet Task Force on Oil Import Control as "late" as the beginning of 1970: "Because demand is growing faster than domestic supply, imports would have to expand by 1980 at the present [i.e., 1970] $3.30 price, and by then could amount to about 27 percent of our demand at that price."[9] In fact, as we have already noted, oil imports had already reached a rate of 35 per cent on the eve of the October, 1973, war.

The second assumption was that large unexplored and unexploited areas, onshore and especially offshore, were waiting for the drill and would supply the incremental quantities to meet rising demand. In fact, however, exploratory and developmental activities in the United States declined, as the industry channeled its funds into overseas ventures that were less beset by constraints and generated higher yields in both oil and profits.

Words of alarm from oil interests regarding the lagging domestic effort failed to strike a responsive chord among persons long turned cynical to what they viewed— rightly or not—as special pleading and poor-mouthing by an apparently prosperous industry. Unfortunately, the same plea (greater financial incentives) could equally well be read as an expression of need for capital to sink into further exploration and development, or as solely a desire for enhanced rates of return. Oil companies have not often been given the benefit of the doubt on this proposition.

Insight through hindsight is appealing. Perhaps the forecasters had shut themselves off too much from the real world—and particularly the political world—when they made their assumptions. For, in addition to erring in economic premises, most planning efforts did not appear to have taken sufficiently seriously threats by the Arab governments to use oil as a political weapon as well as to hold back output for what—to the governments, at least—may have seemed a proper means toward income maximization. Popular recognition that a crisis was approaching and acceptance of the necessity for belt tightening and for policies to expand domestic energy resources to counteract it could not easily have been won in this climate.

Oil Prices and Producer-Country Policies

At the same time that snowballing demand and concentration on oil, on oil imports, and on the Middle East accelerated, resistance to price increases underwent a gradual softening. The evolution of Middle East prices is described in detail elsewhere in this volume. Suffice it to say here that, by October, 1973, a typical grade of Arabian crude oil yielded government receipts of about $3.05—more than a 100 per cent increase over the level prevailing at the start of 1972—and that within half a year that figure would again be more than doubled (see Table 6). In the early stages of these upward price moves, the short-run inelasticity of demand for petroleum products allowed the oil companies to pass on to their customers the amounts they were obliged to pay to the host governments. Since the sharp escalation in prices at the beginning of 1974 coincided with, and probably contributed to, the onset of world-wide recession, it is difficult to determine the extent to which the ensuing drop in demand was a specific response to these higher prices, or simply reflected a slowdown in the economy generally.

Along with the dramatically higher prices and monetary flows to oil-producing governments, these governments exerted pressure to gain control over their own oil deposits and production facilities. The changes that have taken place or are still under way range from outright nationalization in some countries to "participation" agreements in others. The latter arrangements provide for the acquisition of majority—and, ultimately, total—control by host governments over oil-company assets. Initially designed to take place within a decade, the whole process is now clearly being telescoped into a far shorter interval.

To what extent the OPEC moves of 1973-74 reflect the exercise of clear-cut mo-

Table 6
POSTED PRICES AND TYPICAL PRODUCER-GOVERNMENT RECEIPTS PER BARREL OF OIL EXPORTS,
1965-74

	Posted prices, Arabian light crude (34° gravity) f.o.b. Ras Tanura (in dollars)	Saudi Arabia, government take of royalties and taxes[a] (in dollars)
December, 1965	1.80	0.83
December, 1970	1.80	0.88
June, 1971	2.29	1.26
January, 1972	2.48	1.45
January, 1973	2.59	1.52
October, 1973	5.12	3.05
January, 1974	11.65	7.00

[a]The first three lines in this column refer to annual averages rather than to the months shown. In order to relate the figures in the column to world market prices, one must take account of production costs, company profits, and transportation charges. To illustrate: The $7 government take for January, 1974, translates to "tax-paid" costs of, say, $7.15 if one adds production costs; to an f.o.b. price of, say, $7.75 if one allows for an estimated company return of 60¢; and to a delivered price in the United States of around $9 after taking transportation costs into account.

Sources: *Petroleum Press Service*, November, 1973, and February, 1974; and Foster Associates, *Energy Prices 1960-73, A Report to the Energy Policy Project of the Ford Foundation* (Cambridge, 1974), p. 18.

nopoly power that can be expected to endure or simply, in a much more benign interpretation, the momentary exploitation of unique and temporary market conditions continues to be a topic of intense interest and controversy that is not about to be tidily resolved.[10] Those who subscribe to the latter hypothesis cite, for example, the fact that the world-wide economic expansion of recent years had dramatically depleted inventories of crude oil and petroleum products in the industrial countries just before the 1973 war. Indeed, the draw-down of petroleum stocks was quite striking. In the nine-country EEC area, the inventory *accumulation* of 4.2 million tons of crude oil in 1971 gave way to an inventory *reduction* of 3.4 million tons in 1972. In petroleum products, the rate of accumulation fell from 4.2 million tons to 1.9 million tons during the one-year period.[11] The United States experience was similar. Considered in isolation, that would have given OPEC price increases the character of a traditional—but not necessarily lasting—commodity price boom. On the other hand, successive OPEC price increases that began several years prior to October, 1973, have stuck. This is the more remarkable as these prices represent levels reflecting a huge multiple of marginal producing costs (conventionally put at around 20 cents a barrel) and have endured under circumstances of excess productive capacity.

While it is not the purpose of this paper to analyze the conditions and prospects for OPEC monopolistic behavior, one tangential aspect is central to our topic and deserves additional attention. That concerns the shifting role of the United States on world oil markets and in world reserves. This topic has already been touched upon, but it will be taken up somewhat more specifically in the following three sections.

The Background to Rising American Oil Import Dependence

The steadily increasing weight of American imports on world oil markets probably was an important factor in helping to create conditions for the producer-country price and supply moves of 1973-74; for, at considerably lower levels of United States imports than the 35 per cent degree of dependence that we saw developing just before the October, 1973, war, a world-wide supply "overhang" might have been sufficient to preclude the concerted OPEC actions, or at least their success. Indeed, had conditions in 1973-74 not been substantially different from those prevailing in 1967, the American role as a supplier of last resort would probably have critically altered the situation. It is for that reason, rather than our greater familiarity with American developments, that a brief recounting of the circumstances that led up to this sharply increased reliance on foreign sources of supply seems in order. The contributory factors in the United States that are worth highlighting are: accelerated demand for energy in the aggregate; a dramatic falling off of reserve additions of oil and natural gas; severe constraints, largely for environmental reasons, on the use of coal; lags in the scheduled completion of nuclear power plants; and protracted delays in oil and gas leasing.

The mosaic within which these pieces fall can be reconstructed about as follows.[12] To begin with, hindsight provides ample reasons why American concern over the adequacy of fuel and power supplies should have preceded the October, 1973, war. The nub of the problem was that this nation's aggregate consumption of energy resources during the latter part of the sixties continued to grow very rapidly. Real prices of oil in the early seventies had declined, and they remained below their level in the fifties and middle sixties, thus exerting no pressure toward conservation; nor did falling prices create a climate in which public conservation policies could be instituted or, if instituted, be effective. Concurrently, a leveling off in domestic output of oil and gas and partial restrictions on the use of coal led to the need for rapidly rising imports, made feasible by the abandonment of remaining import controls.

To judge the significance of this emerging foreign component of the American energy supply, we need to note that in the early nineteen-seventies oil accounted for around 40 per cent—and, combined with natural gas, for about four-fifths—of United States energy consumption. At the same time as domestic production of both oil and gas had flattened out in the wake of declining levels of exploration and, consequently, declining reserve levels (see Table 7), consumption of oil—the balancing energy source in times of demand pressures—accelerated. Several factors were responsible. For one, oil was rapidly supplanting coal as a power-plant fuel. Having reached a low of 6 per cent as its share in utility fuels in the mid-sixties, oil's share had unexpectedly risen to nearly 16 per cent by 1972, largely on the basis of the high cost or nonavailability of low-sulfur steam coal. To illustrate, in New England, electric utilities had burned 9 million tons of coal as recently as 1966. By 1972, they had reduced their coal consumption to 1.3 million tons, and this happened during a period of steadily rising power generation. Oil made up for both the decline in coal and the rise in power generation. In the Middle Atlantic states, the absolute volume of coal burned remained constant, but oil, and to a much smaller extent, nuclear power, supplied the growth increment. Nationwide, coal had provided the same share of electric generating fuel in 1972 as it had, say, in the mid-sixties, 1.1 million barrels of oil per day would have been "saved," representing some 7 per cent of United States oil consumption or 23 per cent of oil imports in that year. In the main, environmental restrictions operated against the use of coal and foreclosed that possibility; but one must also note that the coal industry had generally been "written off" as declining, that it had severe manpower problems, difficulties in raising capital, and generally showed many signs of being a depressed industry. Environmental problems, both in mining and combustion, greatly aggravated the picture.

The late sixties also witnessed a largely unanticipated burgeoning demand for motor gasoline. This demand rose by 2.8 per cent yearly between 1960 and 1965. After the mid-sixties, it accelerated to more than twice that rate. Continuation of the lower, 1960-65 growth rate would have "freed" another 630,000 barrels of oil per day in 1972—another 4 per cent of American oil consumption, or 13 percent of oil imports. Though much cited as one of the reasons for this development, reduction of automotive efficiency due to pollution control is probably a minor cause when compared with steadily increasing car weight, large engine size, rapid spread of air conditioning, and other extras that lessened fuel efficiency. Failure of annual mileage per vehicle to decline, as had been anticipated by some as a result of more two- and three-car families, is still another. In any case, the high gasoline requirement of large horsepower cars certainly goes far to explain the high *level* (even if it is only one of the reasons for the high recent *growth rate*) of American oil consumption, and, importantly, there appears to have been little anticipation of the compound effect of these elements on gasoline demand.

One could cite still other factors contributing to the tight American oil situation. A halt to the expansion of natural-gas output, not unconnected with the controversies over continued regulation of the natural-gas price, may have added 1½ million barrels per day to 1973 oil demand. The nonavailability of oil from the Alaskan North Slope and from the Santa Barbara Channel, as well as the delay in offshore operations in general, may have "subtracted" about 2 million barrels per day from potential current production. Finally, nuclear power plants were being constructed at a rate considerably below the level foreseen just a few years earlier, owing to design and construction problems, opposition on grounds of safety and environmental hazards, and cumbersome, time-consuming licensing procedures. Various combinations of these possible, but problem-ridden, supply sources would in all likelihood have greatly diminished the level of import dependence the United States was experiencing when the Middle East war began.

Attempts to cope with these problems substantially preceded the war, but, both before and since, short-run success by way of market adjustments has been limited. In the case of natural gas for space and process heat, for example, it is not easy to switch quickly to substitute energy forms. Redesigning automobiles for greater fuel efficiency runs into resistance from manufacturers and from a not insignificant segment of consumers, given the tradeoffs in size and comfort. In a more general sense, demand restraint is not assured unless energy price increases are of such magnitude that demand is suppressed at the cost of increased unemployment and industrial dislocation.

On the supply side, the regulatory lag inherent in federal controls over natural-gas prices insured sluggish action and sluggish production responses. Productive capacity (as in the case of refineries) is not rapidly expandable; accelerated leasing and new reserve development pay off only after a lag of, say, three or four years, and a quick turnaround after years of declining activity (by 1972, the annual number of oil and gas wells completed had declined to less than half the number completed in the mid-fifties) is beset by supply and other difficulties.

In the meantime, for the last few years, annual gross-reserve additions in both oil and gas have been falling short of production, resulting, as shown in Table 7, in net declines in United States reserves. This is in contrast to the period prior to 1968, when additions to reserves had at least kept slightly ahead of annual output. In the case of oil, a key element in this trend was the propensity of the American oil industry during the sixties to divert increasing shares of its capital investment from the United States to foreign producing areas. Thus, foreign sources of American energy supply began several years ago to loom as a major factor on the national scene. Other oil-importing regions of the world, notably Western Europe and Japan, have, of course, long lived with far higher degrees of import dependence; but for the United States, previously confident that its own excess productive capacity could be deployed in times of crisis (see below), the increased reliance on imports signified an unexpected turn of events. It coincided with—if, indeed, it did not pave the way for—suddenly heightened bargaining

Table 7

PRODUCTION AND PROVED RESERVES OF OIL AND NATURAL GAS, UNITED STATES, 1964-73

	Oil (billion barrels) [a]			Natural gas (trillion cubic feet)		
	Proved reserves [b]			Proved reserves [b]		
	At end of year	Change during year	Production	At end of year	Change during year	Production
1964	38.74	0.09	3.18	281.3	5.1	15.3
1965	39.38	0.64	3.24	286.5	5.2	16.3
1966	39.78	0.41	3.45	289.3	2.8	17.5
1967	39.99	0.21	3.68	292.9	3.6	18.4
1968	39.31	-0.69	3.83	287.3	-5.6	19.4
1969	37.78	-1.53	3.93	275.1	-12.2	20.7
1970	37.10 (46.70)	-0.68 (8.93)	4.07	264.8 (290.7)	-10.3 (15.6)	22.0
1971	35.77 (45.37)	-1.34 (-1.34)	4.00	252.9 (278.8)	-11.9 (-11.9)	22.1
1972	33.53 (43.13)	-2.24 (-2.24)	4.04	240.2 (266.1)	-12.7 (-12.7)	22.5
1973	32.15 (41.75)	-1.37 (-1.37)	3.93	224.1 (250.0)	-16.1 (-16.1)	22.6

[a] Including natural-gas liquids.

[b] Figures in parentheses refer to a measure of proved reserves, or change therein, which includes proved reserves ascribed to the Alaskan North Slope—presently unavailable as a source of production. North Slope proved oil reserves are carried at 9.6 billion barrels; natural gas, at 22.9 trillion cubic feet.

Source: American Petroleum Institute, *Reserves of Crude Oil . . . as of December 31, 1973* (June, 1974).

power on the part of the petroleum-exporting countries, raising the prospect of increasingly stiff terms for imported energy even without political upheavals in the Middle East. It was a new and sobering situation.

The Cushioning Impact of American Supplies in Prior Oil Disruptions

A condition for successful monopolistic behavior is tight control over, or concentration of, supply. The existence of surplus capacity in one place immediately dilutes the rewards of holding back supply in another. A distinguishing feature of world oil disruptions prior to the one connected with the October, 1973, Arab-Israeli war was that a much greater supply capability then still existed in the United States. In both 1956-57 and 1967, spare crude-oil producing capacity in the United States stood at around 25 per cent; toward the end of 1973, this figure had moved down close to 10 per cent.[13] The shut-in production capacity, which had prevailed in the United States throughout much of the past two decades and which had occasioned the operation of state prorationing schemes under the protection of mandatory oil-import quotas, had all but disappeared by the early nineteen-seventies.

It is interesting to contrast the different degrees of American oil-import reliance around the time of each of three Middle East crises. This is done in Table 8. For the twenty years as a whole, the oil import trend had, of course, been a sharply rising one. But while at the time of the 1967 war the import share was still kept relatively in bounds (at around 20 per cent), by the time of the October, 1973, outbreak, it had really taken off. Note, moreover, how amidst each of the earlier two disruptions, the

Table 8

U.S. NET IMPORTS OF PETROLEUM AS PER CENT OF CONSUMPTION DURING THREE "CRISES"
(1954-73)

Year	Imports as per cent of consumption
1954	8½
1955	10
1956	11
1957	11
1965	20
1966	20
1967	19
1971	25
1972	28
1973	35½

Sources: Various issues of U.S. Bureau of Mines, *Minerals Yearbook*; and British Petroleum Co., *Statistical Review of the World Oil Industry.*

rising trend in United States oil-import shares dampened or temporarily flattened out; in contrast, in 1973, the trend actually accelerated. There was nowhere else to go.

Geographic Shifts in World Oil Production and Reserves

Even though distribution of petroleum deposits is highly concentrated in limited parts of the globe, oil formations are nonetheless sufficiently widespread to permit production in many different—and sometimes unexpected—areas. To be sure, there is great variability in the size of fields, productivity of the wells, quality of the oil and, hence, costs of production; but exploratory and developmental activities now occurring in such widely separate regions as the North Sea, North American Arctic, and China Sea reflect the geographic breadth of oil occurrence. Nonetheless, the center of gravity both in reserve development and in production has shifted spectacularly and continuously toward the Middle East and North Africa, where yields per dollar of input far outdistance rewards likely elsewhere. As we have noted, the American industry played a major role in this shift.

World-wide shifts in reserves and output are recorded in Tables 9, 10, and 11. The steadily eroding position in United States reserves becomes particularly vivid when related to *changes* in proved oil reserves of the non-Communist countries. Thus, while the United States accounted for approximately one-fifth of non-Communist reserves in the mid-fifties, its share of reserve *additions* during the subsequent eleven- and six-year intervals fell to 2½ and 1½ per cent, respectively. The Middle East and North Africa accounted for nearly 80 per cent of reserve additions throughout both periods. (The North African component figures importantly only in the nineteen-sixties.) As for oil

Table 9
WORLD PROVED OIL RESERVES YEAR END, 1955, 1966, 1972

	1955		1966		1972	
	billion barrels	per cent	billion barrels	per cent	billion barrels	per cent
United States	35.4	18.2	39.8	10.2	43.1	6.4
Middle East/ North Africa	126.2	64.8	263.0	67.2	433.7	64.5
Other non-Communist countries	22.5	11.5	54.9	14.0	97.9	14.6
Communist countries	10.8	5.5	33.8	8.6	98.0	14.6
TOTAL	194.9	100.0	391.5	100.0	672.7	100.0

Sources: 1955 and 1956 data as shown in S. H. Schurr, P. T. Homan, *et al.*, *Middle Eastern Oil and the Western World* (New York, 1971), p. 68; 1972 data from *Oil and Gas Journal*, December 25, 1972.

Table 10
ADDITIONS TO PROVED OIL RESERVES, NON-COMMUNIST REGIONS, 1955-66 AND 1966-72

	1955–66		1966–72	
	billion barrels	per cent	billion barrels	per cent
United States	4.4	2.5	3.3	1.5
Middle East/North Africa	136.8	78.8	170.7	78.7
All other	32.4	18.7	43.0	19.8
TOTAL	173.6	100.0	217.0	100.0

Source: Table 9 above.

production, just before World War II, the United States accounted for 62 per cent of the world-wide total; by 1972, the figure was down to 21 per cent.

What lies behind these statistics? For one, lagging investment by the oil industry in the United States. Overseas markets were expanding more rapidly than was the domestic market in the sixties. United States tax regulations—particularly the foreign tax credit—when coupled with the (then) low foreign production costs insured higher rates of return than at home. Uncertainty about future United States oil-import policy seemed to put investment in foreign reserve development in less jeopardy than domestic ventures. To illustrate what was happening at that time, when corrected for price increase the trend of capital expenditures in the United States approached a flat line. Indeed, between 1968—the peak year in current dollars—and 1971, investment in the United States declined over 10 per cent. Rapid increases in 1972 and 1973 primarily reflect large outlays for lease acquisitions. Within the totals, "geological and geophysical expense and lease rentals" show even less of an increase, and they decline as a share of total investment.

Table 11
WORLD CRUDE OIL PRODUCTION, 1938-72

	1938	1957	1960	1965	1970	1972
	In thousands of barrels per day					
United States	3,475	7,980	7,965	9,015		11,185
Canada	20	510	540	925	1,475	1,830
Venezuela	515	2,780	2,890	3,505	3,760	3,305
Other Latin America	310	670	905	1,190	1,535	1,670
Middle East/North Africa	325	3,585	5,480	10,245	18,755	21,715
Nigeria	—	—	20	275	1,085	1,820
Indonesia	150	320	415	480	855	1,080
Other non-Communist countries[a]	70	385	450	655	1,185	1,510
Subtotal	4,865	16,230	18,665	26,290	39,945	44,115
Communist countries[a]	720	2,265	3,355	5,395	7,845	8,865
TOTAL	5,585	18,495	22,020	31,685	47,790	52.980
	As per cent of world total					
United States	62.2	43.1	36.2	28.5	23.6	21.1
Canada	0.4	2.8	2.5	2.9	3.1	3.5
Venezuela	9.2	15.0	13.1	11.1	7.9	6.2
Other Latin America	5.6	3.6	4.1	3.8	3.2	3.2
Middle East/North Africa	5.8	19.4	24.9	32.3	39.2	41.0
Nigeria	—	—	0.1	0.9	2.3	3.4
Indonesia	2.7	1.7	1.9	1.5	1.8	2.0
Other non-Communist countries[a]	1.3	2.1	2.0	2.1	2.5	2.9
Subtotal	87.1	87.8	84.8	83.0	83.6	83.3
Communist countries[a]	12.9	12.2	15.2	17.0	16.4	16.7
TOTAL	100.0	100.0	100.0	100.0	100.0	100.0
	As per cent of non-Communist[a] total					
United States	71.4	49.2	42.7	34.3	28.3	25.4
Middle East/North Africa	6.7	22.1	29.4	39.0	47.0	49.2
All other	21.9	28.7	28.0	26.7	24.8	25.4
TOTAL	100.0	100.0	100.0	100.0	100.0	100.0

[a] 1938 tabulation follows present classification of Communist and non-Communist countries.
Source: British Petroleum Co., *Statistical Review of the World Oil Industry,* various issues.

In contrast, investment outside the United States rose 150 per cent in current dollars from 1963 to 1971. Between 1963 and 1973, it more than tripled for the rest of the world, while it less than doubled for the United States. The 1973 figures are particularly affected by heavy investment in foreign-flag tankers, but the basic differences remain. The shift away from United States sources is best illustrated in Table 12, condensed from a tabulation made by the Chase Manhattan Bank that was released in December, 1974. The stagnation, even in current dollars, in capital spending on explo-

Table 12

CAPITAL AND EXPLORATION EXPENDITURES BY THE WORLD PETROLEUM INDUSTRY, 1963 AND 1973 (IN MILLIONS OF CURRENT DOLLARS)

	1963				1973			
	United States	Middle East	Rest of world	WORLD	United States	Middle East	Rest of world	WORLD
Crude oil and natural gas	3,525[a]	150	1,290	4,965	7,290[a]	850	4,275	12,415
Natural gas liquids plants	125	0	80	205	150	5	355	510
Pipelines	375	20	230	625	450	130	650	1,230
Tankers	40	—[b]	900	940	100	—[b]	6,450	6,550
Refineries	325	55	1,355	1,735	1,050	300	3,515	4,865
Chemical plants	275	5	350	630	425	30	720	1,175
Marketing	650	25	1,060	1,735	850	25	1,605	2,480
Other	160	20	135	315	325	50	395	770
Total capital expenditures	5,475	275	5,400	11,150	10,640	1,390	17,965	29,995
Geological and geophysical expense and lease rentals	600[c]	30	420	1,050	850[c]	50	800	1,700
COMBINED	6,075	305	5,820	12,200	11,490	1,440	18,765	31,695

[a] Of which $575 represents lease acquisitions in 1963 and $3,600 in 1973. The 1972 figure was $2,475. Prior to that year, annual figures ranged between $500 and about $1,500. This row includes producing wells as well as dry holes.

[b] Non-U.S. tanker investments are geographically unallocable.

[c] Of which $440 represents geological and geophysical expenses in 1963 and $675 in 1973.

Note: Total annual U.S. and world-wide capital and exploration expenditures during 1963-73 were:

(in million current dollars)

	United States	World-wide
1963	6,075	12,200
1964	6,750	13,405
1965	6,985	14,355
1966	7,775	15,785
1967	8,265	16,765
1968	9,065	19,230
1969	8,900	19,755
1970	8,890	21,465
1971	7,965	23,195
1972	9,790	26,490
1973	11,490	31,695

Source: Chase Manhattan Bank, *Capital Investments of the World Petroleum Industry* (New York, December, 1974).

ration and development in the United States is evident. If one deducts the heavy outlays for offshore lease acquisitions in 1972 and 1973 and allows for inflation, the flat trend has continued unabated.

Apart from the shift of oil-industry investment away from the United States, a related point worth noting concerns the composition of the world-wide allocation of oil-industry investment and its trend from 1963 through 1971. One finds about a 15 per cent increase under the oil-and-gas-production heading, a doubling in pipeline investment and in marketing, a tripling in tanker investment, and a near-tripling in refinery and chemical-facility capital expenditures. As a consequence, by 1971, about 28 per cent of all investment was being made in crude oil and gas development compared to 40 per cent in 1963.

Hindsight and Foresight

The economic background against which the events beginning in mid-October, 1973, unfolded seems to go a long way toward explaining the oil-producers' success; one might say that the oil consumers had so boxed themselves in that no short-term escape routes were available, and that oil producers had become sufficiently aware of consumer "softness" to perceive correctly that the time was right to close the lid on the box. Having recognized and correctly analyzed the conditions preceding the embargo-cum-price-boost events, oil-consumer nations might be expected to cope better with similar episodes in the future. Reviewing the areas in which the major oil-importing countries have sought to build defenses or provide contingency remedies provides the elements for an answer. In our judgment, the answer varies among principal areas of concern.

1) *Managing transfers of funds from consuming to exporting nations:* Here early fears of fatal shocks to trade and the international monetary system were replaced in late 1974 by indications that the system might weather the storm. The exporters' capacity to absorb goods and services has been larger than anticipated; oil exports have declined, and the ability of the oil importers to reach a minimum level of agreement on handling balance-of-payments problems has mitigated the original panic.

2) *Reducing the rate of growth in energy consumption:* Efforts in this direction have been made in all consuming countries. Although in large part the result of economic recession, sharply higher prices, and supply stringency early in the year, the oil surpluses developing during 1974 also reflected conservationist policies. Ironically, their very success made the individual consumer doubt the validity of such policies. Few significant policies are firmly installed, and demand trends, after having retreated a bit, could reassert their previous upward strength, especially as the lagging economies of the consumer countries revive.

3) *Halting or reversing the shift to oil:* It has become obvious that adequate growth of alternate energy sources to lessen the role of oil is not a short-term proposition, especially as each of these sources is constrained by environmental issues or technological barriers. No significant early success is likely.

4) *Lessening dependence on imported and especially Middle Eastern and North*

African oil: No substantial replacements have come into view. North Sea oil will somewhat relieve reliance on imported oil for the United Kingdom and obviously for Norway, but it does not appear to be heading for a larger role. Periodic reports of new oil finds elsewhere in the world, even if verified by subsequent drilling, merely indicate relief in a distant future. In the meantime, dependence might increase (as it has done in the United States) rather than decline.

5) *Insurance against supply interruptions:* Due in part—perhaps even largely—to the recession, commercial crude-oil stocks have increased substantially in most importing countries, with storage space scarce early in 1975. Deliberate attempts to stockpile are more halting. In Western Europe, the OECD, as long ago as 1962, urged its European member countries to adopt sixty-day stockpiles. This program seemed adequate for a time. But in 1971, the OECD recommended early adoption of ninety-day stocks for major petroleum products. This target has slipped. Indeed, the newly created International Energy Agency in mid-1975 was still deliberating a time schedule for meeting the ninety-day objective. Japan's record in security stockpiling does not appear to have been notably more successful than Europe's.

6) *Mutual assistance:* Efforts have been most successful in the financial field (see 1 above). Work on a tentative mutual-aid pact among OECD countries to share scarce supplies in specified emergencies was begun in mid-1974 and continued into 1975. Achieving domestic acceptance and implementation for the contemplated system presents problems; but the effort has gone faster and farther than skeptics would have believed possible.

A summary judgment would be that the importing countries are now in a somewhat better position to cope, should a situation arise similar to that of the 1973-74 winter, but, at the same time, so are the exporters. They have as a group, in the meantime, accumulated very large financial resources that could see them through a long period of confrontation. What looms ahead, therefore, are probably the most important unanswered questions: Will the cohesiveness of OPEC be sufficient to stand up during a prolonged period of excess oil capacity and will that excess capacity be slowly eliminated with a resumption of upward demand in the importing countries, and a reduction or cessation of investment in productive capacity? If what we have described in this paper as explanatory economic variables emerge as a product of hindsight, the importing countries might learn enough from the experience to benefit from foresight. A look at the post-1973 trends suggests that at best the shift is proceeding slowly.

REFERENCES

[1] A portion of what follows in this section and the three subsequent sections has been adapted, in revised form, from J. Darmstadter and S. H. Schurr, "The World Energy Outlook to the Mid-1980s: The Effect of an Alternative Supply Path in the United States," *Philosophical Transactions*, 276 (Series A) (London, Royal Society, 1974).

[2] In all energy statistics, only energy moving through commercial channels and statistically recorded is accounted for. Undoubtedly, much energy consumption in Africa and Asia escapes the record and thus makes per capita consumption in those areas appear lower than it is.

[3] Real GNP growth rates shown here and in the following paragraph are taken from U.S. Agency for

International Development, *Gross National Product—Growth Rates and Trend Data*, Release RC-W-138, May 1, 1974.

[4]Japan, Bureau of Statistics, Office of the Prime Minister, *Statistical Yearbook 1973/1974*.

[5]"Indirect" shipments refer to Arab crude oil processed in third-country refineries (e.g., in Italy or the Netherlands Antilles) prior to shipment as petroleum products to the United States.

[6]These estimates are contained in a paper by the Petroleum Industry Research Foundation, "U.S. Oil Imports and Import Dependency," New York, December 19, 1974. Similar estimates were presented by G. M. Bennsky in testimony before the Subcommittee on the Near East of the House Foreign Affairs Committee, November 29, 1973.

[7]See Joel Darmstadter, "Energy Consumption: Trends and Patterns," in Sam H. Schurr, ed., *Energy, Economic Growth, and the Environment* (Baltimore, 1972), especially pp. 168-71. While, since 1970, the U.S. energy-GNP ratio seems once again to have reverted to its long-term downward trend, it does seem, from the rather erratic energy-GNP elasticities tabulated for a number of countries, that this familiar (and perhaps over-used) relationship may have limited utility as a basis for forecasting energy consumption, particularly at a time when volatile price trends, with or without the reinforcement of purposeful conservation policies, may quite markedly alter historical relationships.

[8]Hans H. Landsberg, "Learning From the Past: RFF's 1960-1970 Energy Projections," in Milton F. Searl, ed., *Energy Modeling: Art, Science, Practice*, RFF Working Paper EN-1 (Baltimore, 1973), pp. 416-36.

[9]U.S. Cabinet Task Force on Oil Import Control, *The Oil Import Question* (Washington, D.C., February, 1970), p. 125.

[10]We might note in passing that, in what was probably a misguided legacy of the "Limits to Growth" debate, skyrocketing oil prices and shortages were deemed by a few as symptomatic of approaching resource exhaustion.

[11]European Economic Community Eurostat, *Energy Statistics Yearbook, 1969-1972* (Brussels, 1973)

[12]Adapted from "Energy in Crisis," *Resources*, No. 45 (Washington, D.C., 1974), pp. 5-6.

[13]Derived from data appearing in the American Petroleum Institute publications, *Petroleum Facts and Figures* (Washington, D.C., 1971), and *Reserves of Crude Oil, Natural Gas Liquids, and Natural Gas in the United States and Canada and United States Productive Capacity as of December 31, 1973, 28* (June, 1974).

EDITH PENROSE

The Development of Crisis

THE CHRONOLOGY OF EVENTS in the international oil industry from January, 1970, to the present shows clearly the progressive interaction of three separate historical developments. Each of these developments had its origin in circumstances unrelated to the others and long preceding the period under discussion. Each developed independently, but each was approaching its own "crisis," or "turning point," at the beginning of the decade. As it moved toward its own crisis, each interacted with and reinforced the others and produced a concatenated series of events that culminated in what we loosely call the "oil crisis."

The three historical developments were, first, the rising bargaining power of the governments of the oil-exporting countries of the Middle East vis-à-vis the international companies that had discovered, developed, and long controlled Middle Eastern oil; secondly, the growing dependence of the United States on Middle Eastern, and specifically Arab, oil; and thirdly, the establishment and expansion of Israel in Palestine against the bitter opposition of the Arab countries, but with the strong support of the United States. A fourth development, which intensified the effects of the others, was the increasing confusion in international monetary relationships combined with the rising rates of inflation in the world's industrial economies.

Many of the other papers in this volume deal with one or another aspect of these developments. In this paper, I want to analyze their interaction as the circumstances leading to the crisis emerged, slowly at first and then accelerating in the winter of 1973.

Although the establishment of the Organization of Petroleum Exporting Countries in 1960 marked the beginning of increasingly effective cooperation among the crude-oil exporting countries, it was the reopening of price negotiations with the new Libyan government in 1970 that set in motion a chain of events, the direct repercussions of which were rapidly to overturn most of the established arrangements between companies and governments in the Middle East. The Teheran and the Tripoli price negotiations in 1971 stemmed directly from the Libyan agreements in 1970. The success of the former from the point of view of the exporting countries, together with growing political pressures, paved the way for participation by governments in the equity of the companies and to eventual government control.

Immediate Background

In the year or two preceding the reopening of the Libyan negotiations, many of the major oil-exporting countries were becoming increasingly militant in their determination to achieve their objectives. Their actions reflected the OPEC "Declaratory Statement of Petroleum Policy in Member Countries," adopted in June, 1968, at the XVI Conference. This had not been taken very seriously by the oil companies at the time, but the objectives that the member countries were subsequently to pursue had been plainly set out.

Stating that "changing circumstances" justified revision of existing as well as future contracts and concession arrangements, the member countries asserted their right to participation in ownership. They agreed that posted (tax reference) prices should be determined by the government and should ensure no deterioration in the price of oil relative to the prices of imported industrial products. Even more important, however, they resolved that their foreign oil operators should not have the right to "excessively high net earnings after taxes"; excess profits, as determined by the government, "shall be paid by the operator to the Government" (Resolution XVI.90).

By the end of 1968, Algeria was beginning to show dissatisfaction with the 1965 Franco-Algerian agreement, the rate of investment of the French companies in the country, and the slow rise in discovered reserves. The entry of the Getty Petroleum Company in October, an independent American company willing to offer better terms than either the majors or the French companies, intensified Algerian dissatisfaction with existing arrangements. During 1969, Algerian militancy grew, and increasing pressure was exerted on all foreign companies in the country.[1]

In May, 1969, Iran succeeded in forcing the companies to increase payments substantially by mounting a large-scale international publicity campaign and threatening to take over a 50 per cent equity participation in the consortium oil fields if its demands were not met. The Shah insisted that the consortium must meet the deficit in the financial requirements of the country's development plan. To increase output sufficiently to provide the required revenues would have created difficulties for the companies in view of their interests in the Arab world; and the Arab governments had warned them against increasing output from Iran at the Arabs' expense. Other means of providing funds to Iran were worked out that avoided a rise in posted prices and an extraordinary increase in output.

In June, 1969, the Saudi Arabian oil minister, Shaikh Zaki Yamani, opposed to the increasingly strident public demands for nationalization of the oil companies in the Middle East, which were encouraged by the more radical regimes in the Arab world, countered with a strong statement in favor of equity participation or "partnership."[2] A month later, at the Vienna meeting of OPEC, the question of participation dominated the agenda, and a "high powered" technical committee was set up. At the Doha meeting in December, 1969, OPEC members offered strong support to both Algeria and Libya, which were in conflict with the companies, and set in motion further work on participation. The pace of events was clearly accelerating.

Throughout the Middle East, the companies must have known that they were fighting only a rear-guard action, and they must have deliberately decided that it paid them to do so. The longer-term necessity for changes in the control of the industry was evident to most, even then. In view of this, negotiations with individual countries would only have been a form of playing for time and testing temporary breaking points; the companies would eventually have to give in unless their own home governments decided to intervene.

Libya, 1970

The Libyan negotiations opened in January, 1970, and by September the companies had conceded the Libyan government's demands, which involved substantial increases in both posted prices and tax rates. Government pressures on the companies were extremely severe and included threats of nationalization and enforced reductions in the output of almost all companies.[3] Pressures were concentrated on Esso and on Occidental Petroleum, one of the leading independent American companies, which relied on Libyan oil for its European refineries. Esso's parent, Exxon, is the largest of the major companies, and had extensive oil resources in Iraq, Iran, and Saudi Arabia, as well as in Libya. The company very much wanted to prevent the acceptance of Libyan demands, which would have extensive repercussions on the terms of its Middle Eastern concessions.

Libya was strongly supported by OPEC and especially by two other Mediterranean exporters—Algeria and Iraq[4]—both "revolutionary" regimes and both engaged in outright confrontation with their concessionaire companies. Algeria in particular maintained extremely close contact with the Libyan government, advising it during the negotiations and coordinating policies with it. In June, 1970, Algeria took over the operations of a number of foreign oil companies, including Shell and Phillips, which had been under "sequestration" since 1967, when they had refused to sign new agreements more advantageous to the government. In July, it unilaterally raised the tax prices applicable to the French oil companies by 77 cents a barrel and insisted that the companies reinvest $1.80 a barrel in the country, an action that was called at the time a "decisive turning point in the dramatic process of change which is evidently in prospect for the structure of the international oil industry."[5] The companies resisted strongly, but Algeria's actions could only serve to strengthen the will of Libya.

The Libyan government, thus supported and willing to risk extreme action, forced a settlement on the companies that irrevocably broke the pattern of the past. The benefits of the settlement would have to be generalized to other OPEC countries. The primary problem facing the major companies was to work out appropriate settlements with the countries delivering oil to the Persian Gulf and, at the same time, to prevent the development of a situation in which Libya and the Mediterranean exporters, on the one hand, and the Gulf exporters, on the other, tried to outdo one another, "leapfrogging" over each other with continuously escalating demands.

The Libyan government's success can be attributed to a combination of factors. First of all, the negotiations took place in a period when the demand for oil was rising rapidly and oil in the Mediterranean was in short supply, due to the closure of the Suez Canal and other temporary interruptions. Prices and freight rates were already climbing steeply, and the reductions in output that had been enforced on the companies had an exaggerated effect. Europe feared a bleak winter. Secondly, Libya had the vociferous support of other OPEC countries, and the companies could not know how widespread and damaging the reaction might be if negotiations broke down and Libya took unilateral action.

Thirdly, the companies were divided—the majors being unwilling to go to great lengths to protect those independents who had no other source of oil for their European operations. Moreover, the Libyan government, although ill-prepared to carry through the actions it threatened if its hand had been called, could nevertheless have greatly damaged the existing companies. In any case, some other new companies might also have stepped in.[6] At the same time, the failure of the governments of importing countries, and particularly that of the United States, to react strongly to the Libyan threats implied that the risks to the Libyans of serious retaliation were not great.

In spite of its strong position, however, the Libyan government did not simply legislate to achieve its ends, it argued its case.[7] Although some of the arguments were plainly disingenuous and were backed up by severe economic pressures, it was clear that the government did not want to risk forcing the companies to leave the country, and it was equally clear that the companies did not want to go. Many observers did not think that the Libyan demands were unreasonable, even if they deplored some of the tactics used to achieve them.

It is, I think, significant that extended negotiation, argument, and economic pressures were preferred by the government to unilateral, non-reversible legislative action in order to force compliance on the companies, for this attitude indicates their uncertainty. It continued to characterize the negotiations between most of the Middle Eastern oil-exporting governments and the companies right up to the outbreak of the October, 1973, war. The war itself, however, marked a decisive turning point.

The Price Agreements

The prices settled on in the Libyan negotiations of 1970, though appearing excessively high to some, were not in fact really disruptive. The oil companies and the governments of the consuming countries had long known that there would have to be some rise in the cost of oil. It was clear that similar increases would have to be given to other oil-exporting countries, but even more important for the companies was the necessity of making arrangements that would stabilize the situation and prevent excessive demands for increased prices by producing countries in competition with each other. The OPEC countries, meeting in Caracas in December, 1970, agreed on the minimum terms they would accept from the companies, and they laid down a rigid timetable and procedure for obtaining them, promising "concerted and simultaneous action by all Member Countries" to force compliance, should the negotiations prove ineffective (Resolution XXI.120).

In order to protect themselves against escalating demands, the companies (those from the United States having obtained permission from the American government to negotiate collectively) tried unsuccessfully to persuade OPEC to negotiate as a group, but OPEC preferred to negotiate separately for oil shipped from the Persian Gulf and that from the Mediterranean terminals.[8] This put the companies in a vulnerable position: they had to recognize that Libyan oil could command a higher price than oil from the Gulf because of its quality and its proximity to Europe. Moreover, this price differential, justified by these inherent advantages, was further heightened by temporary advantages arising from the closure of the Suez Canal and the currently high freight rates. If Libya, in separate negotiations, were able to extract a price that exceeded the Gulf prices beyond the amount justified by the recognized differentials, then any settlement with the Gulf countries would clearly be unstable.

Fundamentally Libya wanted to establish Mediterranean oil as the base to which oil prices everywhere would have to be related. But the Gulf, particularly Saudi Arabia, was the source from which additional supplies of oil would largely come. Hence the underlying problem in the negotiations (which opened in Teheran in January, 1971) with the states producing oil in the Gulf was to resolve this conflict in a way that would permit some continuing stability of oil prices. The negotiations were stormy, and agreement was reached only after a special OPEC conference was called and had passed a resolution[9] providing for the terms agreed on in Caracas to be incorporated in the legislation of the countries concerned, and for a concerted embargo by all nine members of companies that refused to accept the legislation.

The Caracas resolution provided some comfort to the companies, however, in that it did not ensure *carte blanche* support from OPEC for Libyan and Algerian demands outside the terms of the resolution passed. Libya and Algeria were much more extreme in their approach than were Iran, Kuwait, and Saudi Arabia, and they felt that the price demands of the latter were "much too low."[10] They were also much opposed to accepting the companies' demand for a five-year agreement. Nevertheless, after only a month of negotiation, a five-year agreement was concluded with assurances from the Gulf governments against "leapfrogging," should better terms be obtainable elsewhere. The posted price on Arabian light crude rose from $1.80 to $2.18 a barrel and the tax rate to 55 per cent. But, so far as the Gulf countries were concerned, a period of peace seemed within reach, provided that acceptable arrangements could be made with the Mediterranean countries.[11]

Negotiations opened in Tripoli in February, 1971, with Libya acting as negotiator for the Mediterranean countries. Agreement was reached by April with great difficulty. Libya got most of what it wanted, but the companies also obtained the five-year commitment they wanted. The prices agreed upon were higher than was justified by the underlying factors mentioned above and were reported to have aroused some resentment in the Gulf states, but they were the result of compromise,[12] so there were grounds for hope that both agreements would hold.

The Teheran and Tripoli agreements were designed not only to raise the revenues per barrel for the oil-exporting countries, but also to stabilize revenues in real terms. Thus, they provided for a 2½ per cent annual escalation in posted prices to offset the effects of inflation in the prices of imported commodities. The Gulf countries' share of the proceeds from the sale of their oil implicit in the Teheran agreement was estimated to be around 20 per cent. As we shall see, however, the assumptions about the course of inflation, market prices (as contrasted to tax [or posted] prices), and monetary parities (particularly the value of the United States dollar), which were held by both the companies and the governments and upon which the agreements were constructed, proved to be unjustified. As events continued to reveal the weaknesses of these assumptions, the price agreements themselves became progressively more inappropriate.

Devaluation of the Dollar

As early as the summer of 1971, the continued depreciation of the United States dollar in relation to other currencies began to cause trouble, for the posted price of oil was stated in dollars and the Teheran agreement had not provided for a fall in its value. When, in August, 1971, the United States announced the suspension of the convertibility of the dollar into gold, the secretary general of OPEC warned that the "Teheran and Tripoli price agreements did not deal with the question of the parity of money, and, therefore, should the US dollar be devalued, the gains achieved by the Teheran, Tripoli and related agreements would be substantially eroded." OPEC spokesmen referred to Resolution XXI.122 adopted at Caracas, which expressly provided for adjustments of posted or tax reference prices, if changes in the parity of monies adversely affected the purchasing power of oil revenues.[13]

By December, 1971, OPEC was expressing great impatience with the lack of progress on the monetary issue, which was of course complicated by the disorderly international monetary conditions brought about by floating rates of exchange. Because of this disorder, the companies had no international monetary frame of reference within which they felt able to negotiate with the governments. The Smithsonian Agreement signed by the ten major industrial countries on December 18, 1971, in Washington, supplied the necessary framework. It provided for a formal devaluation of the dollar against gold and for a revaluation of leading currencies against the dollar. Subsequently, in January, 1972, an agreement was reached in Geneva with the OPEC countries for an increase in posted prices of 8.59 per cent to take account of these monetary adjustments. Although some companies insisted that the Geneva agreement was a violation of the terms of the Teheran agreement, it was treated as a supplement to it and was to run for the same five-year period. Posted prices were to change quarterly in accordance with an agreed formula that produced an index of exchange-rate movements. The price of Arabian light rose from $2.285 to $2.479 a barrel.

Participation

In the meantime, OPEC's long-standing demand for equity participation in the major concessions began to gain momentum. The major companies, having access to "owned" or "cost" crude oil, which gave them oil at a cost far below the prices paid by companies that had to buy their oil, had a corresponding competitive advantage. The American companies also obtained tax advantages from producing their own crude oil. Continued access to this low-cost crude oil was one of their primary objectives in negotiations with the governments, and it was one of the chief reasons why the companies clung so tenaciously to their position in the Middle East. On July 1, 1971, Shaikh Zaki Yamani, the Saudi Arabian oil minister, stated that participation in the ownership of concession-holding companies "has become a national demand, especially after the Algerian action nationalizing 51 per cent of the assets of the French companies operating in Algeria." He asserted that, due to the coming energy shortage, a sellers' market had arrived.[14]

In Geneva, directly after the conclusion of the monetary agreement, preliminary discussions on the participation issue were held, with Yamani delegated as leader of the OPEC countries. By March, 1972, Aramco had accepted 20 per cent participation in principle, and by April serious negotiations over an agreement had opened. The crucial issues for the companies were the terms of compensation, guaranteed access to oil on preferential terms (that is, oil on terms better than those at which other companies that had not had concessions could buy it in the market), and the rate at which the governments' share would rise from the initial 20 per cent to the 51 per cent they ultimately had in view.[15]

Additional complications were introduced in the summer, when Iraq finally nationalized the Iraq Petroleum Company in the north of Iraq, and the Shah announced that Iran was not a party to the OPEC bid for participation—Iran had already nationalized its oil in 1951 and now only sought arrangements with the consortium more suited to its special position.[16] Nor was Libya interested, since in any event it intended to take at least 50 per cent immediate participation on much less favorable terms for the companies.

On October 5, 1971, it was announced by Shaikh Yamani, after meetings in New York, that a general agreement had been reached, subject to its acceptance by other governments. This agreement would provide a framework for each government to negotiate the details of its own participation, and in particular the price at which oil that the government became entitled to but could not sell would be bought back by the companies. In the subsequent negotiations over the individual agreements the level of "buy-back prices" caused considerable difficulty. Because of continually rising market prices, all governments had increased the prices for the oil sold back to the companies. By the end of December, only Saudi Arabia and Abu Dhabi had signed agreements. Qatar had not yet done so, because it was not satisfied with the companies' evaluation of its crude oil in comparison with that of Abu Dhabi. Kuwait's reluctance was largely due to internal political disagreements. Iraq, deeply involved with the consequences of the nationalization of the Iraq Petroleum Company, was still reserving its position, especially with respect to the methods of calculating compensation.

Market Reactions

As was to be expected, the oil market was not unaffected by the negotiations, which were rapidly changing the organization of the industry and the control over crude-oil supplies. Although output continued to rise, growing demand from the United States and the effort of each of the individual companies, majors as well as independents, to secure its own supplies of oil led to steadily increasing market prices. Crude-oil imports into the United States rose by 26 per cent in 1971, 17.5 per cent of it coming from the Middle East, as compared with 10.3 per cent in 1970. Every forecast of American imports presented during 1972 revised upward all previous forecasts, and imports rose by 32 per cent. *Petroleum Intelligence Weekly (PIW)* reported in November that contracts for major crude purchases were being signed "without regard for economics"; overseas crude was a "necessity" for American inland refiners.[17]

At the same time, the major companies became increasingly cautious about outside sales as uncertainty increased regarding the effects of participation on the amount of oil they might have for their own use and on its cost. *PIW*, reporting on the rising prices, noted: "Attesting to the current crunch are indications that a traditionally crude-long major international oil firm is ready to post a sold-out sign on its Persian Gulf crudes for the rest of 1972, and another is active in the market now as a net buyer."[18]

To conserve its resources, Kuwait imposed a ceiling on output in April, 1972, and British Petroleum, which usually lifted less than its entitlement of oil, began lifting its full share, leaving its partner, Gulf, normally a large crude-oil seller, with less for its outside sales.[19] Again, in December, it was reported that prices were rising in advance of the expected cost increases and ". . . many suppliers have recently been reluctant to sell crude, due either to their own internal needs or to awaiting a clear determination of participation costs."[20]

During the first nine months of 1973, the oil-importing countries saw themselves in the midst of crisis, as both they and the exporting countries struggled to adjust to the new situation created by the price and participation agreements, by the rapidly changing market conditions, as each of the oil companies attempted to secure its own supplies, and by the continued monetary instability and inflation. Nevertheless, there were a number of hopeful developments. Saudi Arabia, Abu Dhabi, and Qatar signed participation agreements, although ratification by the Kuwaiti Assembly met with difficulties. Iran came to an agreement with the companies of the consortium. An overall settlement was reached between Iraq and the Iraq Petroleum Company group that ended more than a decade of confrontation and disruption.

On the other hand, the devaluation of the United States dollar at the beginning of 1973 threw the monetary problem once more into confusion, while rising United States demand for imported oil continued to push up market prices. In the background, but moving to the foreground within the year, were the political implications of the increasing dependence of the United States on the Arab oil-exporting countries.

Second Devaluation of the Dollar

The devaluation of the dollar by 10 per cent in February, 1973, gave rise to a sharp controversy. There is no need here to go into the complexities of the matter, but briefly the problem was that the formula in use since the "Geneva I" agreement did not completely offset the effects of the second devaluation of the dollar on the revenues of the exporting countries. These countries now insisted that the spirit of that agreement required that they should be "compensated" in full for the new changes in monetary parities. Shaihk Yamani argued that OPEC was not trying to obtain an increase in prices, in violation of the earlier agreement, but simply a "correction" of the evident unfairness of the way in which the old formula worked in the light of the dollar devaluation. The companies were willing to make some modifications, but they strongly re-sisted the specific OPEC proposals, which would have produced a large increase in posted prices.

In the end, OPEC had its way, but not before Algeria had made an effort to get the Teheran and Tripoli price agreements repudiated, and Libya and Iraq had urged the governments unilaterally to announce higher posted prices if the companies continued to resist OPEC demands. The threat to call an extraordinary conference of OPEC was used several times to enforce compliance. But Iran and Saudi Arabia would not sanction a unilateral breach of their earlier agreements, and no drastic steps were taken by any of the OPEC countries. In the final, or "Geneva II," agreement signed on June 2, posted prices in the Gulf were "adjusted" to a level nearly 12 per cent higher than they had been in January; government "take" rose by slightly over 12 per cent and tax-paid cost by 11.4 per cent.[21] The posted price of Arabian light had now reached $2.90 a barrel.

Supply, Demand, and Prices

A more serious threat to existing agreements arose from the impact of increasing market prices, largely as a result of the scramble for supplies by American companies. Early in 1972, the Bureau of Mines reported that United States crude-oil output was expected to fall for the third year running by over 2 per cent, and that imports would rise rapidly to meet the consequent abnormally high demand. Some predicted that im-ports would reach 35 per cent of the total United States supply, but that even this amount would be insufficient to prevent shortages of gasoline in the summer and of home-heating fuels in the winter. At the same time, however, the Middle Eastern out-put rose rapidly to a level nearly 38 per cent higher in the first six months of 1973 than it had been in the same period in 1972.

Between the spring of 1970 and the summer of 1973, market prices for crude oil more than doubled measured in dollars, and nearly doubled in terms of other cur-rencies, but the fiscal income of the host governments accounted for less than half of the increase. The rest can be attributed to a "tighter fit between import demand and export capacity,"[22] in other words, to "market forces." There is a clear record in the trade press of the short-term pressures of demand on market supply and of the increasing prices offered by buyers—mostly American, but also Japanese and European, indepen-dents.

In addition, the large, integrated major companies became increasingly concerned about the extent to which the participation of host governments in their major con-cessions would affect their own supplies. In particular, they were uncertain about the terms on which they would in fact have access to government-owned crude oil; uncer-tainty about prices naturally induces some caution with respect to future sales com-mitments.[23] As a consequence, the integrated companies began reducing their outside sales of crude oil, forcing new buyers and new demands into competition in an even tighter market. The situation was summarized by Dillard Spriggs, a New York in-vestment consultant:

1973 was the time when the push by independents for participation crude was on, when everyone was seeking assured sources of supply, when the U.S. Government freed American refiners to go abroad for foreign crude, and when Libya was threatening to nationalize the American companies.[24]

President Nixon's energy policy, announced in a message to Congress in April, 1973, reinforced the continuing rise in international oil prices which had become one of the central considerations in the prevailing atmosphere of crisis, because it freed American importers from the mandatory quota system without imposing any restrictions on the freedom of either established or new refiners and marketers to bid against each other in a tight international market. The supply situation encouraged marketers to enter into refining, but, in order to get oil, they had to bid for the participation crude available from the governments of the exporting countries. In other words, although United States import demand was instrumental in creating a continuing crisis, the United States did nothing either to control it or to manage the way it hit the market.[25] Instead, the government concentrated its diplomacy on attempting to organize a cooperative program with other consuming countries, thus, in effect, shifting the problem to the political arena. In that arena, however, the problem could not be dissociated from the broader question of American foreign policy in the Middle East.

The rapid inflation of prices and radical changes in market conditions created a situation between the companies and their host governments that became increasingly unfavorable to the latter, a result which had not been anticipated by the companies any more than it had been by the governments. This inadvertent effect was due to the way in which profits and taxes in the oil-exporting countries were calculated. During the nineteen-sixties, the market prices of crude oil and oil products had been falling, but OPEC had prevented further cuts in posted prices; in that period, therefore, the government share in realized prices rose. This mechanism for increasing the government share was basic to the Teheran and Tripoli agreements; the *rate* of the "profit" tax rose to 55 per cent, and the price used to calculate the profits to which the new rate was applied was the higher posted price rather than the lower market one. Clearly, however, if market prices came to exceed posted prices the government share in "realized profits" would fall. Thus, the Teheran agreement protected the governments against *falling* market prices, but not against *rising* market prices, the expectation shared by both companies and governments being that posted prices would remain above market prices. It could only have been on this basis that the governments (or even, presumably, the companies) expected the agreement, in its own words, to constitute a "fair appropriate and final settlement."[26]

Changed market conditions similarly affected the terms in the participation agreements that applied to oil sold back to the companies by the government. It was assumed that the governments would not quickly find markets for their share of the oil at prices higher than the agreed buy-back prices, and certainly not at or above posted prices. Market developments in the early nineteen-seventies were seriously undermining both expectations. As early as May, 1972, Saudi Arabia sold all of its participation crude available for 1973 and part of that available for 1974 and 1975 (admittedly very small amounts) at "record prices."[27] In June, Qatar was reported to have initialed an agreement with an American independent company for all of its participation crude for the period 1973-75 at a price in excess of the posted price.[28] By August, market prices in the Mediterranean generally had gone to "well above" posted prices,[29] and *PIW* reported that the demand from American refineries for low-sulfur fuel was pushing prices in Latin America and in Nigeria to $5 a barrel and over.[30] In September, Algeria, which was not a party to any of the agreements, raised its prices to $5 a barrel and imposed an exploration tax on the companies of 35 cents a barrel.

Not only were the price expectations of the preceding year thus being everywhere exceeded, and market prices beginning to rise above posted prices, but the governments were having no difficulty in selling their share of the oil. If such conditions continued, there would be no necessity for a "phase-in" period, during which the governments could seek market connections, and no need, therefore, for a gradual increase in the governments' equity entitlement to a greater share of the oil.

It is no wonder, then, that the more "radical" members of OPEC began to press harder. At the XXIV Conference in Vienna on June 17 and 18, 1973, Iraq led a move to scrap the Teheran and Tripoli agreements and to adopt instead the Venezuelan policy under which governments would unilaterally set their own prices. The attempt failed, but all members felt that at least the annual escalation of prices by 2.5 per cent, which had been provided for in the agreements, should be reconsidered in the light of the 6 to 7 per cent rise in the prices of goods imported by the oil-exporting countries. At the same time, the members were reluctant to advocate a unilateral abrogation of their agreements.[31]

As crude-oil costs rose, the oil companies raised prices in product markets to an extent greater than would have been justified by the rise in crude costs alone. The exporting countries interpreted this as a strategy for enlarging company profits over and above the increases already resulting from the higher prices of crude oil. Since the governments' own opportunities for profit were restricted by their price agreements,[32] this was an additional reason to subject these agreements to a reexamination.[33] The preferential access of the major companies to crude oil gave them an increasing competitive advantage as market prices rose in relation to costs, including taxes paid to the exporting governments. *PIW* calculated in August that, if a major company met all of its refinery needs from its own sources, the cost advantage on Arabian light crude from Saudi Arabia delivered to Rotterdam was some 93 cents a barrel over independent companies, with a transport cost advantage ranging from 30 to 62 cents a barrel.[34]

By September, 1973, all the OPEC countries were prepared formally to request a revision of the price agreements, for they could no longer accept the steady decline in their share of the profits from oil. It was calculated that the "national profit split on realized prices between governments and companies may have changed from 80-20 in the governments' favor at the time of the Teheran agreement to around 64-36 in the governments' favor now."[35] Negotiations with the companies to revise the Teheran agreement opened on October 8, in Vienna. On October 12, in spite of the outbreak of war on the sixth, the companies asked for a two-week adjournment, since they felt that the increases in price being asked (reportedly around two to three dollars a barrel) were larger than they could agree to without further consultations.

Four days later, on October 16—the day before output restrictions were decided upon by the Arab countries in connection with the war—the ministerial committee representing the six Gulf states of OPEC, including Iran, met to consider the status of the negotiations. The ministers decided to negotiate no further: the governments would post their own prices, thus completing once and for all the transfer of control over tax prices. New posted prices were announced, representing a 70 per cent increase for Gulf crudes (from $3.01 to $5.12 a barrel for Arabian light) and raising the government's "take" from $1.77 to $3.05 a barrel.

The general policy adopted was to maintain posted prices at 40 per cent above the applicable market prices—that is, the prices at which governments sold crude to third parties. On this basis the new posted prices were represented as an increase of 17 per cent over prices realized in recent sales. In the future, the posted price would change as the market price changed. The intention was "to re-establish the same relationship between posted and realized prices as existed in 1971 before the Teheran agreement and, by extension, a roughly similar profit-sharing ratio as well."[36] Libya raised its tax-reference price by 94 per cent.

Even at this point, the Arab-Israeli conflict, which by now had erupted into full-scale war, had not influenced the posted price of oil. Although the Arab countries had long warned that United States policy in the Middle East could have serious effects on supplies of oil, and these warnings had been intensified during the spring and summer of 1973, as other papers in this volume show, Middle Eastern output continued to increase rapidly. The October, 1973, increases in posted prices were not related to the war, but to the fact that the assumptions underlying the Teheran agreement had proved unjustified. The exporting countries therefore felt that the tax prices agreed upon in Teheran required adjustment to the new market and monetary conditions. These conditions were not, in their view, of their own making, since they had not cut back supplies, although they had increasingly emphasized the possible effects on supply of United States policy toward Israel.

Prices and Output Restrictions

The output and destination restrictions imposed by the Arabs, on October 27, 1973, to support their war effort further narrowed the market for oil, and, in particular, for oil free of destination restrictions. Hence, the prices bid in the free market by independent companies scrambling to obtain supplies rose even further, and decisions had to be made regarding posted prices. On November 17, the oil companies, at their own request, met with the six Gulf states concerned with the Teheran agreement to discuss future arrangements for setting posted prices. The discussions made little headway, for the companies were strongly opposed to the pricing decisions taken by the ministers on October 16. But they presented no alternative proposals acceptable to the ministers, who argued that future price should be "market oriented." Given the supply situation as it had developed in 1972-73, together with the restrictions imposed to support Arab political objectives, this position was clearly unacceptable to the companies. There was, however, considerable difference of opinion among the exporting countries. Some wanted to "play the market for all it was worth right now," and some preferred a slower and more moderate approach.[37] This difference was to come to the surface in December.

The National Iranian Oil Company (NIOC) had called for bids by December 10 for the 470,000 barrels a day of crude oil that it intended to market in the first six months of 1974. This was the first major government sale of crude oil since the oil restrictions had been put into effect. NIOC departed from its previous policy of simply announcing the price at which it would sell; for the first time, it went out deliberately to test the market. Bids, mostly again from United States independents (many to their subsequent regret) but including at least one from a European company, reached from $16 to over $17 a barrel.

These prices, together with reports of bids for Nigerian and Libyan crude of over $20 a barrel, led the OPEC ministers—particularly the Iranian minister—to demand, at their meeting in Teheran on December 22-23, that the posted price to become effective on January 1, 1974, be in the region of the bids Iran had received. (It will be recalled that OPEC had earlier decided that market prices, as determined by direct sales, should be the chief guide for the setting of posted prices.) Saudi Arabia resisted the increase on the ground that the high bids reflected the embargo and destination restrictions on Arab oil and thus did not reasonably represent appropriate market prices. On the other hand, the Iranians had long maintained that the price for oil should equal the cost of the nearest substitutes; it could be argued that buyers would be willing to pay any price that left the cost of oil below the cost of substitutes. A compromise was reached on a posted price of $11.65 a barrel for the "marker crude" (Arabian light 34° API) yielding a government "take" of $7 a barrel.

Thus, the posted price prevailing before the outbreak of war was nearly quadrupled. The OPEC communiqué announcing this shock to the world, with its enormous implications for both monetary inflation and increased industrial costs, ended with a sentence of alarming simplicity: "Considering that the government take of $7 a barrel is moderate, the Ministers hope that consuming countries will refrain from further increases of their export prices."

The oil minister of Saudi Arabia subsequently announced that in the opinion of his government a "lower price would have been more equitable and reasonable," but the Shah of Iran, in a press conference in Teheran, insisted that a seven-dollar-a-barrel government take represented the minimum cost of developing alternative energy, notably the extraction of oil from shale or the liquefaction or gasification of coal. He argued that such costs were the only proper basis for pricing oil. Export prices all over the world, from Indonesia to Bolivia, were immediately raised in line with OPEC's decision.

Post-War Problems

The "oil crisis" proper was now over, for a "crisis" cannot by definition be a long-continuing state of affairs. The "turning point" had been reached in October-December, 1973. During 1974 and 1975, the problems facing importing countries were essentially of two kinds: the financial and economic implications for their individual balances of payments and economic policy of the accumulating financial assets in the hands of oil-exporting countries, on the one hand, and the implications for their economies of the higher cost of energy, on the other. The amount of crude oil available rose quickly as the Arab countries (with the exception of Libya) lifted the embargo against the United States in March, 1974, and later removed other restrictions as well. Posted prices were not further increased, but, during 1974, average revenues per barrel of oil rose enough to offset, and probably more than offset, the effects of inflation on the prices of the imports of the oil-producing countries.[38]

At the same time, signs of excess supply appeared as the underlying surplus of productive capacity began to make itself felt and the demand for oil declined, primarily as a reaction to recession and high prices, but also in response to the policies adopted by governments in the importing countries. The attempts of the latter under strong pressure from the United States to organize a response to the new situation had not, by the summer of 1975, met with notable success.

Nevertheless, the decline in demand forced some exporting countries to reduce output; they found themselves unable to sell as much as they had planned to produce at the prices they were prepared to accept. The output reductions fell most heavily on Kuwait and Libya, but, by January, 1975, the unrealistic differentials that Libya and Abu Dhabi were attempting to sustain for the price of their oil had reduced sales from those countries to unacceptable levels, and both reduced their prices. Libya took similar action in June. Whether these reductions could be looked upon as forerunners of a general softening of exporters' prices was not yet clear, but plainly problems were arising for the OPEC countries for which solutions might not be easy.

In December, 1974, OPEC made an attempt to bring some order into the pricing system and to offer a measure of stability to the consuming countries. The different price arrangements for crude of different categories—bridging crude, phase-in crude, equity crude, marker crude, etc.—were considered to be excessively complicated and confusing. Proposals to establish a unified price, although allowing a discount in some form for the major companies, were discussed, along with proposals to relate the price of crude to some import-price index. At the XLII Conference of OPEC, on December 12, a new pricing system was adopted. Prices were based upon an average government take from the operating oil companies of $10.12 a barrel for the "marker crude." This price, effective January 1, 1975, was to be maintained for nine months. In the face of inflation, this implied a fall in real prices. The major companies would still be able to obtain crude at a price somewhat lower than the price at which market sales would be made by the government companies. The margin, however, would not be high enough to permit the companies to compete with government sales in third-party markets.[39] All exporters would adjust their prices to that of the marker crude at appropriate differentials.

In January, 1974, Kuwait took 60 per cent of the equity of the Kuwait Oil Company, thus scrapping the earlier framework for the participation agreements. Talks leading to a "100 per cent participation" in Aramco for the Saudi Arabian government had made considerable headway by the spring of 1975, and Kuwait had also decided to take over its industry completely. It seemed likely that most governments would soon take over their own industry, leaving the companies as contractors and operators with little, if any, equity holding.

Conclusion

With the establishment of OPEC in 1960, the governments of the oil-exporting countries effectively constrained the actions of the oil companies with respect to changes in tax prices. They were unable to obtain more than relatively minor gains in other directions, although they did develop an efficient forum for cooperation and for the formulation of common objectives to be pursued, as it became possible for them effectively to do so. The emergence of "radical" regimes in Iraq, Algeria, and Libya, which were willing to take the lead in challenging the companies and in running risks in their negotiating tactics, put the more moderate regimes in the area on the defensive each time a victory was won. The willingness of the companies to accept defeat by these governments and to "roll with the punches" steadily increased the confidence of all OPEC countries. Unlike the legendary giant, the companies lost strength every time they hit the ground. By 1973, the takeover of power by the host governments was complete.

At the same time as the bargaining power of the individual countries was being put to the test, the growing dependence of the United States on imported oil changed the demand and supply conditions against which the bargaining was conducted. There were numerous reasons for the tightness of the market for oil in the period before the reemergence of "surplus" in mid-1974, but the most important by far was the rapidly growing United States demand for Middle Eastern oil, which reached a peak in 1973. This was the result of a number of long-standing trends in the United States economy (including, as an important element, growing concern for the environment). In any case, increasing pressures of supply on demand were created, which gave the advantage to suppliers and provided favorable conditions for the formation of an effective suppliers' cartel, especially given unrestricted competition by buyers for supplies that could not be quickly expanded. Thus, the bargaining strength of the OPEC countries was reinforced by the interaction between this growing market pressure and their own growing solidarity and confidence. In particular, the influence of the "radical" countries was enhanced in OPEC.

Early in the period considered here, agreements on prices were reached between the exporting countries and the oil companies, and, later, arrangements were made to give the governments a share in the ownership of the companies producing their oil. I think the evidence indicates that, although power had shifted, most of the governments wanted in good faith to reach agreement with the companies; they made concessions to do so and did not lightly abandon the agreements reached. But the fall in the value of the United States dollar, in terms of which prices had been set, combined with unexpectedly high rates of international inflation and unexpectedly rapid rises in the market prices of oil in 1972 and 1973, created circumstances that undermined the basis of the agreements by vitiating the expectations that were held by *both* companies and governments at the time when they were made. Renegotiation became essential if the agreements were reasonably to serve the mutual interests of the parties.

The political response of the United States to the increased influence of the exporting countries served only to strengthen the confidence of these countries. The growing dependence of the United States was presented as an emerging and imminent "global" energy shortage resulting from the depletion of world oil reserves, thus reinforcing the belief of the exporting countries that their long-run, as well as their short-run, position was strong. It was also presented as a political crisis arising from the implications of such dependence for American national security. This was variously represented as a danger to the "free world," because of the feared Russian influence in the Middle East, as a force weakening the ability of the United States to exercise its "responsibility" for the security of that world, and as a clear threat to the existence of Israel.

So long as the Arab countries had little cohesion and were in a weak position to exercise market control over oil supplies, there was little fear that they could effectively "use the oil weapon" in the conflict with Israel. But, as their position changed in the ways we have been describing, it became clear that, if the deteriorating relations between Israel and the Arabs should come to open conflict, it would very likely bring a genuine crisis in the international oil industry as well. I say a "genuine crisis" because I do not see that there were strong reasons, apart from the Arab-Israeli problem, for treating the changing relationships between oil companies and their host governments, or even the increasing cost of oil following on the price and participation agreements, as constituting a "crisis" in any meaningful sense. There was a sharp shift in the terms of trade affecting oil; events underlined the urgent necessity, already recognized, of overhauling the institutional framework of the international monetary system; and the oil-exporting countries did acquire enormous monopoly power. All of these would create problems, some of them very difficult, but they would not necessarily constitute a "crisis."

The October war, which was the culmination of another independent course of events, brought all the separate developments to a single head. The open clash occurred, and the Arab exporting countries imposed their oil restrictions; they also abandoned all pretense of negotiation over prices, and they imposed the new price levels they had earlier put before the companies. The level of realized prices reflected the effect on the market before the October war of the hectic bids of oil companies, largely American independents. This, in turn, reflected the extraordinarily rapid increase of American dependence on oil imports from the Middle East. The really startling price increases at the end of December, 1973, adopted largely under pressure from the Shah of Iran, were also in part a reaction to this same demand, but its impact was enormously enhanced by the effect of an embargo and of output restrictions.

The efforts of national economies, as well as of the international economy, to adjust to the new situation and the efforts of the OPEC countries to determine their own future policies constitute the continuing oil problem for 1975 and the continuing challenge to national and international authorities, including those of the oil-exporting countries. Looming as always in the background is the specter of another war between the Arabs and Israel. This would bring another oil "crisis."

REFERENCES

[1]The French state oil group, Elf-Erap, argued editorially in its monthly bulletin (May 20, 1969) that Europe, being so dependent on Middle Eastern oil, should recognize the political as well as the economic aspects of the situation and work out a plan for long-term economic cooperation and collaboration with the Arab oil producers generally.

[2]See the Supplement to Middle East Economic Survey (hereafter cited as MEES), 12:33 (June 6, 1969) for an authorized summary of Zaki Yamani's remarks.

[3]Only the operations of British Petroleum-Bunker Hunt, which produced some 400,000 b/d from the Sarir field in southeast Libya, were exempt from the cutbacks. In December, 1971, the government nationalized British Petroleum's interests in the Sarir concession as a political act in retaliation for the seizure by Iran of three islands in the Persian Gulf. Colonel al-Qadhafi held Britain responsible for Iran's action, since the islands were nominally under British protection. In June, 1973, Bunker Hunt's assets were nationalized as a "slap in the face" for the United States, in the words of Colonel al-Qadhafi, because of its support of Israel, and as a warning to other companies that were resisting the government's demand at the time for 100 per cent participation with compensation at net book value.

[4]In 1961, Iraq had expropriated over 99 per cent of the concession territory of the Iraq Petroleum Company and ever since had been engaged in a bitter dispute with the company. The IPC still controlled almost all of Iraq's oil production and used its control of oil exports as a means of bringing pressure on the government to change its policy. Exports had been severely held down, although in 1968 and 1969 they were substantially increased after the closure of the Suez Canal, when supplies in the Mediterranean commanded higher prices.

[5]MEES, 13:39 (July 24, 1970).

[6]A senior American official has indicated that United States policy was influenced by the possibility that others might step in if the American companies had not reached a settlement. See James E. Akins, former Director of the Office of Fuels and Energy of the U. S. Department of State, in "The Oil Crisis: This Time the Wolf Is Here," Foreign Affairs, April, 1973.

[7]According to one of the outside (though Arab) advisers, the Libyan government began the negotiations with the presentation of a case that it thought was technically sound, justified, and reasonable. When the negotiations got nowhere, it decided to adopt militant tactics. See Taki Rifai, The Pricing of Crude Oil: Economic and Strategic Guidelines for an International Energy Policy (New York, 1974), pp. 251-52.

[8]It is ironic to note that when OPEC was formed it wanted to negotiate as an organization with the companies, but the companies resisted, even refusing to "recognize" the organization for bargaining purposes. This was widely interpreted as an attempt by the companies to maintain a "divide and rule" attitude toward the producing countries. Ten years later, attitudes were reversed, with the governments wanting "separate negotiations" and the companies insisting that they would only negotiate with OPEC as a group! Libya, at least, appreciated that "divide and rule" was an effective policy for it to adopt.

[9]Resolution XXII.131.

[10]MEES, 14:15 (February 5, 1971).

[11]However, the Shah stated in an address to an open session of this conference that if the governments agreed to a five-year price freeze, the companies should agree not to raise product prices, or at least should agree to tie product prices to a reasonable index. This can be interpreted as an additional warning that OPEC would continue to use increases in product prices (and company profits) as an argument for raising taxes. No provision of this sort appeared in the Teheran agreement, the demand having been dropped by the governments as the various compromises were worked out. On the other hand, partly because of the steep rise in product prices, a number of balanced (though not necessarily unbiased) observers did not believe that the demand of the Gulf countries for a 35-cent increase in posted prices, or less than half the recent price rises in the product markets of Europe, in exchange for a five-year agreement, was unreasonable. Ian Seymour of MEES, for example, held "that they made use of their extraordinary new bargaining power with considerable restraint, responsibility and reasonableness." Ibid. Nevertheless, the increases demanded would have meant a very large rise in the cost of oil to the companies.

[12]It has been reported that Libya was feeling isolated at the time of the Tripoli negotiations (see Taki Rifai, *op. cit.*, p. 274). Libya realized that the Gulf countries would be unwilling to support it if its demands went beyond the "Caracas terms," and it also knew that they did not want the Teheran agreement upset. Hence the government felt constrained to compromise with the companies.

[13]See *MEES*, 14:43 (August 20, 1971). Company spokesmen held that the reference in the Teheran agreement to the Caracas Resolution XXI.122 was intended to cover changes in monetary parities. OPEC spokesmen held that the reference was only to the "inflation part" of the Resolution, not to the "parities part."

[14]Zaki Yamani, interview with *MEES*, 14:36 (July 2, 1971).

[15]In the general agreement as it was finally worked out, the government share was to start at 25 per cent, rising to 51 per cent only by 1983. The original OPEC demand was for 20 per cent rising to 51 per cent by 1980. The additional 5 per cent initial share was a *quid pro quo* for the additional three years given the companies.

[16]Press conference, London, June 24, 1972.

[17]*Petroleum Intelligence Weekly* (hereafter cited as *PIW*), 11:45 (November 6, 1972).

[18]*PIW*, 11:33 (August 14, 1972).

[19]*Ibid.*

[20]*PIW*, 11:52 (December 25, 1972).

[21]*Ibid.*

[22]J. E. Hartshorn, "Oil Diplomacy: The New Approach," *The World Today*, 29:7 (July, 1973), pp. 282-83.

[23]See, for example, the reports in *PIW* for January 8 and February 19, 1973. By July, *PIW* reported that the majors were no longer suppliers of participation crude and that new independent refiners had to rely on the market (July 16, 1973).

[24]Testimony of Dillard Spriggs, Executive Vice-President of Baker Weeks & Co., Inc., in U.S. Senate, *Hearings before the Subcommittee on Multinational Corporations* [Church Committee] *of the Committee on Foreign Relations on Multinational Petroleum Companies and Foreign Policy*, 93rd Congress, second session, January 30, 1974, Part 4, p. 61. In his testimony, Mr. Spriggs stated that the companies' response to the new situation was "to cease making long-term supply or sales commitments to third-party purchasers. They could not know whether they would be able to honor any commitments they might make. The name of the game in international oil became 'hold on to supply' . . ." (*ibid.*, p. 57). "This situation has had a significant impact on what the independents have done. Purchasers of crude oil . . . bought increasing amounts of spot oil because they couldn't get long-term supply contracts from the majors. That forced the price up in the market, and the independents began to line up to buy participation crude from the government-owned companies. . . . On top of this situation the United States removed all quantitative controls of crude-oil imports May 1, 1973. Because of declining production of domestic crude oil and its availability, most U. S. refiners were at that time having trouble meeting operating schedules and, as they looked to the future, they faced a widening gap between their needs and what they could obtain domestically. Most of these companies joined the Europeans and Japanese, all seeking to buy participation crude from the producing countries. Even certain American utility companies sent missions to the Middle East looking for supplies" (p. 60).

[25]In contrast, Japan did make such an effort. In February, 1973, the Japan Lines made a contract for large supplies of crude from Abu Dhabi at very high prices (reportedly taking most of the participation crude Abu Dhabi had available). MITI immediately took steps to control the bidding of Japanese companies, thus incurring much criticism from Abu Dhabi; see *MEES*, 16:16, 21, 22 (February 9, March 16, and March 23, 1973).

[26]The companies were, in principle, protected against falling prices by the effect on their costs of taxation based on posted prices, since such a tax, acting as a tax per barrel of exports, set a floor to prices.

[27]*MEES*, 16:29 and 31 (May 11 and 25, 1973).

[28]*MEES*, 16:36 (June 29, 1973).

[29]*MEES*, 16:41 (August 3, 1973).

[30]*PIW*, 12:34 (August 20, 1973).

[31]See, for example, reports in *MEES*, 16:49 (September 28, 1973).

[32]Dillard Spriggs calculated that the profit margins on crude oil for the companies had, by the spring of 1973, gone up to around 90 cents a barrel from the slightly over 30 cents a barrel of 1971. He also testified that the companies were increasing product prices disproportionately in order to shift profits from crude-oil production to their refining and distributing operations, in view of the possibility that their owned crude-oil production might eventually be lost. *Hearings, op. cit.*, p. 61.

[33]It will be recalled that the Shah of Iran had originally insisted during the Teheran negotiations in 1971 that, in return for a five-year agreement on prices, the companies should agree not to raise petroleum-product prices. This issue had been dropped, but not surprisingly it was raised again when product prices were seen to rise at a greater rate than crude prices.

[34]*PIW*, 12:34 (August 20, 1973). There was also some evidence that, when prices sagged, at least some companies took action to prevent the resale of oil at lower prices by purchasers. In July, an easing of prices in the United States was reported by *PIW*, and some sellers emerged who were offering to resell Saudi Arabian participation crude at lower prices than they had paid. "One refiner . . . sought to sell a few summer cargoes of the crude he received on a long-term contract from a major international oil firm. The supplier learned of it and squelched the offering. The refiner was told his contracted volumes would be cut if he was unable to take the oil for himself." *PIW*, 12:29 (July 16, 1973).

[35]*MEES*, 16:48 (September 21, 1973). *PIW* calculated that the real cost of importing oil for several major European countries and Japan in August, 1973, was still below 1957 levels. For countries with depreciating currencies, this was not true (see *PIW*, 12:39 [September 24, 1973]).

[36]*MEES*, 16:52 (October 19, 1973). The same issue reprints the text of the OPEC communiqués on the subject.

[37]See *MEES*, 17:5 (November 23, 1973), which contains extensive documentation on the issues.

[38]In November, 1974, it was estimated that the average government revenue per barrel for marker crude had risen from $9.31 in January to $10.14 in November, the revenue being calculated to include "government take" (royalty plus income tax) and revenue from sales (assuming 40 per cent equity crude and 60 per cent buyback crude at 93 per cent of posted price). See Shell International, *Shellbrief*, November 15, 1974.

[39]For a discussion of the new system and its expected interpretation, see *MEES*, 18:8 (December 13, 1974), Supplement.

Part 3: THE STATES AS ACTORS

GEORGE LENCZOWSKI

The Oil-Producing Countries

IN THE NINETEEN-SIXTIES, a semblance of balance still existed in the relations between the oil-producing countries and the international oil companies, but the scale began to tilt in favor of the producers in the nineteen-seventies, a process that reached its culmination in the crisis of 1973-74. A heterogeneous group in ethnic and political composition, the oil producers displayed certain common traits side by side with substantial differences in perceptions, attitudes, strategies, and behavior. Both the similarities and differences merit identification and analysis.

The Common Denominators

The producers, grouped since 1960 into the Organization of Petroleum Exporting Countries (OPEC) and embracing Asian, African, and Latin American states of different cultures, political complexions, and levels of development were at the same time linked by a number of common bonds. Chief among them was a simultaneous striving toward decolonization and modernization. Full national control of their oil resources, including the control of transportation, refining, and distribution, claimed high priority as one means of achieving these goals. Using the slogans "from negotiation to legislation" and "from concession to service contract," the producing countries declared the passing of the old order, in which vital matters had been either decided by the companies outright or subject to bargaining with them.

This new attitude of assertiveness can be illustrated with a few examples. Between 1951 and 1953, Iran had attempted to take unilateral action against the then concession holder, the Anglo-Iranian Oil Company. It met with resistance from the major oil companies which, in a show of solidarity, instituted a virtual boycott on Iranian oil and thus largely thwarted Iran's designs.[1] The conflict ended in a formal recognition of the nationalization, but the basic pattern of control and profit-sharing did not differ from that already in force in the neighboring countries. Similarly, when Iraq nationalized most of the Iraq Petroleum Company's concession area in 1961 and subsequently rejected offers of settlement, the companies penalized it by restricting production in favor of a higher output in other producing states of the region.[2] Even with the advent of OPEC in the nineteen-sixties—a move indicating the producers' resolution to strengthen their ranks by coordinating their efforts—the emphasis was for a long time

59

defensive and protective and only gradually did it take on a more militant flavor. OPEC's initial emphasis was on the preservation of the posted-price level and, in due course, on production increases, for which it tried—fruitlessly—to find a satisfactory prorating formula.[3]

The oil countries' position of weakness and the corresponding strength of the companies could be attributed to three interrelated factors: the existence of a buyers' market, the self-sufficiency of the United States as a producer and consumer of oil, and the dominant position of the major oil companies. All three factors were undermined or reversed in the early seventies. The OPEC members as a group could consequently abandon their concern with production increases and concentrate almost exclusively on prices. In fact, their former preoccupation with the growth of production, which was a potentially dividing factor within the organization, gave way to the acceptance of output restrictions as a legitimate and viable policy alternative for individual members or their groups. Depending on the country and its particular economic position, the policy of output restrictions could be based on concern for conservation or on the inability to use effectively the revenue surpluses that even before 1973 had been accumulated by some states.

Until the outbreak of the October war, it was thus possible to find a number of common denominators that enhanced the solidarity of OPEC members. This community of interests was likely to prevail over potential differences so long as major political issues did not become involved. The Arab group within OPEC, however, differed from others in that it was directly concerned with the unsettled state of Arab-Israeli relations and in that it admitted (and occasionally practiced) the use of oil as a weapon in the political struggle. Furthermore, even the Arab group itself was not fully unanimous as to the degree and manner in which this weapon should be used.

The Policies of Saudi Arabia

The Arab-Israeli war began on October 6, 1973. On October 17, a conference of Arab oil ministers in Kuwait resolved to use the oil weapon in the struggle for the Arab cause. They agreed in a binding decision to reduce their exports by a minimum of 5 per cent as compared with the average September output and to follow this reduction with further cuts of 5 per cent per month until the completion of the total withdrawal of Israel from the territories occupied in 1967 and the restoration of the rights of the Palestinians, "or until the production of every individual country reaches the point where its economy does not permit of any further reduction without detriment to its national or Arab obligations."[4] The consumer countries were to be treated according to their stand on the Arab-Israeli issue: the friendly ones were to be exempted; others—unfriendly and neutral—were to be subject to varying degrees of restriction. To this basic decision was added a recommendation to apply an embargo on the United States due to its unfriendly policy toward the Arabs. Iraq opposed the decisions and withdrew. The remaining nine states—Abu Dhabi, Algeria, Bahrein, Egypt, Kuwait, Libya, Qatar, Saudi Arabia, and Syria—approved the actions unanimously. Despite this apparent solidarity, however, their underlying attitudes were not identical.

The key position in the Arab group was held by Saudi Arabia on both economic and political grounds. With an output of 8.5 million barrels a day on the eve of the war, the Saudi kingdom accounted for 40 per cent of the total Arab production, hence its voice in the policy of sanctions was bound to be decisive. Furthermore, as a country of wealth and as the center of traditional Islam, it not only wielded major political influence in the Middle East and beyond, but was an acknowledged leader in the conservative camp of the Arab world. Under King Faisal, after the Khartoum Conference of 1967 and especially after the advent of President Sadat to power in Egypt, Saudi Arabia maintained close relations with Cairo and, through its policy of financial aid, helped Egypt to remain solvent. In the nineteen-seventies, Saudi Arabia cooperated with Egypt in developing a general Arab strategy in relations with the outside world. It also responded with increasing sensitivity to the trends in Arab public opinion, where three elements were especially important: Saudi domestic opinion, the militant revolutionary regimes in certain Arab countries, and the Palestinians.

In spite of its authoritarian system, the Saudi ruling establishment had been paying attention to the moods and expectations of its steadily growing intelligentsia, many of whom were American-educated. These men expected their country not only to modernize itself but also to assume responsibilities in Arab affairs more generally, including the problem of unsettled Arab-Israeli relations. Moreover, its high-ranking technocrats were part of the ruling elite and, although subordinate to the leading princes of the royal family, they played an increasingly influential role.

The influence of the radically inclined Arab governments was less tangible, partly because of their frequently aggressive attitude toward the Saudi monarchy, especially in the case of Libya and Iraq. But again, their militancy could not altogether be disregarded even in traditionalist Riyadh. A similar militancy was present among various Palestinian groups. Despite the ideological extremism and excesses of some of them, Saudi Arabia broadly supported the general Palestinian struggle for the recognition of their rights and provided subsidies to the Palestine Liberation Organization. King Faisal's dedication to the Palestinian cause was, moreover, strengthened by his interest in Jerusalem and his often cited determination to see its Arab part restored to Arab rule so that he might make a pilgrimage to its holy places.

Pressures from these quarters conflicted with the traditional Saudi policy of "not mixing oil with politics." If, during the previous crises (1956, 1967), the Saudis had resorted to the use of the oil weapon, they had done so reluctantly and only for brief periods.[5] But by the spring of 1973 the continuing deadlock in the Arab-Israeli conflict caused King Faisal to accept with greater enthusiasm the idea that oil might be used as an instrument in Arab-American diplomacy. On at least three occasions, the King expressed his concern and issued warnings to the United States: first through his oil minister, Shaikh Ahmed Zaki Yamani, to certain American cabinet members in April; then through Aramco's parent companies in May; and finally through *Newsweek* and the BBC in September.[6] The idea that the oil weapon would have to be employed by adoption of restrictive measures was further emphasized by a series of consultations King Faisal held throughout the summer of 1973 with President Sadat of Egypt. But the basic Saudi "philosophy" in this respect was to avoid mentioning, if possible, any specifically punitive actions.[7] The original intent was rather to alert world public opinion to the gravity of the Arab-Israeli conflict should a new war break out. This explains why Saudi Arabia favored adopting a *decision* on production cutbacks at the Kuwait meeting of October 17 while supporting only a *recommendation* with regard to the embargo. Prior to the meeting, King Faisal had sent an urgent message to President Nixon to cease assisting Israel in the war. When, on October 19, he learned of the President's decision to request Congress for an authorization to provide $2.5 billion for arms assistance to Israel, he decided to impose an embargo.

Attitudes of Other Gulf States

The smaller Arab states of the Gulf—Kuwait, Qatar, Abu Dhabi, and Bahrein—have traditionally followed Saudi leadership and in this case also conformed to the Saudi line, both at the ministers' meeting in Kuwait and after it, when King Faisal decided on an embargo policy. Of the four states in question, three not only had intimate political relations with Riyadh but also were broadly similar to it in their political structures and tendencies, which could be described as conservative, prudent, and in favor of cooperation with the West. These were Qatar, Abu Dhabi, and Bahrein. As for Kuwait, it followed a more independent policy because it was richer and it consequently carried more weight in inter-Arab councils than its minor Gulf neighbors. Moreover, it differed from them in that it was a constitutional monarchy with a parliament, had greater freedom of speech, and a much larger and more heterogeneous population, composed mainly of immigrants from other Arab states, including some 150,000 rather articulate Palestinians. In addition, its close proximity to Iraq and Iran compelled it to pay close attention to the trends and moods in these two comparatively powerful countries. Consequently, despite its basic conservatism, Kuwait was more sensitive to domestic and external pressures than were either Saudi Arabia or the smaller Gulf states.

The Lone Course of Iraq

If Saudi Arabia stood at one end of the Arab spectrum, Iraq could be regarded as standing at the other. Anti-imperialist and radical in its domestic and foreign policies, Iraq—since the revolution of 1958—not only had a history of tense relations with Washington and London but also regarded itself as outside the free world. Linked by a treaty of friendship with the Soviet Union and with a Baath-dominated[8] cabinet, which included two Communist ministers, Iraq was largely isolated from the Arab community, its only friendly ties being with revolutionary Libya and the Marxist-oriented regime of South Yemen. Under the circumstances, it was not surprising that Iraq displayed skepticism or outright hostility to the Saudi attitudes on oil and foreign policies. At the October Kuwait meeting, the Iraqi delegation resisted the decision on production cutbacks while accepting the idea of selective embargoes. It argued that the cutback policy did not distinguish sufficiently between the friendly and unfriendly consumer states; instead, it proposed a more radical policy of promptly nationalizing foreign oil companies as well as banks and any other assets or enterprises of hostile countries that might be located in the Middle East.[9] In support of this stand, the Iraqis invoked their own decision, enacted on October 7, immediately following the outbreak of the war, to nationalize American holdings in the Basra Petroleum Company (BPC). These consisted of shares owned by Exxon and Mobil.

The official Iraqi argument revealed only part of the real Iraqi motivation. Two other more weighty reasons lay behind Iraq's opposition to the majority scheme: first, as a result of the dispute with the IPC parent companies, following General Kassem's[10] nationalization decree of 1961, Iraq fell behind in exploration, development, and production, with attendant financial consequences. To accept the cutback policy would mean to perpetuate and aggravate its position of inferiority as a producer. The second reason could be ascribed to the socialist nature of Iraq's regime: a policy of punitive nationalizations, if adopted by the Arab oil ministers, would give greater legitimacy to Iraq's own economic practices. Even though the policy of gradual participation leading to full takeovers was already adopted and was being practiced by other Arab producers, the sweeping abruptness of the Iraqi proposal, which reached beyond the oil industry, appeared too radical in its social implications to be acceptable to the more conservative regimes. As a result, the Iraqi-Saudi confrontation at the Kuwait meeting was of considerable intensity. Finding itself a minority of one, the Iraqi delegation withdrew. From then on, Iraq absented itself from all meetings of the Arab oil ministers held through the period of crisis, 1973-74.

Having rejected the majority policy on cutbacks, Iraq designed its own. It joined in the embargo against the states that were designated in successive meetings by the oil ministers, but instead of restricting exports to other states, it supplied them according to its own productive capacity. In the initial stages of the October war, these exports suffered some decline as a result of the Israeli bombing of the Banias terminal in Syria. After the terminal was repaired, exports were not only resumed, but even increased in comparison with the prewar period. (A good part of the increase could be attributed to the Basra region's production.) Iraq's defection from the common Arab front did not, however, make an appreciable difference in the overall Arab export situation.

The Positions of Algeria and Libya

Of the other participants in the Kuwait meeting, two deserve particular attention—Algeria and Libya. Algeria shared the characteristics of both the moderates and the radicals in the broader Arab community. Issuing from a nationalist struggle for independence, Algeria's regime was known for its militancy in pan-Arab causes, including the question of Palestine. If this was the feature linking it to the radical camp, it nevertheless refused to criticize President Sadat for his policies of curbing Soviet influence and restoring workable relations with Washington. On the oil issue, Algeria was therefore prepared to accept the Saudi and Egyptian idea of a partial boycott based on production cutbacks, in spite of the fact that, as a middle-ranking producer hungry for oil revenue, it could be expected to suffer financially as a result. Furthermore, if an embargo were to be applied to the Netherlands as well, Algeria's discomfort was bound to be acute, considering the volume of its exports directed to Dutch ports for local consumption, processing, or transshipment.

Libya's position during this period was characterized by contradictory trends and pressures. A militant revolutionary country, it was sympathetic to the position of Iraq, although—because of Colonel al-Qadhafi's Islamic orientation—it did not share Iraq's proclivity for cultivating friendly relations with Moscow and tolerating the local Communists. Furthermore, after abortive attempts at union with Egypt in 1972, Colonel al-Qadhafi felt estranged from Cairo and clearly unhappy at having been excluded from the planning and waging of the October war. Yet to reject the cutback policy followed by the majority would have further aggravated Libya's actual or potential isolation. On the other hand, if Tripoli had adopted Iraq's policy of punitive nationalizations of American interests, it would have produced very different results. In Iraq, such a move could be viewed as no more than an accounting operation because Exxon and Mobil, the two American companies affected, were only remote shareholders in an operating company—the Basra Petroleum Company—whose production process would continue uninterrupted. In Libya, American companies, with a few exceptions, held direct concessions by contrast, and these accounted for a great part of Libyan production. To nationalize them, especially under the impact of the Arab-Israeli war, would be to cause serious disruption of Libya's production and this would result in a denial of supplies to many consumers in excess of the cutbacks proposed in Kuwait. To this, two other considerations might be added: first, having nationalized British Petroleum's interests in 1971 for political reasons connected with Britain's Persian Gulf policy, Libya would have found it difficult to secure a suitable replacement for the American operators of its oilfields. Secondly, in contrast to Iraq, Libya was not lacking in revenue and had, by October, 1973, accumulated sizable financial reserves. This financial strength had in preceding years allowed it to apply certain curbs on production that had been officially explained by conservation policies. Consequently, a moderate cutback such as that adopted at the Kuwait conference was not at all incompatible with Libya's interests. Libya, like Iraq, was in full agreement with the embargo policy. Interestingly enough, however, it formulated its own conditions for lifting the embargo. These were the supplying of modern weapons and technology by Europe to the Arabs, and the denial of arms and economic assistance to Israel. The conditions were not so sweeping as they appeared, since only two European countries were being embargoed, the rest being subject merely to cutbacks.[11]

The Boycott Crisis: The Saudi Leadership

Although the embargo provision figured among the Kuwait resolutions only as a recommendation, it soon became a stated collective policy of the Arab producing countries. The initiative came from Saudi Arabia, whose government in two successive proclamations, on October 19 and 20, formally announced a whole range of restrictive measures. These included both general cutbacks and an export embargo to specified countries, including the United States. Cutbacks were decreed at 10 per cent of the September production level (instead of the mandatory minimum of 5 per cent enjoined by the Kuwait meeting) with anticipated further monthly reductions. A number of friendly countries, including such industrial states as Britain and France, were exempted from cutbacks. Other Arab countries soon followed Saudi Arabia's example. Within a few days, the Netherlands was also subjected to the embargo.[12]

As mentioned earlier, placing the United States under embargo was the direct result of the American decision to supply arms to Israel on a major scale. The Netherlands was included in the hostile category for several reasons, including the Dutch offer to replace Austria as a relay center for emigrating Soviet Jews, various pro-Israeli pronouncements by Dutch leaders, recruitment on Dutch territory—with permission from the authorities—of volunteers for service in Israel, and the use of the official Dutch airline, KLM, for chartered flights to Israel in connection with the Israeli war effort. The embargoes against the United States and the Netherlands were to apply to direct as well as indirect exports, i.e., to those initially shipped to third countries in which oil was processed or transshipped for final American and Dutch destinations, and for supplies to the United States Navy. The Saudi proclamation listed in this third-country category: Trinidad, the Bahamas, the Dutch Antilles, Canada, Puerto Rico, Guam, Singapore, and Bahrein (whose refinery was processing 50,000 barrels a day for the United States Navy). Similarly, certain European refineries—in Italy, Greece, and France—which supplied the American markets or navy were also subjected to full or partial embargo.

All in all, the Saudi proclamation constituted a document revealing that the Saudi authorities possessed sophisticated and accurate information about the destinations of their oil. As for enforcement, Oil Minister Yamani stated that all movements of exported oil were being watched with the aid of a computer.

Refining the Boycott Policy

Although the Saudi decree thus set a pattern for collective Arab action, the effects of the boycott policy were not identical for each producing country. Those producers that used to ship large volumes to the Netherlands and the United States saw their exports reduced much more than those who supplied other, non-embargoed, markets. Moreover, no uniform method was adopted as to whether the embargoes and cutbacks should be treated incrementally or not, that is, whether the decreed cutbacks should be applied before or after deducting the embargoed volumes. Saudi Arabia applied the incremental method, with the result that the Saudi exports fell by about 36 per cent, an amount the Saudis themselves found disquieting.[13]

It was with an eye to the clarification of some of these uncertainties that the second conference of Arab oil ministers was held in Kuwait on November 4. The decisions of the conference can be summed up under four points: 1) The November production was to be reduced by a uniform 25 per cent from the September level, with a further 5 per cent planned for December. 2) This uniform cutback was to include the embargoes to the United States and the Netherlands (a non-incremental method), which meant that, for other consumers, reduction in supplies would be less than 25 per cent. 3) Consumer countries were to be warned that a more positive attitude toward the Arab cause would have to be taken before they could qualify for the exempted category. 4) Saudi Arabia's Yamani and Algeria's Abdelsalam Belaid were to tour major consumer countries to explain these Arab restrictive measures.

The Kuwait announcement on November 4 evoked a prompt response in Europe. Two days later, at a meeting in Brussels, the European Economic Community published a declaration urging both parties in the Arab-Israeli war to withdraw to the positions occupied on October 22 (i.e., the initial United Nations-sponsored cease-fire line) and calling for a peace settlement which would conform to the Security Council Resolution 242 of November 22, 1967.[14] Because the latter was favored by the Arab countries, EEC's statement clearly marked a point scored by collective Arab diplomacy. As such, it was received with satisfaction in the Arab world.[15] However, the Arab governments also wanted the EEC countries to insist on Israeli troop withdrawal from all territories occupied in 1967. Because the EEC as an organized body would not issue such a statement, the two Arab ministers, Yamani and Belaid, sought such a declaration from individual EEC members. Such additional statements paved the way in several cases for a reclassification of a country into the exempt category.

The two Kuwait meetings thus launched the basic boycott policy. What followed was, on the one hand, as elaboration of the initial broad principles into a more detailed set of measures and, on the other, a series of adjustments on the part of the Arab governments in response to the changing attitudes of the consumers. This "refining" of the boycott policy was accomplished at the Arab oil ministers' meetings in Vienna on November 18, in Kuwait on December 8-9 and again on December 24-25, and at the Arab summit meeting in Algiers on November 26-28.[16]

Among its numerous resolutions, the Algiers summit meeting decreed severance of all diplomatic and economic relations (including exports of oil) with Portugal, South Africa, and Rhodesia.[17] It was, however, at the successive oil ministers' meetings that the policy guidance was set in detail. By the end of December, 1973, the consumer countries were divided into four categories, as follows:

1) The most-favored countries, i.e., those entitled not merely to the September level of supplies but actually to as much as their own requirements dictated. Among these were Britain, France, and Spain among the European states, Arab importing countries, African states that had broken off diplomatic relations with Israel, and Islamic countries.

2) Friendly countries, i.e., those that modified their policies in favor of the Arab cause. To this category belonged those countries previously in the neutral group which, due to the change of their attitudes toward the Arab cause, became eligible for supplies at the September level (for example, Belgium and Japan).

3) Neutral countries: those to which general export cutbacks applied. Because of their pro-Arab statement of November 6, the EEC countries (with the exception of the Netherlands) were temporarily placed in a more favored position within this category inasmuch as they were exempted from the further 5 per cent general cutback promised for December. Later, after individual EEC governments—following Yamani's and Belaid's visits—gave additional assurances to the Arabs, they became qualified to be treated as "friendly."[18]

4) Hostile countries to which full embargo applied. These included the United States, the Netherlands, and as mentioned earlier, Portugal, South Africa, and Rhodesia.

Initially, the Arab producers were committed to successive cutbacks of 5 per cent per month in addition to the basic 25 per cent decreed at the second Kuwait meeting. If unchecked, these progressive reductions would in due course bring production to zero, which some producers—those with smaller financial reserves—could obviously ill afford, even temporarily. For this reason, a secret resolution at the Algiers summit meeting decreed that successive cutbacks would be applied only to the point at which the reduction of a producer's revenue reached the maximum of 25 per cent of its income for 1972.[19] It has been estimated that, because the price increased by 70 per cent as compared with 1972, the Arab producers could decrease their output to approximately 45 per cent of the 1972 production level on the average before their revenue would fall to 75 per cent of the 1972 revenue level.

Relaxation and End of the Boycott

As late as the end of November, the Arab mood was still one of defiance. In response to Secretary Kissinger's reference to possible American counter-measures, Yamani declared in Copenhagen on November 22 that any military intervention would simply induce the Arabs to destroy their oil facilities.[20] Beginning in early December, however, Arab producers, with Saudi Arabia as a leader and spokesman, began to give signs of relaxing their conditions. This new stance could be attributed to Yamani's visit to the United States on December 5 and to the active role as mediator that Washington had assumed in the Arab-Israeli conflict by that time. The Arabs' retreat from their position on the issue of boycott was effected in several stages: In the first, on December 9, the Arab oil ministers—accepting a new Saudi formula—decided to link the lifting of embargo to the adoption of a timetable for Israeli withdrawal from occupied Arab territories, including Jerusalem. In the second, on December 25, the ministers decided to decrease the 25 per cent production cutback to 15 per cent and cancel the 5 per cent reduction scheduled for January. In the third stage, President Sadat began pressing for an early end of the boycott policy as a result of the conclusion, under American auspices, of the Egyptian-Israeli disengagement agreement on January 17, 1974. Because the terms of the agreement fell short of the Arab conditions formulated in December (a timetable for full Israeli withdrawal), Sadat encountered temporary resistance from Presidents Assad of Syria and Boumedienne of Algeria, as well as from King Faisal of Saudi Arabia. He was also angrily denounced by Libya's Colonel al-Qadhafi, who, in protest, boycotted certain joint meetings of the heads of states.

On March 18, however, the Arab oil ministers, at their meeting in Vienna, decided, over Libya's and Syria's negative votes, to remove provisionally the embargo against the United States until June 1, 1974. The conclusion, on May 29, 1974, of the Syrian-Israeli disengagement agreement opened the way for two final acts in the boycott drama: on June 3, in Cairo, the ministers tentatively agreed to remove all restrictions except for the embargo on the Netherlands, and on July 11, this decision became formal and the ban on the Netherlands was lifted as well. The boycott crisis came to an end; nevertheless, in their June and July statements, the ministers warned that embargoes could be reimposed if a new war broke out.

The Price Crisis

While the price crisis partly overlapped the boycott crisis, it differed from the latter in two respects: first, the reasons for raising prices were primarily economic, as opposed to the political reasons for the boycott; secondly, it was OPEC as a whole rather than the Arab oil-producing group that initiated the price increases.

On October 16, 1973, the ministerial council of six Gulf OPEC member-states decided in Kuwait to boost the posted price of the Saudi light oil ("marker crude") by 70 per cent from $3.011 to $5.119 per barrel. The decision was subsequently approved at the thirty-sixth OPEC conference held in Vienna on November 19-20. And at their meeting in Teheran, on December 22, the Gulf group of six raised the posted price for the same type of crude to $11.651 per barrel, effective January 1, 1974. This amounted to an increase of 130 per cent from the October level and about 300 per cent from the pre-October prices. The new, quadrupled price was to assure the producing countries a seven-dollar "take" per barrel. If the September production rate of 25 million barrels a day was to be maintained for the Middle East-North African region, the revenues to these governments would reach $62 billion a year. The drain on the finances of certain consumer nations, not to speak of many other dislocations, was thus bound to be of unprecedented proportions.

The first price boost—that of October 16—came on the very eve of the Arab boycott decisions. As a unilateral action of the producing countries (rather than as a result of a negotiated agreement with the companies), it certainly broke new ground. It could, however, partly be explained by the expected decision of the Arab states to impose oil sanctions, which would cause shortages and provide a realistic basis for the increased prices. What was in October a mere expectation became in December an accomplished fact. By the time the Gulf Six met in Teheran, the Arab boycott had been in operation for two months and the world-wide shortages clearly enabled the entire OPEC community to cast off restraints in boosting the prices.

While the anticipated or actual shortages help explain the ability of the producers to increase prices, their reasons for doing so should be traced to different origins. First of all, the producers felt that for a number of years their oil had been underpriced and their share of profits had been inadequate. The inflationary spiral for goods imported from the industrial states only aggravated this feeling. The immediate reason for the decision, however, was the failure to reach agreements on prices with the companies at meetings held in Vienna.

On the price issue, OPEC as a group demonstrated outward solidarity, although substantial differences in approach emerged among its members. Of particular importance were the distinctive positions of Iran and Saudi Arabia.

Iran's Position on Prices

It was Iran that both proposed and secured acceptance, over Saudi objections, of the elevated Teheran price level. Iran justified its stand by saying that the hike in prices was a long-delayed act of justice to compensate for the depressed price levels; that it was high time for producers to start practicing conservation and rational use of their depletable resources; that the West should develop alternative sources of energy rather than squander the "cheap" energy obtained from the producers; that the price of oil should follow the price of the nearest substitute; and that, contrary to Western complaints, the oil price boost did not account for more than 2 per cent of the world-wide inflation—a phenomenon that in any case, antedated the oil price rises.[21]

These official explanations should be supplemented by a few other clarifications which, by their very nature, could not be formally articulated. One was undoubtedly political: assertion of Iran's sovereign will on the price question and a spectacular increase of the state revenue for development and defense were of tremendous importance to Iran's energetic ruler, Mohammad Reza Shah Pahlavi. Attainment of these objectives was a clear vindication of the Shah's national policies and was bound to provide a contrast with Mossadegh's abortive attempts twenty years earlier. By securing higher revenues, the ruler of Iran was not only increasing his and his country's prestige but also paving the way for the entrenchment of Iran's influence in the entire Persian Gulf region. Perhaps the move could also be viewed as a psychologically compensatory act for the position, resented by some groups in Iran, of previous dependence on the West. When queried about the ill effects of the drain of currencies upon certain European countries vulnerable to social unrest and a possible leftist takeover, the Shah acknowledged the existence of the danger, but he rejected the idea that it was his responsibility to watch over the political welfare of such states.[22] As for the less-developed lands, the Shah recognized their difficult plight but pointed to his policy of giving generous loans and grants to a variety of countries, ranging from India and Pakistan to Egypt and certain African states, designed to alleviate their difficulties. Similarly, examples of major deals with France and Britain, each involving at least a billion-dollar credit for industrial goods and services, were invoked as evidence of unselfish and imaginative policy.[23] Moreover, the Shah himself and his spokesmen repeatedly offered to enter into a dialogue with the industrial countries. Its purpose would be to examine the prices of major industrial commodities imported by the OPEC members and to compare them with the price of oil with an eye to arresting the inflationary spiral. In short, while admitting the reality of complications likely to be generated or aggravated by the price increases, Iran's ruler perceived that the major advantages to be gained from the new price levels would outweigh their negative consequences.

The Saudi Price Policy

At the December Teheran meeting, the Saudi oil minister initially opposed the Iranian price proposal but, following intense Iranian urging and guided by last-minute instructions from his government, accepted it and thus assured the unanimity of the decision. But, despite this agreement, Saudi Arabia did not cease to voice criticism of the high price level and to strive for its reduction or, failing this, for at least its stabilization to prevent further substantial increases. Certain smaller Arab producers in the Gulf area tended to follow the Saudi lead, though their spokesmen avoided publicity on this issue. As for Kuwait and Iraq, both—despite their somewhat strained mutual relations—inclined toward the Iranian position.

Saudi Arabia was critical of the high prices for several reasons. King Faisal was concerned about a possible collapse of the West European and Japanese economies with resulting harm to their political strength and their ability to resist international Communism. While this concern was present in Iran as well, Saudi Arabia's record in refusing to have diplomatic relations with Russia and the Soviet bloc countries as well as its opposition to Communism in all its manifestations could hardly be equaled anywhere in the world. Linked with this was the Saudi desire to revert to its traditional cooperation with the United States. The embargo notwithstanding, King Faisal never ceased to consider this cooperation as a cornerstone of his foreign policy. By March of 1974, as we have seen, the embargo had been removed and the Saudis had restored friendly relations with the United States. Prince Fahd's visit to Washington in June continued this trend through the inauguration of a comprehensive program of American technical assistance to Saudi Arabia.

Two other considerations, mutually interrelated, might be mentioned in explaining the consistent Saudi opposition to high oil prices. The Saudis felt that the radical price boosts were apt to nullify the advantages gained from the boycott policy. This policy was essentially Saudi-designed and, as its designers saw it, its merit lay in its sophisticated differentiation between friends, neutrals, and enemies. Now, by contrast, the price policy hurt friend and foe alike and antagonized those on whose support the Arab world was counting. The Saudis feared that this situation would provide a fertile ground for Israel's political activity in the Western world directed against the Arabs and especially against Saudi Arabia as a key oil exporter. Linked with these apprehensions was the Saudi awareness of hints and discussions in the United States about a possible military intervention to seize Arab oil fields.[24]

With all this, the choice for the Saudis was far from easy because it involved a decision whether to maintain or break OPEC's solidarity. The least the Saudis could do was to go on record with a formula of their own while still broadly adhering to the OPEC-decreed level of prices. At the conference of the Gulf Six in Abu Dhabi on November 10, 1974, Saudi Arabia rallied to itself Qatar and the United Arab Emirates to announce the decrease in the posted price of the marker crude by 40 cents a barrel to $11.25, while raising the royalty to 20 per cent of the posted price and the income tax on profits from 65.75 to 85 per cent. Although the net result was even an increase in the actual payments by the oil companies to the three governments in question, the decision to come forth with one's own formula independent of OPEC signified Saudi Arabia's inclination to resist the will of OPEC's majority.

This brings us to the concluding point of this essay—that of the reality and durability of collective action by the producers. The creation of OPEC had been a first step in the direction of solidarity, but it was a modest beginning. Harmony prevailed so long as interests appeared common, especially in protecting the posted prices and the governments' share of profits in the nineteen-sixties. On another issue—that of prorating the desired production increases among the members—no agreement was reached and the matter was abandoned. However, OPEC experienced a revival in the nineteen-seventies when the sellers' market provided a more solid ground for cooperation. But it was not until the October war of 1973 that the apogee of cooperation was reached. Even then it was primarily due to the catalytic impact of Arab solidarity, itself generated by political causes which, by inducing production shortfalls, enabled OPEC to maximize its advantageous position. In the short run, OPEC members derived obvious financial benefits from their cooperation and firmness on the price issue. In the long run, political and economic factors could not be ignored in their calculations. Because each individual producing country had its own political priorities dictated by the domestic requirements of the group or individual in power and the broader concern for external security and survival, the solidarity based on economic interests was bound to find its limits imposed by broader fundamental considerations.

REFERENCES

[1] The companies were unwilling to face lawsuits threatened by the AIOC if they were to buy and sell oil of contested ownership.

[2] Actually, Iraq penalized itself automatically because, by seizing over 99 per cent of the concession area, it precluded further exploration and expansion by the company.

[3] This was primarily a reflection of the rivalry for higher production levels between Iran and Saudi Arabia. It caused the companies to engage in a delicate "balancing act" to keep both of these countries satisfied.

[4] Quoted from the official statement issued in Kuwait, October 17, 1973; courtesy of the Government of Saudi Arabia.

[5] For more details, see George Lenczowski, "Arab Bloc Realignments," Current History, December, 1967.

[6] For King Faisal's interview with Nicholas C. Proffitt, see Newsweek, September 10, 1973. Based also on the author's audience with H. M. King Faisal, September 1, 1974.

[7] Based on the author's interviews with Shaikh Ahmed Zaki Yamani, Minister of Oil, in Jeddah, January 13-14, 1975.

[8] The Baath is the abbreviated name for the Arab Socialist Renaissance Party, known for its militant nationalism and hostility toward conservatism and imperialism.

[9] For an official explanation of Iraq's oil policy, see Iraq News Agency release of November 6, 1973, summarized in Baghdad Observer, November 6, 1973, and the Arab World (Beirut), November 8, 1973, p. 7. See also the Arab World, November 19, 1973, p. 2, and December 18, 1973.

[10] Major General Abdul Karim Kassem, "the Sole Leader" of the Iraqi Republic, 1958-63.

[11] For Premier Abdul Salam Jalloud's statements on this subject, see Al-Anwar (Beirut), November 13, 1973.

[12] Technically, Abu Dhabi was the first to proclaim an embargo against the United States on October 18, 1973, while Algeria was the first to impose a ban on exports to the Netherlands over the weekend of October 20-21.

[13] Saudi concern centered largely on the ill effects of cutbacks on the economy of Japan with whom Saudi Arabia maintained increasingly friendly relations.

[14] Text of EEC resolution in The Times (London), November 7, 1973.

[15] For a typical positive comment, see Michel Abu Jawdeh's editorial in An-Nahar (Beirut), November 7, 1973.

[16] The chronology of meetings is listed in the Appendix at the end of this volume.

[17] In the case of South Africa and Rhodesia this, however, was merely a reaffirmation of the earlier decisions taken by the United Nations.

[18] Based on interviews with Shaikh Zaki Yamani; see note 4 above.

[19] For details and text of the Algiers resolutions, see The Arab World (Beirut), November 28, 29, and December 4, 1973.

[20]*New York Times*, November 23, 1973.

[21]Based on the statements of H. I. M. the Shah of Iran, *Kayhan International*, December 29, 1973; November 9, 1974; and December 14, 1974; as well as on the statements of Jamshid Amuzegar, Minister of Finance, *ibid.*, January 26, 1974, and March 16, 1974.

[22]Based on the author's personal interview with H. I. M. the Shah, August 24, 1974.

[23]Iran's commitments to aid-and-trade with less-developed and industrial countries amounted to a total of $7 billion in 1974. Personal communication from Mr. Ardashir Zahedi, Iranian Ambassador in Washington.

[24]Based on the author's interviews with Shaikh Ahmed Zaki Yamani; see note 4 above.

JAMES W. McKIE

The United States

THE OIL EMBARGO OF 1973-74 PRESENTED some extraordinarily difficult political and economic problems to the United States. It struck two practically simultaneous blows, the ultimate consequences of which have gradually come to light during the year or so since the end of the winter crisis. One blow was political: The American public became aware for the first time during the embargo that the country was now vulnerable in a vital matter—its energy supply. The other impact was primarily economic: The very large increases in the price of oil, coming on top of some considerable increases in prices generally between 1971 and 1973, confronted the oil-importing countries with an essentially new situation. In the United States, the higher prices of imported and domestic petroleum worked their way through the chain of economic interdependencies until they appeared in the form of increased prices for other kinds of energy and for energy-intensive goods and services, such as transportation. They raised the cost of living, particularly in their massive impact on the heating and cooling bills of householders. They also may have contributed to the onset of recession and unemployment in 1974.

The problems did not originate in 1973, but had grown gradually over the preceding decades. There is no doubt that the United States could have dealt with them much more successfully had it earlier understood their implications and perceived the need for coping with them.

The Economic Roots of the Crisis

The United States became a net importer of petroleum (crude and products) in 1947. Thereafter, total consumption in the United States rose from approximately 6 million barrels per day in 1948 to a bit under 17 million just before the embargo in 1973.[1] Production of domestic crude oil and liquids rose from 5.9 million barrels per day in 1948 to 10.8 million in 1973, but it had peaked in 1970 at 11.2 million. Imports represented over 30 per cent of consumption by 1973; midway in that year, direct imports of Arab oil were running over one million barrels per day, up from less than half that amount a year and a half earlier.[2]

The reasons for the rapid growth in the demand for oil in the United States through the nineteen-sixties were many, but the most important was probably the fall in domestic real prices for oil combined with a high rate of growth in the economy. In Janu-

73

ary, 1969, the price of oil, when compared with the price of other products at wholesale, was 10 per cent lower than it had been eleven years earlier. There were similar
drops in the real prices of gasoline and fuel oil.[3] Energy was becoming cheaper relative
to almost everything else, and demand both in the United States and in the world reacted accordingly.

The principal factors that led to a rapid increase in American imports during the
early nineteen-seventies were the curtailment of natural-gas supplies under regulation,
the disappointingly low rate of growth in nuclear power, the environmental restrictions
on strip mining and on the burning of coal, on the development of domestic oil supplies, and on motor vehicle emissions, and conflicting government policies. These deserve a brief recapitulation.

The Convergence of Demand and Supply Factors

In the United States, consumption of oil and its products grew at an annual rate of
over 4 per cent during the nineteen-sixties, rising to a 5.4 per cent annual growth rate
during the period 1967-72.[4] The supply from domestic sources failed to keep pace.
One factor depressing the long-run development of domestic crude-oil reserves was the
severe restriction on output that state prorationing controls imposed during the fifties
and early sixties. In 1963, for example, the large, efficient, low-cost fields in Texas
were cut back to under 30 per cent of maximum efficient rate, while the high-cost
stripper wells were allowed to produce without restriction. The result was a dampening of profit incentives for further exploration and development.

One ordinarily expects, when demand presses upon capacity, as it did upon domestic capacity after 1970, that the price will rise and the market will go in search of
cheaper alternatives. As matters worked out in the early nineteen-seventies, the cheap-
est alternative was imported oil. Until 1973, quantitative restrictions on imports prevented the wholesale substitution of foreign for domestic production, but when increasing demand for lagging domestic supply began to push prices up, the import
restrictions gave way. Meanwhile, the alternative domestic energy sources, instead of
absorbing part of the incremental demand as oil prices rose, actually contributed elements of their own to the growing energy shortage.

1) *Natural gas.* At present, this fuel is in even shorter supply than domestic oil. It is
preferred to all other fuels for environmental protection. It has, in fact, become rather
precious of late. A large part of the difficulty with the natural-gas supply, of course,
stems from the imposed ceiling-price regulation.

Natural gas was a cheap fuel before World War II. It was virtually a by-product of
oil, and was priced accordingly. But the development of long-distance pipeline transmission after the war introduced gas into vast new markets, and its value began to rise
as consumption accelerated. Then, in 1954, the Supreme Court decided that the Natural Gas Act of 1938, regulating the prices of gas sold by pipelines, also applied to natural-gas sales in the field. The Federal Power Commission then began its several attempts to hold field prices of gas down to "costs" rather than permitting it to find its
own value in the market.

The culmination of this approach was a decision in 1965 that established a two-level price ceiling: 16.5 cents per million cubic feet on gas not yet committed or sold, including that still to be developed, and a lower ceiling restricting escalation of prices on gas already committed under contracts containing escalator clauses.[5] It sharply limited windfall gains for producers, but it also set the price of "new" gas below its long-run marginal cost. Price ceilings stimulated low-value uses for what was becoming a premium fuel. Between 1950 and 1970, gas provided more than half of the total growth of energy in the United States.[6] It displaced coal from many industrial uses. It came into use as boiler fuel in electric power-generating plants. Surging demand and lagging supply caused the ratio of proved gas reserves to annual production to fall, from 27 to 1 in 1950 to 20 to 1 in 1960 to 13 to 1 in 1970. By then the reserve ratio had reached levels that began to cause difficulty with deliverability, led to spot shortages, and required most gas pipeline companies and gas utilities to impose a moratorium on additional customers.

By 1973, the Federal Power Commission had raised price ceilings on gas considerably. The Administration was soon calling for removal of regulation of the field price. (By December, 1974, the ceiling price on new gas had been raised to 51 cents per MCF, still considerably below its free-market value.) In the meantime, demand deflected from gas by shortages fell on fuel oil; but since the domestic oil industry was operating at capacity, the demand for more fuel oil had to come from the total import demand. From 1.5 to 2 million barrels per day had been added to imports by 1973 owing to the shortage of gas.[7] Domestic gas supplies had not yet responded to the higher prices. Natural gas, like oil, requires a long time for increases in production to respond to price changes.

2) *Nuclear power.* Seldom in human affairs do all of the influences on a complex problem become unfavorable simultaneously, but that is what seems to have happened to energy sources in the United States. The nation might legitimately have hoped that the great efforts devoted to the development of nuclear energy would have begun to pay off before the other fuels went into short supply, and that it could confidently look forward to transferring a rapidly growing share of energy demand from fossil to nuclear fuels. But in 1973 nuclear power accounted for less than 5 per cent of total generation of electricity—far less than had been projected for it ten or fifteen years earlier.[8]

The reasons for the delays are easy to identify. First, there are the unforeseen difficulties with the development of nuclear technology, including problems with reliability, maintenance, and cost overruns. Second, opposition from environmentalists, and from other citizens who distrust the safety of nuclear plants and who are concerned with disposal of radioactive wastes, has assumed formidable proportions, especially in the past few years. Third (partly as a result of the second), government inspection and licensing procedures, which now include a requirement for environmental impact statements, have slowed the investment process down considerably. Fourth, an unfavorable turn in electric-utility economics, owing primarily to rising interest rates, has caused difficulties with financing.

3) *Coal.* This is the most abundant source of energy available to the United States,

but unfortunately it is regarded as an inferior fuel to gas and fuel oil for most uses. Production of coal fell from a high of nearly 700 million tons in 1947 to less than 420 million in 1961, notwithstanding a fall in its real price throughout the nineteen-fifties. The real price decline also continued throughout the nineteen-sixties,[9] although demand began to revive during those years. However, the substitution of coal to take care of increments in oil and gas demands was restrained by the environmental-protection movement as well as by rapidly escalating costs and prices after 1969.

The environmental issues are two: pollution of the air by the burning of coal with high sulfur content, and destruction of the natural landscape by strip mining. Efforts to limit environmental damage by requiring the scrubbing of stack gases and restricting the burning to low-sulfur coal have elevated the cost of using coal. No comparable environmental restrictions were imposed on strip mining before May of 1975, though Congress was readying some regulations that would limit strip mining and increase its costs. Prices were also pushed upward by the Mine Safety Act of 1969, which increased labor costs and led to the closing of many substandard small mines. Booming foreign demand, chiefly from Japan, added to the upward pressure on prices. The average price for domestic coal in the United States rose from $4.58 per ton in 1968 to $7.66 in 1972[10]—over 66 per cent—and surged upward another 20 to 30 per cent in 1973.[11]

A combination of environmental constraints (notably the Clean Air Act of 1970), coal-supply scarcities, and escalating coal prices actually deflected some energy demand away from coal to oil in 1970-73, though a national-security interest in reversing this trend in the future, subject to environmental protection limits, has since been widely proclaimed. During the embargo itself, little could be done to substitute coal for oil in existing installations. The long-run allocation of industrial energy requirements among coal, oil, and nuclear power is still in debate.[12] The prices of fuel oil *after* December, 1973, tended to turn energy demand back toward coal in planning for the near future, during which time nuclear power will be unable to absorb much of the load.

Exotic energy sources, such as solar energy, and "synthetics," such as shale oil, contributed nothing to alleviating the energy crisis of 1973-74; the chief effect of the crisis in that quarter was to accelerate research and development.

Other Environmental Restrictions on Supply

We should note two other governmental policies, or rather inactions, that had a major effect on the domestic oil supply in 1973, reducing it well below what might have been available. The first was the delay in building the Alaska pipeline, originally scheduled for completion in 1973 with a capacity of as much as 2 million barrels per day—a volume that would by itself have exceeded the total imported directly from Arab countries in mid-1973. The environmental crusade stopped the Alaska pipeline for four years. Environmentalists feared the ruin of the Arctic ecosystem from breaks in the line and massive oil spills. But, in the end, the oil embargo did clear the way for the start of construction in 1974.

After the well-publicized "blowout" off Santa Barbara and the consequent pollution of nearby California beaches in 1969, the government, under severe pressure from public opinion, imposed a moratorium on off-shore drilling. But the Outer Continental Shelf in several zones is the most promising area (along with Arctic Alaska) for discovery and development of major new domestic supplies of oil and gas. Insofar as rational arguments entered into the decision to suspend off-shore drilling, the decisive one was the belief that protection of the shoreline environment took absolute priority over development of new oil supply. That attitude also did not survive the embargo.

Price Controls of Oil

The prices of oil and all other fuels were rising during the early nineteen-seventies on both domestic and international oil markets. In August, 1971, however, the domestic prices suddenly encountered the barrier of price controls, which held them down until early 1973. During the first seven months of that year, the index of American prices for refined petroleum products rose by over 30 per cent. Domestic crude prices climbed by more than 50 cents per barrel. Meanwhile, in April, 1973, import controls had been suspended, and the rapidly escalating import prices of crude oil and petroleum products had an increasingly strong effect on domestic consumer goods.

On August 17, 1973, the Cost of Living Council imposed a two-tier price ceiling: "old" oil was to be sold at prices prevailing in March, 1973, plus 35 cents per barrel, while "new" oil (that produced in excess of 1972 levels) and imports were free to sell at the uncontrolled market price. The purpose of the two-tier approach was similar to the policy imposed on natural-gas prices: to prevent windfall gains on oil already developed, while encouraging the production of more "new" oil.

It is hard to say whether price controls had had any substantial negative effect on domestic production and capacity on the eve of the embargo in October, 1973. The period of rigorous control was short, and prices did move ahead explosively when controls were relaxed. But what price controls did *not* do was to discourage and restrict demand, an effect that substantial price increases might well have had in the early nineteen-seventies.

By the fall of 1973, the United States energy economy had become heavily dependent upon imported oil supplies. Its domestic petroleum prices were increasingly influenced by the rising world price. For various reasons—imperatives of regulation, of politics, and of environmental protection, as well as technological obstacles—the various components of the domestic energy economy had for several years increasingly fallen short of meeting the demands upon them, forcing the nation to turn to imports in ever larger amounts. But that was not the result of a conscious policy decision taken with the demands of energy security in mind. As it was, prices on oil and products were rising fast enough in 1972 and 1973 to alarm those governmental offices responsible for price stabilization. The government decided to abolish limits on oil and product imports in April, 1973, and to replace them with small tariffs.

Just six months before the Arabs initiated their embargo, what remained of security policy toward oil had finally been eclipsed by the need for price stability and by the continuation of consumption as usual. Let us now look back over the development of that policy.

The Quest for Energy Security

As imports of oil into the United States continued to grow after 1948, demands for restriction also increased. Although arguments were usually couched in terms of national security, they were not always believed. When, for example, the Eisenhower Administration, in March, 1959, placed quota limits on oil imports, many observers were convinced that their *primary* purpose was economic protection for the American oil industry. Ten years later, Milton Friedman, among others, still held that view: "The political power of the oil industry, not national security, is the reason for the present subsidies to the industry. International disturbances simply offer a convenient excuse."[13]

Events have since shown that, whatever the motives may have been in imposing the controls, there *is* a security problem. Although the impact of any "peacetime" embargo would fall primarily on the civilian economy and not on military requirements, both civilian and military demands must be protected against a cutting off of imports, whether it is the result of accident or of design. That the United States avoided a total failure of security policy during the 1973-74 embargo was only providential—the dependence of the United States on Arab oil was not yet sufficiently great nor was the embargo sufficiently prolonged to cause severe damage.

In the nineteen-sixties, on the other hand, when practically no dependence on insecure sources appeared likely, the severe restrictions on imports did seem disproportionate to any probable threat to security, raising suspicions about the true objectives of the program. Imports into the United States east of the Rocky Mountains were restricted by quota, eventually set at the equivalent of 12½ per cent of production in that area. Imports to the Western region were adjusted to equal the difference between "demand" (at a stable price) and domestic supply in that area. For most of the period 1959-72, the import rights, which were distributed to refiners in proportion to refinery inputs, had a cash value arising from the difference between the price of imported oil and the domestic price. The total distribution, at the ultimate expense of consumers of course, amounted during the nineteen-sixties to many hundreds of millions of dollars a year.

Canadian oil was exempt from quota allocations, though such imports were subtracted from the overall import quota, making that much less available from other sources. But many exemptions and exceptions to the quota were granted in the course of time. By the time the quota system ended, the Eastern coastal region of the United States was almost entirely dependent on imported crude and products, especially on heavy residual fuel oil. By 1972, in fact, residual fuel oil accounted for almost one-third of total petroleum imports to the United States.

In 1973, when the world price moved above the United States price, import quotas were abandoned. By that time, imports of crude and products exceeded, even under the quota, 30 per cent of American consumption.

The Need for Protection: The Outlook in 1969

In 1969, OPEC had been in existence ten years, but it was still an ineffectual cartel. Though it had frequently announced intentions to rationalize the world oil industry, allocate output, and raise prices, it had found no effective means for doing so. In the mid-nineteen-sixties, the world oil industry was producing an excess, and prices in fact fell somewhat. However, control of output was still in the hands of the international oil companies; though they were not able to prevent some decline in prices, they did manage to maintain fairly stable conditions in the market. Producing-country governments sensed the revenue bonanza that could be theirs if they could jointly and effectively limit output, but they were unable to contrive an effective world-wide prorationing system.

As late as 1969, there seemed no reason to suppose that this situation would not continue indefinitely.[14] Costs were still very low and supplies abundant.[15] It appeared that the United States did not need to fear interruption or manipulation of supply or prices by OPEC for economic reasons. As for a politically inspired embargo, opinion was also justifiably skeptical.[16] It was thought that the Arab states, the most likely perpetrators of such an embargo, could not cooperate among themselves with sufficient effectiveness to enforce one. In any event, the United States itself did not need protection against such an embargo because it imported very little oil from Arab countries, and any boycott by one or a few producers could be offset by enlarging purchases from other sources. Its principal sources of supply were still in the Western Hemisphere. The best insurance against a calculated embargo, or against accidental interruption such as that caused by the Nigerian civil war of the period, lay in maintaining the availability of multiple sources of supply.

A storage program was considered in 1969 as an added protection against possible future concerted interruptions of imports. The costs were high, but they were within reason as an insurance premium.[17] A recent authoritative estimate of the full annual costs of sufficient steel-tank storage to supply 2 million barrels per day for six months—more than our total daily imports from the Eastern Hemisphere in 1972—is about $730 million per annum, calculating the acquisition cost of the stored oil at $7.50 per barrel, or $820 million at $10 per barrel.[18] The industry, however, preferred to stick to the existing quota system. Besides, those making energy policy in the United States government were not at all convinced that a "crash" program to build up storage capacity was necessary.

The Events of 1969-73

The assumptions underlying the prevailing views of the American position in the world oil system eroded very quickly during the next few years, though the foundations were already shaky by the end of the nineteen-sixties. Excess capacity disappeared in the United States, and domestic supply peaked and began to decline, while domestic demand zoomed. Foreign sources of supply on which the United States had relied failed to meet its expectations. A revolution in Libya replaced the compliant pro-Western government with a hostile and erratic nationalistic one. The Trans-Arabian pipeline was shut down for an extended period, causing pressure on available transportation capacity and prices. Kuwait and Saudi Arabia adopted a conservationist position, announcing that in the future they would not allow the rate of development of capacity to go as high as the oil companies and the consuming countries would have pushed them in an unrestricted market. Finally, the international oil companies continued to lose position and control of output in many producing countries.

While these events were occurring, Canadian policy underwent a major reversal, imperiling another source of United States supply. In view of the importance attached by government policy-makers to maintaining and strengthening the Canadian connection, this was a particularly dismaying development. The government actually imposed quantitative limits on Canadian oil imports for a short time in early 1970—one of the most mindless acts of economic policy on record—and followed this with some general trade restrictions in August, 1971. Meanwhile, an outbreak of Canadian nationalism resulted in a rebellion against American "exploitation" of Canadian resources, especially oil and gas. Most Canadian petroleum deposits were owned by American companies; they left only the local costs of labor and materials plus a small tax and royalty in Canada when they produced their oil and shipped it across the border. The Canadian government decided in the early seventies to conserve petroleum for Canadian needs and to sell only the "surplus" to the United States. The Canadian Energy Board placed export restrictions on natural gas, and, in 1973, it advised the United States not to make any plans that assumed the availability of Canadian petroleum in the future; it would probably all be needed at home. With these changes in policy, North American security of supply for the United States went out the window.[19]

Political Economy

In view of the growing dependence of the United States on imported oil and the evident development of insecurity in supply and of cartel control of world prices in the early nineteen-seventies, one may well ask why United States policy did not react more decisively. Why did it not take steps to counter these economic and political threats? What could or should it have done about both the causes and effects of its new insecurity?

One factor that seems to have affected both governmental policy and public opinion was a simple inability to believe that what eventually did happen could happen. Since no adequate survey of public attitudes on these matters was made before the embargo, this judgment is admittedly subjective. But considering the long American experience of economic power, its general feeling of invulnerability to economic pressure that has been manifest since World War II, and its century or more of essential self-sufficiency before the war, it should not be surprising that the American public and most of the policy-makers in the government did not take foreign threats of economic reprisal very seriously. The image that most Americans had of the national economy was of a mighty structure for national production, the most efficient in the world, buttressed by the best science and technology, the most skilled work force, and the most imaginative and resourceful management, and solidly based on an abundance of natural resources. Damage to the United States by a group of small Third World countries wielding an economic weapon was inconceivable.

We cannot say that everyone was indifferent to or unaware of the developing predicament. The oil industry itself had long been quite outspoken about the threat that the growth of oil imports posed for national security. The National Petroleum Council (an industry advisory board) had produced a series of reports in 1971 and 1972 carefully analyzing the energy position in the United States and predicting a dangerous degree of dependence on imported oil before 1985, if policies toward domestic energy development were not changed.[20] Many individual oil companies had published similar projections, though their recommendations on policy did not adequately specify the security issue or distinguish between more and less secure sources of imports. The domestic "independent" companies had vehemently called for a reduction rather than an increase in import quotas.[21]

The major international companies conveyed several Arab warnings to the United States government concerning the drift of policy toward the Arab-Israeli conflict.[22] The Mobil Oil Company actually went to the length of publishing advertisements calling public attention "to the U.S. stake in the Middle East" and for "justice and security to all the people and all the states in that region." California Standard soon followed its example: The consequence was a consumer boycott of Chevron stations in California, and it abandoned the campaign.[23] The American-based companies with stakes in the Middle East anticipated a renewal of Arab-Israeli hostilities, and they were fearful that American support of Israel would make their position untenable. They were also no doubt genuinely better disposed toward the Arab claims than was the average American—certainly they were better acquainted with the situation in the Middle East. But their advocacy of a change in policy to forestall a possible embargo might also have been regarded as self-serving.

Federal Energy Policy and Its Constituencies

The federal government also included some offices and bureaus that were aware of the emerging problems, and they were not without influence on Administration policy. But the organization of energy policy was diffuse and haphazard. The office that was supposed to have the major responsibility for emergency planning—the Office of Emergency Preparedness—was without a public constituency and virtually without power to remedy any of the fundamental causes of the problem. It could only make recommendations to the President, and, in any case, it was dismantled in 1973. Its energy-planning functions were transferred to the energy office of the Department of the Treasury, which later became the Federal Energy Office, and then (after the embargo) the Federal Energy Administration. This office was only one of several in the government interested in energy policy.

By the middle of 1973, before the embargo appeared as a tangible threat, the United States was already in the midst of a mild shortage of petroleum products caused chiefly by a shortage of domestic refining capacity that had developed while imports of foreign crude were restricted, and by the trends on the world markets mentioned earlier. Other sectors of the energy economy, such as gas and nuclear power, were also causing governmental concern. Along with these problems, the government continued to study means for protecting security. For example, the energy office of the Treasury issued "Emergency Energy Capacity: An Interim Report" on October 18, 1973, the day before the Arab producers announced their embargo against the United States. This report is a good survey of the protective and precautionary measures, including storage and reserve capacity, that might have, but were not—and still have not been— taken. As the acuteness of the energy problem grew, it tended to move toward the center of federal policy planning; it elicited struggles among bureaucrats, agencies, and legislators for control and influence, but it produced very little effective action. Beginning in the late spring of 1973, the centers of power in the Administration were preoccupied by Watergate.

The cross-purposes and indecision evident in the petroleum policies of the United States government before and during the embargo may be better understood if they are considered against the background of the highly diverse constituencies involved, which had different interests and stakes in energy decisions, including both their political and economic aspects. The interests of those groups are reflected in conflicts among the governmental agencies and the monitoring Congressional committees that look on them as constituents. No policy or mixture of policies on energy, or on oil specifically, will satisfy all these constituencies or resolve their conflicting aims and pressures.

The producers of energy—oil, gas, coal, nuclear power—are primarily interested in higher prices, better investment incentives, and unfettered development and production of fuel. The coal industry and its associated United Mine Workers, the most powerful labor union in this sector of the economy, both want restrictions on imports of heavy residual fuel oil. Some federal government agencies, such as the Atomic Energy Commission (recently succeeded by the Nuclear Regulatory Commission and the Atomic Energy Research and Development Administration) and the Office of Oil and Gas of the Department of Interior, have the energy industries as their clients.

The principal industrial and commercial users, on the other hand, primarily want assured supplies and stable prices (or alternatively the power to pass cost changes on to the consumer through price increases) as well as freedom from restrictions on their particular uses of energy. The industrial users—most importantly in petrochemicals, metals, glass, and ceramics—are probably less politically powerful than the transportation interests: the automobile industry, airlines, the trucking industry, railroads, and the highway lobby. The Department of Transportation and certain regulatory commissions (the Civil Aeronautics Board, the Interstate Commerce Commission) look after the interests of these groups. Farmers, who need motor fuel for machinery and LPG for drying crops, are well protected by the Department of Agriculture.

Ultimate consumers, householders, and motorists frequently find themselves in opposition to the energy producers and the principal industrial and commercial users as well. They are not organized for self-protection, but since they include most of the population their unhappiness is heard by Congress. They want supplies in customary volume through customary distributive channels at low prices. They are represented in some respects by price stabilization offices in the government, including state and federal agencies regulating energy utility rates.

Environmentalists, with some government representation (Environmental Protection Agency) and a strong influence through public opinion, have taken a generally hostile stance to expansion of energy production and consumption, especially by industry, since these so often degrade the environment. There are also small but audible groups of futurists and no-growth believers, the latter having little or no organized influence in the government but considerable moral and intellectual impact outside it.

The government includes other divisions with various responsibilities that are powerfully affected by events in the international economy and polity. The United States Treasury and its outposts are acutely concerned with the balance of payments and the fate of the international monetary system, which has received some heavy shocks since the October embargo and the oil price increases of December, 1973. The military services and the foreign-policy establishment have to confront the principal American policy problems in the Middle East that impinge upon petroleum, and they must consider the uses and effects of petroleum as a geopolitical weapon—an exceedingly complex problem.

Finally, we should recognize the influence of an amorphous pro-Israel constituency among the general public, having powerful support in Congress, and represented by smaller activist Zionist organizations. There is no corresponding group of pro-Arab organizations.

Clashes among domestic economic interests in energy have been more significant in the past than domestic conflicts arising out of events in the Middle East; not until the embargo was the public confronted with the new dimensions of the energy crisis and its capacity for generating conflicts of interest in American politics. Security policy in energy somehow has had to find its way through this tangle of aims and purposes.

American Attitudes Toward the Middle East Before and After the Crisis

It seems clear that the Arab attack of October 6, 1973, came as a tactical surprise both to the Israelis and to the United States government. Thereafter, events moved too rapidly for planning; the embargo began two weeks later, destroying the preconceptions on which our policies had been based.

The Arabs directed their embargo as a political weapon against the United States because they strongly believed that Israel was a creation of the United States and that its policies and actions, notably the 1967 war and its refusal to withdraw from the occupied territories, could not have been undertaken without American support. The American public was not fully aware that the Arabs would hold them entirely responsible for Israel's actions. On the whole, public opinion in the United States has been highly favorable to Israel, but not to the extent of supporting direct American involvement in clashes with its Arab neighbors.

It may be fair to say that for the average American the modern history of Palestine began in 1948. The enormous outwelling of contrition after the Holocaust predisposed the public to sympathize with the idea of a national homeland for displaced Jews, whereas it remained ignorant about the plight of displaced Arabs and their grievances (at least until they received more attention in the press following the October war of 1973). After 1948, the State of Israel appeared as a stable, ordered, progressive, democratic island besieged by hostile and (to most Americans) alien peoples.

In July, 1967, a month after the Six-Day War, the Gallup survey found that only one American in seven thought that Israel should be required to return the lands seized from the Arabs in that war. But in 1968, most Americans hoped that the United States would not get involved should hostilities break out again. A large majority was against sending either arms or troops to the Middle East. In 1970, 62 per cent of those surveyed were sympathetic to Israel and 9 per cent to the Arabs, while 29 per cent had no opinion. By October, 1973, the proportion favorable to Israel had fallen to 48 per cent, and only 37 per cent favored supplying it with arms and material. But Arab sympathizers accounted for only 6 per cent.[24]

Certainly Congress was sympathetic to Israel, because it had a strong American constituency while its opponents had none. It was widely believed that Israel could count on a majority of seventy-five or eighty senators. Congress was unfailingly generous with financial aid and expressions of support.[25] Up to 1973, few members of Congress seemed to realize that some peril for the United States lay in that course.

The United States made some feeble efforts after 1967 to bring about a political settlement between Israel and its neighbors, including Secretary Rogers' abortive "peace plan" of 1970. But sympathy for Israel was not seriously diminished. Official strategic thinking in the State Department's public statements (whatever some of its country desks might have thought) and in the Pentagon focused not only on preserving Israel as a strong independent state but on countering Soviet designs in the Middle East. During the period 1958-73, official policy seems to have leaned toward the view that a "balance of power" was necessary in the Middle East, that Israel (along with Iran, perhaps) could be a military "bulwark" or anti-Communist "bastion" for United States policy there, and that some neighboring Arab nations had become tools of the Soviet Union and were "fronting" for its designs. Thus the American commitment to sustain Israel and back its policies was related to its concern for resistance to Communist aggression and subversion everywhere in the world. During the October war, this attitude was vehemently expressed by some military leaders and by sympathizers with Israel as a reason for massive military assistance. "Like the Korean aggression of 1950," said one commentator, "the Arab attack of October 6 was a carefully planned Soviet move, threatening fundamental security interests of the United States and its allies."[26]

Though the opposing view had few supporters before 1973, there were some skeptics even then. Some of them thought that American policies had in fact caused Soviet penetration of the Middle East, and that an uncritical commitment to the policies of Israel might itself be damaging to the "fundamental security interests of the United States and its allies"; advice to this effect from allies had not been lacking. Senator Fulbright, one of the few doubters in Congress, attacked the emergency appropriation of $2.2 billion in military aid for Israel after the October war on the ground that "promoting the military interests of Israel" would encourage "further intransigence" on its part and would be seen by the Arabs as a "re-affirmation of the inability of the United States to pursue an even-handed policy."[27]

Events and Perceptions During the Embargo

As in all crises, the public looked for scapegoats, and it found them primarily in the oil companies and in the federal government's lack of preparedness and its ineffectual response. Again we call on Dr. Gallup: In answer to the question, "Who or what do you think is responsible for the energy crisis?" asked on December 7-10, 1973, the following answers were offered (including some multiple answers):[28]

	Per cent
The oil companies	25
The federal government	23
The Nixon Administration	19
U.S. consumers	16
Arab nations	7
Big business	6
Leaders "playing politics"	4
U.S. exporting	3
There is *no* shortage	6
Misc./no opinion	19

There were some well-publicized incidents of speculative middlemen holding oil off the market during the embargo in hopes of getting very high spot prices, but the public fury against the oil companies was more fundamental: Many Americans suspected the international oil companies of having *instigated* the embargo and the subsequent price increases for the industry's own benefit. While basically irrational, this suspicion fed on reports of profit bonanzas that began to appear in the press in December, 1973, at a time when the average American motorist was beginning to feel a severe gasoline-supply pinch and householders were paying sharply higher prices for petroleum products. The profit bonanza for the internationals was caused primarily by the high posted prices set by OPEC, and it has since turned out to be short-lived, but it made its impression nonetheless. Moreover, the profits were shared by domestic oil companies, as prices of American-produced "new" oil followed the international prices up. Total profits for seventy oil companies operating in the United States rose 53 per cent from 1972 to 1973, while total manufacturing profits rose by only 33 per cent.[29]

Independent, non-integrated oil companies in the United States were suddenly confronted with shortages of supply, which, some of them asserted, were due to the efforts of the "major" companies, i.e., the larger integrated oil companies, to take advantage of the crisis in order to extinguish competition in refining and marketing. Such charges fed public suspicion, and they ultimately led to the imposition of a system of mandatory allocations of crude oil among refiners. However, available data on gasoline marketing do not support that phase of the charge: Independent marketers of gasoline with non-major brands had a market share of 12.9 per cent in February, 1972, and 12.8 per cent in February, 1974, while the independent distributors of non-branded gasoline increased their share from 9.5 per cent to 11.5 per cent during the same two years.[30]

At all events, the public did not at first reveal any hostility toward the Arabs. Later (after September, 1974), talk of a "military response" began to surface, notably in the famous Kissinger interview with *U.S. News and World Report*, but not as a result of public jingoism. In the meantime, the embargo had ended, and the petroleum problem had metamorphosed from a problem of shortages into a problem of high prices (and of balance of payments). But all OPEC countries, not just the Arabs, participated in the price increases, and even Canada, following the lead of OPEC, imposed limits on exports of oil to the United States and added an export tax of $5.20 per barrel. Economic exactions by the OPEC cartel became an issue no less important than the continuing Arab-Israeli political confrontation. Public opinion tended to confuse the two issues, but by early 1975 resentment against all oil-producing countries had risen remarkably in the public opinon polls. It was, however, directed especially against the Arabs because of their use of oil as a political weapon and because of their key role in the cartel.[31]

One serious consequence of the embargo was a breakdown of cooperation between the United States and its principal Western allies, as a *sauve qui peut* atmosphere seemed to take over in the Western alliance. The United States followed its policies of resupplying Israel during the October war without consulting its NATO allies beyond informing them of its actions; the EEC nine reached agreement on the Middle East Resolution of November 6, 1973, without consulting the United States.[32] The consuming nations scrambled for oil supplies without any attempt to form a common front. By February, 1974, however, the OECD group had unified sufficiently to meet in Washington and examine the question of consumer-country cooperation. In November, 1974, the United States and fifteen other countries (France stayed out) had created an International Energy Agency to plan OECD counter-pressure against OPEC and to arrange for the sharing of oil supplies in the event of another embargo. The United States government took the lead in these moves, though it is less dependent on imports than its allies. It was apparently prepared to promise a share of its own supplies in return for avoiding future diplomatic isolation on Middle Eastern issues. It also urged a general cutback of oil consumption by OECD countries.

Energy Policy Since October, 1973: The Brokerage of Claims

In view of the sudden and unanticipated onset of the oil embargo and of the huge price increases in oil that it precipitated, the general lack of attention to security planning in the preceding period, the loose organization of responsibility for energy in the federal government, and the conflicting aims and demands of politically significant groups detailed above, it is not surprising that energy policy in the United States during the crisis was confused and ineffective. At the beginning, the Federal Energy Office (FEO), which was created after the embargo began, did not have a clear idea of how much oil and petroleum products were needed for inventories; it lacked control over movements of foreign supply to the United States; it had no means of controlling domestic supply of oil and other fuels. It did not succeed in eliciting any appreciable emergency output from American oil fields (not even from the Elk Hills Naval Reserve), though some extra short-run capacity to produce was undoubtedly present.

Attempts by the FEO to cope with the oil shortage were also handicapped by the price controls that had been forced on it by political considerations, and the consequent system of direct allocations. Restricting the prices on "old" oil to $5.25 per barrel while allowing the price on "new" oil to rise to nearly $10 meant that oil companies would not produce a barrel of old oil (from wells unable to exceed 1972 production levels) whenever it would have cost them more than $5.25. Moreover, beginning in February, 1974, the FEO ordered refiners with more than the industry's average of crude-oil supplies on hand to share their "excess" with those having less. However, they could only charge a price equal to the average of crude costs from all sources. Hence crude-rich refiners would lose money on every barrel of oil imported at, say, $10, which could now only be sold to a crude-poor refiner at $7. Imports during the later stages of the embargo probably fell below what they might have been, in response to these obvious disadvantages.[33]

The FEO apparently made some mistakes in inventory management and control of product mix; it constricted gasoline supplies on the market unnecessarily while accumulating too large inventories and over-producing fuel oil.[34] It was unable to work out an equitable plan of sharing gasoline supplies among states and cities, although it announced several times that it would do so. Shortages remained particularly severe on the Eastern seaboard, the area most dependent upon imports.

The demands of various groups for more favorable treatment and their political stratagems to get more or to resist giving up what they had were far too numerous to catalog here, nor was the Federal Energy Office the only theater for such conflicts. But we might just mention a few examples of what the FEO, other administration offices, and Congress had to deal with almost daily:

—In December, 1973, Energy Administrator Simon moved to take away 1.5 million barrels of jet fuel from the military (principally from Air National Guard training flights) and allocate them to civilian airlines, which were faced with a severe shortage that threatened curtailment of service. Secretary of Defense Schlesinger protested; the result was a "compromise" permitting the Pentagon to keep 600,000 barrels.[35]

—Intercity truck operators demanded preferential allocations, higher speed limits, and a price rollback on diesel fuel, and they tried to enforce their demands by blockading and disrupting highway traffic, necessitating intervention by the National Guard in some cases.

—Environmental protection suffered some reverses, as the government, under the visible impact of the shortage, decided to reopen offshore oil leasing and to clear the way for the Alaska pipeline. The Clean Air Act and automobile emissions standards came under heavy attack by industry lobbies. The automobile industry asked for postponement of the deadline for Environmental Protection Agency (EPA) standards under the act. The EPA countered by asking for comprehensive plans for automobile transport control. Later in 1974, President Ford vetoed a bill providing for a coal tax to pay for reclamation of land that had been subjected to strip mining.

Power brokerage still goes on. The gains and losses of the various interested parties may be seen in the record and guessed for the future. Public policy, recovering from its crises of the last five years, is beginning to focus on the key issues of the next decade or two. Environmental restrictions are to be somewhat relaxed without being abandoned; foreign policy will propose a more balanced stance between Israeli and Arab interests and will attempt to create a cohesive strategy vis-à-vis OPEC; the United States will try to limit the growing rate of demand for oil, to stimulate supply, and to develop alternative fuels as a means of lessening dependence on foreign sources and the political vulnerability that goes with it.

The government can choose between a policy of rigorously excluding all "insecure" imports—which seems to mean most oil imports nowadays—and severely curtailing oil consumption, or a policy of controlled importing, while protecting acceptable consumption levels with a program of standby reserve and storage.[36] The decision between them has not been made. "Project Independence" seems to imply the former; but it is more probable that the Administration, with Congressional backing, will eventually choose the latter. It has signaled its intention to follow a high-price policy to attain its ends—high prices for imported oil and (probably) a minimum-price-guarantee approach to preserve domestic incentives if world prices should again fall below acceptable levels. Along with this will go massive research and development expenditures on new forms of energy and new ways of economizing energy. The prospect is vastly favorable for energy producers, but less so for consumers, who will pay much higher real prices for energy than they have been used to in the past.

Whether the political mechanism can sustain the load that energy policy is now putting on it is open to doubt. The disarray of policy has not ended; no consensus exists for solutions to a number of the most serious problems. Conflicts over foreign policy in the Middle East, over price stabilization and equitable distribution of energy costs, over environmental protection against production and use of energy, and over energy-demand curtailment, supply incentives, and the rate of economic growth are substantially unresolved, notwithstanding the elevation of the energy problem in all its guises to the leading issue of the mid-nineteen-seventies.

REFERENCES

[1]Details of the change during the last decade will be found in the article by Joel Darmstadter and Hans Landsberg in this issue.

[2]See also the article by Robert Stobaugh.

[3]Cf. *Petroleum Facts and Figures* (Washington, D.C., American Petroleum Institute, 1971), pp. 448, 452, 469.

[4]British Petroleum Company, *Statistical Review of the World Oil Industry*, 1972.

[5]*Area Rate Proceeding* 61-1, 34 FPC 159 (1965).

[6]*Exploring Energy Choices*, A Preliminary Report of the Ford Foundation Energy Policy Project (Washington, D.C., 1974), p. 7.

[7]Cf. Paul W. MacAvoy, "The Separate Control of Quantity and Price in the Energy Industries," Sloan School of Management and Energy Laboratory, Massachusetts Institute of Technology (manuscript, 1975), p. 19. See also the article in this issue by Darmstadter and Landsberg.

[8]*Exploring Energy Choices, op. cit.*, p. 8.

[9]National Petroleum Council, *U.S. Energy Outlook*, December, 1972, pp. 150-51.

[10]Thomas D. Duchesneau, *Competition in the Domestic Primary Energy Industry* (Washington, D.C., Ford Foundation Energy Policy Project, 1975), Table 3-13.

[11]*Wall Street Journal*, November 1, 1974.

[12]Cf. R. L. Gordon, "Coal: Our Limited Vast Fuel Resource," in E. W. Erickson and L. Waverman, eds., *The Energy Question: An International Failure of Policy*, 2 (Toronto, 1974), pp. 49-75.

[13]*Newsweek*, June 26, 1969. See also the testimony of Walter Adams in U.S. Senate, *Hearings before the Subcommittee on Antitrust and Monopoly of the Committee on the Judiciary, Governmental Intervention in the Market Mechanism: The Petroleum Industry*, 91st Congress, first session, April 2, 1969, Part I, pp. 304 ff.

[14]*The Oil Import Question*, A Report on the Relationship of Oil Imports to the National Security by the Cabinet Task Force on Oil Import Control (Washington, D.C., February, 1970), p. 53.

[15]M. A. Adelman, *The World Petroleum Market* (Baltimore, 1972), Chapters II, VI.

[16]This paragraph summarizes the prevailing (though by no means unanimous) views in the government in 1969, as the author was able to gather them during his work for the Cabinet Task Force on Oil Import Control.

[17]*The Oil Import Question, op. cit.*, p. 53.

[18]Douglas R. Bohi, Milton Russell, and Nancy M. Snyder, *Oil Imports and Energy Security: An Analysis of the Current Situation and Future Prospects, Report of the Ad Hoc Committee on the Domestic and International Monetary Effect of Energy and Other National Resource Pricing of the Committee on Banking and Currency, House of Representatives*, 93rd Congress, 2nd Session, 1974, p. 156.

[19]"The future does not hold any remote possibility of a continental energy policy." Leonard Waverman, "The Reluctant Bride: Canadian and American Energy Relations," in *The Energy Question, op. cit.*, 2, p. 218.

[20]The definitive summary report was the National Petroleum Council's *U.S. Energy Outlook*, December, 1972.

[21]Cf. Minor S. Jameson, Jr., "Present and Future Import Policies" in I. J. Pikl, Jr., ed., *Public Policy and the Future of the Petroleum Industry* (Laramie, Wyoming, 1970), pp. 71-83. Mr. Jameson was the executive vice-president of the Independent Petroleum Association of America.

[22]See the article by Robert Stobaugh in this issue for details of these communications.

[23]*Wall Street Journal*, August 7, 1974.

[24]*Gallup Opinion Index*, No. 61 (July, 1970) and No. 103 (January, 1974).

[25]Total U.S. governmental grants and loans to Israel totaled over $3 billion from 1949 to mid-1973. U.S. Senate, *Hearings for the Committee on Foreign Relations, Emergency Military Assistance for Israel and Cambodia*, 93rd Congress, first session, December 13, 1973. p. 55. The estimate of 75-80 votes is that of Senator Fulbright (p. 130).

[26]Eugene V. Rostow, "The Middle East Conflict in Perspective" (October 28, 1973), *Vital Speeches*, XL:4 (December 1, 1973), p. 107. See also the speech by Admiral T. H. Moorer, "The Middle East: A Recognized Military Beachhead" (May 3, 1974), *Vital Speeches*, XL:17 (June 15, 1974), pp. 517 ff; and the statements by Deputy Secretary of State Kenneth Rush and by Senators Jackson, Dole, and others, U.S. Senate, *Special Hearing before the Committee on Appropriations, Emergency Military Assistance*, 93rd Congress, first session, November 5, 1973. Senator Jackson added a special warning: "[A] conclusion to the present hostilities that left the Soviets in possession of the Suez Canal would do great harm to the position of the United States in the Persian Gulf and the Mediterranean. . . . Israel must win this war and it must win decisively."

[27]U.S. Senate, *Providing Emergency Security Assistance Authorizations for Israel and Cambodia, Report No. 93-657 of the Committee on Foreign Relations*, 93rd Congress, first session, December 19, 1973, p. 9.

[28]*Gallup Opinion Index*, No. 104 (February, 1974), pp. 4-5.

[29]First National City Bank, *Monthly Economic Letter*, March, 1974, p. 7.

[30]"Petroleum Market Shares, A Progress Report on the Retailing of Gasoline," Washington, D.C., Federal Energy Administration, August 6, 1974.

[31]Louis Harris Poll, as reported by the Chicago Tribune News Service, February 11, 1975. Seventy-six per cent of those polled believed that, if the U.S. yields now to "Arab demands," the Arabs will eventually "dictate" much of our foreign policy. The Harris Poll also reported some resurgence of sympathy for Israel since November, 1973.

[32]Brig. General Richard C. Bowman, "Effects of the Middle East War and the Energy Crisis on the Future of the Atlantic Alliance," *Proceedings of the National Security Affairs Conference*, The National War College, July 8-10, 1974, p. 180.

[33]Richard B. Mancke, *The Performance of the Federal Energy Office* (Washington, D.C., 1975), p. 15; Paul W. MacAvoy, Bruce E. Stangle, and Jonathan B. Tepper, "The Federal Energy Office as Regulator of the Energy Crisis," Sloan School of Management and Energy Laboratory, Massachusetts Institute of Technology (manuscript, 1974). MacAvoy *et al.* estimate that domestic crude production in 1974 was .58 million barrels per day less than it would have been in the absence of price controls on "old" oil.

[34]See note 33 above.

[35]*Time*, January 21, 1974, p. 23.

[36]William Nordhaus argues persuasively that a combination of tariffs, storage, and reserve capacity in domestic fields gives optimum protection (at lowest cost) and permits continued low-level dependence on insecure foreign sources of petroleum. "Project Independence is poor economics: it always pays to take a risk if the odds are in your favor." "The 1974 Report of the President's Council of Economic Advisors: Energy in the Economic Report," *American Economic Review*, September, 1974, p. 563.

ROMANO PRODI AND ALBERTO CLÔ

Europe

It is not an easy task to give a brief description of the problems that Europe had to face during the petroleum crisis, nor of the reactions they elicited. Not only did the crisis itself unfold rapidly, but national interests differed, and national responses therefore varied accordingly, precisely at the moment when a unified response was most urgently needed. Moreover, the crisis arose just as the long and slow process of forming the European Community had reached its most precarious stage. The power of the supranational institutions in Brussels to direct or even coordinate action was still minimal, while national governments lacked adequate institutions to handle a problem of that magnitude autonomously. Part of the reason for this weakness was that six of the Community's nine members had only recently joined. While this expansion of the European Community may increase Europe's importance and power on the international scene in the long run, at that moment it raised new obstacles to the formation of a common strategic policy. This situation is not yet fully understood outside Europe, especially in the United States, where the twenty-year-old treaty of Rome still gives the false image of a Community which, solid in its principles and in its historical mission, heads steadily toward a united Europe.

Because the process of achieving European unity is slow and the vitality of the national decision-making centers is undiminished, European energy policy has still to be examined in terms of the similarities and differences of individual national policies, an approach that would have been unnecessary had a common strategy existed.

One basic thesis of this paper, then, is that the institutions of the Community entered the crisis period in a weakened state and were further enfeebled by it. Though the point of no return for these institutions was not reached, restoring vitality to them has been—and will continue to be—delicate and may easily be endangered by outside factors. A second thesis is that the energy crisis arose almost completely from political and economic conditions extraneous to Europe. That is to say, Europe was a protagonist of the second rank, a position consistent with the decline of its political and military influence, especially with respect to the Middle East, during the previous two decades.

National Energy Policies Before the Crisis

A fundamental shift took place in Europe's energy position after the Second World War. Its coal-mining industry shrank drastically, and it lost the self-sufficiency in ener-

91

Table 1
PRIMARY SOURCES OF ENERGY IN WESTERN EUROPE, 1955 AND 1972*

	1955	1972
	per cent	
Use:		
Coal	75	23
Petroleum	22	60
Natural gas	1	9
Other	2	8
Produced in Europe	78	35
Imported from non-Europe, net	22	65

*For EEC countries, see Table 4.
Source: ENI, *Energia e Idrocarburi, Sommario Statistico*, 1955-73.

gy that it had previously enjoyed (see Table 1). While it is true that Europe may well have derived major economic advantages from the shift to the extent that it released manpower and capital to other productive sectors, the economic system became extremely vulnerable in its supply of this vital resource.

The substitution of petroleum for coal in the many uses where such a substitution is possible cannot be attributed simply to price competition, though that of course had its effect; numerous other factors were operative, and some of them were more important than relative prices.[1] One of these was the different set of rules of the market by which the petroleum and the coal-mining industries operated. From the very start of the development of the European Community's organization, coal and oil were treated as separate commodities. The treaty that instituted the European Coal and Steel Community (ECSC) in 1951 placed the policy for coal with that organization. Atomic-energy policy was left primarily in the hands of the individual states, with some power granted to Euratom. Other energy sources were entrusted, in 1959, to the European Economic Community (EEC). No provision was made for a common energy policy, and those of the various Community institutions were extremely diverse. For example, the ECSC imposed special rules for the disclosure of price and commercial practices that had no parallel with respect to other energy sources in the EEC. For this reason, as well as others, production and marketing were much less flexible for coal than for petroleum.

From time to time, particularly in the mid-fifties, some European governments expressed concern over the possibility of an international energy shortage and the risks posed by an excessive reliance on foreign energy sources.[2] Nevertheless, steps to reactivate coal production never materialized. By the end of the fifties, its production in Europe had "ceased to be an industry and became only an instrument for reaching objectives of social security."[3] The virtual abandonment of the industry was largely taken for granted; the main preoccupations of policy-makers became those associated with conversion from coal to oil. During the nineteen-sixties, the number of coal miners in the nine countries of the present European Community declined by more than 60 per cent, from 1,600,000 to 615,000.[4]

From the early nineteen-fifties, the policies of European governments toward the energy market were based, at least in part, on the conviction that the cost of energy represented an important variable in the costs of industrial production. Low-cost energy was considered decisive in determining the position European industry would have in the international market. As a result, the price of coal was held down to the point where it would amount to no more than 5 per cent of total industrial costs. Control over the price of coal, however, prevented the coal industry from accumulating reserves sufficient to deal with periods of crisis.[5] When the demand for energy suddenly increased in 1956-57, foreign coal had to be imported at prices 40 to 50 per cent higher than those in Europe.[6] The decline in the price of oil in the late fifties dealt the final blow to the European coal industry.

Another aspect of European policy had contributed to the decline of Europe's coal industry. Fearful of a return to the monopolistic practices of the pre-war period, the ECSC maintained a rigid control over the coal market. It enforced the principles of market disclosure and non-discrimination in sales to customers, and it limited the use of restrictive agreements or mergers among enterprises. As a result, prices in the various countries were held down without regard for the production costs of the various coal deposits and the competitive situations in the various regional markets for alternative fuels. This leveling of prices without regard for the regional circumstances in the market "denied the miners the benefits of a market economy while taxing them solely with its costs."[7]

On the other hand, no rules existed for limiting the entry of the multinational oil companies into the European market. These already enjoyed various advantages over the coal industry, including a greater operating flexibility that derived from their vertical integration and their international structures, as well as the numerous advantages that petroleum as a source of energy enjoyed compared with coal.

The strategy of the oil companies to penetrate the market was based on the establishment of refineries in Europe and the relative pricing of their various oil products. Consequently, in the nineteen-fifties, with occasional help from government subsidies and tax concessions, the refinery capacity of the oil companies in Europe grew over five times, reaching 220 million tons. Supporting this huge capacity was a price structure that favored heavy products from the processing of crude oil—that is to say, those products in most direct competition with coal. The products were underpriced, while light products, such as gasoline, that were destined for uses that were not exchangeable with coal were priced much higher.[8] Fuel oil was being dumped, while gasoline was being monopolized. As Table 2 shows, gasoline prices were consequently two to three times higher than fuel oil prices in Great Britain and France, and 3½ to 7 times higher in Germany and Italy. These different price levels bore no relationship to the relative costs of production.

Beginning in 1957, the price of fuel oil in Europe began to drop simultaneously with the rise in the price of European coal, resulting in a progressive deterioration in the competitive position of coal and the financial collapse of the coal industry. In Germany, for example, as Table 3 shows, the relation between the prices of coal and fuel oil for industrial uses (reduced to the same thermic value) rose between 1957 and 1960 from 0.55 to 1.35; in household uses from 0.67 to 1.21.

Table 2
AVERAGE PRICES OF GASOLINE AND FUEL OIL
(in dollars per ton; net of taxes)

Country	Gasoline			Fuel Oil		
	1956	1960	1970	1956	1960	1970
France	74	67	69	35	30	25
Germany	96	81	51	26	12	16
Great Britain	68	70	50	29	30	28
Italy	80	57	108	22	16	16
United States	78	77	89	14	14	25

Source: ENI, *Energia e Idrocarburi, Sommario Statistico,* 1955-71.

Table 3
RELATION BETWEEN THE PRICE OF COAL AND OF FUEL OIL
(per thermal unit)

	1957	1960	1970
France			
Industrial Use	0.91	0.98	2.08
Household Use	1.40	1.99	2.20
Germany			
Industrial Use	0.55	1.35	1.45
Household Use	0.67	1.21	2.02

Source: Eurostat (Statistical Office of the European Communities), *A Comparison of Fuel Prices, Oil-Gas-Coal, 1955-1970.*

To bring the prices of the two fuels into line during the nineteen-sixties, the European countries could either have altered their tax structures or restricted the importation of oil—as the United States was doing during the same period. However, they preferred to follow a policy compatible with the pricing strategy of the petroleum companies, placing the highest taxes on the light products, such as gasoline, and the lowest taxes on the heavy products, such as fuel oil. In 1965, for instance, Great Britain taxed gasoline at about $100 per ton, and placed no tax on fuel oil. In Italy, the gasoline tax was about $200 a ton, with a $6 tax on fuel oil. Moreover, the tax on fuel oil was not connected with the price of coal; it was a simple consumption tax, not a protectionist measure. In this respect, at any rate, the European states behaved in the same way, though they did not regard this uniform pattern as a common policy.

Where European states did display some interest in the security of their energy supply, the problem was seen as a long-range issue of national prestige, rather than as an immediate and pressing problem of supply security. France's aspirations to control the Near East and North Africa, for example, were not connected with concern over the security of its energy supplies, nor did the crisis of 1956 alter this perspective. Events such as the founding of OPEC in 1960 were considered to be of minor impor-

Table 4
PRIMARY SOURCES OF ENERGY IN SELECTED EUROPEAN COUNTRIES, 1973
(in per cent)

Country	Petroleum	Natural Gas	Coal	Others*
Belgium-Luxembourg	62.1	13.8	23.7	0.4
France	72.5	8.1	16.1	3.2
Germany	58.6	10.1	30.1	1.3
Great Britain	52.1	13.2	33.6	1.2
Italy	78.6	10.0	8.1	3.2
Netherlands	54.2	42.3	3.4	0.1

*Including nuclear, hydroelectric, and geothermic energy
Source: British Petroleum, *Statistical Review of the World Oil Industry, 1973*

Table 5
DOMESTIC PRODUCTION AS PROPORTION OF DOMESTIC CONSUMPTION, BY ENERGY SOURCES,
SELECTED EUROPEAN COUNTRIES, 1972

Country	Coal	Petro-leum	Natural Gas	Hydro-elect.	Nuclear	TOTAL
Belgium	65	—	—	2	—	16
Denmark	—	—	—	—	—	—
France	69	1	54	30	9	23
Germany	115	7	64	5	3	50
Great Britain	98	2	97	2	11	51
Italy	5	1	93	33	3	15
Luxembourg	—	—	—	42	—	2
Netherlands	63	6	171	—	1	70

Source: OECD, *Long-Term Energy Assessment, Preliminary Draft Report*, Paris, September 11, 1974

tance. Throughout the nineteen-sixties, Europe continued to ignore the growing power of that tightening organization. As a result, by the time of the oil crisis, dependence on foreign sources of supply was enormous in all European countries. Excluding the Netherlands, the external sources of energy supply in these countries reached or passed 50 per cent of their energy consumption.

Most of this dependence was on petroleum, although the percentages varied from 52 in Great Britain to 79 in Italy. Some also did much better than others in terms of total self-sufficiency, and these differences are likely to increase in the future as new discoveries of oil and natural gas are made in the North Sea. Indeed, it is probable that self-sufficiency in energy will shortly be achieved in Northern Europe (Great Britain, the Netherlands, Norway, and, to a limited extent, Germany). Southern Europe, including France, Italy, and Spain, have much more limited prospects for autonomy, at least in the short and medium terms.

National Differences in Institutions and Policies

In this situation of general energy dependence, each nation has responded according to its philosophy, its institutions, and its interests. Not the least of these philosophical differences have to do with the national approach to the regulation of materials. The various approaches range from French-style *dirigisme* to the German free-market approach. The other countries in Europe fall between the two models, but most are closer to the German pattern. The French regulation of its national oil market, beginning in 1928, coupled a strong measure of public intervention in the market with a system of guaranteed market shares for the eight main companies (among which three are under direct government control) that account for 90 per cent of the market. The 1928 policy apparently tolerated and even authorized agreements among the eight to share markets and prices.[9] As part of the same approach, the French government participated directly in the oil market through its ownership of 35 per cent of the shares of the Compagnie Française des Pétroles (CFP) and its 100 per cent ownership of Elf-Erap. In short, French policy was an intricate marriage between *dirigisme* and monopoly. Later on, it would prove an efficient tool to support a policy of national independence, to foster cooperation with the oil-producing countries, and to contain the role of the international oil companies in the French market.

French government intervention took several different forms, but all of them were designed to give to the CFP the task of defending the national interests among the "club" of international companies, and of allowing Elf-Erap to act as agent for French policy in negotiations and relations with the producing countries.

Indeed, all the European countries tended to use their owned enterprises as agents in any efforts to enlarge the European presence in the oil-producing countries of the Middle East and Africa. They normally operated in conjunction with the publicly owned enterprises of the oil-producing countries, as was the case, for instance, in 1957 when the government-owned Agip of Italy and the government-owned NIOC of Iran entered into their initial agreement. A similar pattern was followed in other arrangements including those of French Elf-Erap with the national oil companies of Iran and Iraq. This interaction was a factor in the development of attitudes in the producer countries with regard not only to the oil situation but even to the whole process of development.

It cannot be said, however, that the interaction was especially helpful in the end to the European countries that promoted it. The French, along with everybody else, experienced four-fold increases of oil prices. In the end, Italy's ENI—for many years the symbol of Europe's effort to achieve independence from the international oil companies—fell into a pattern of cooperation with them. Eventually, the international companies' complaints in Italy were directed not against ENI but against the structure of the administrative controls that insulated the Italian market from the rest of the world and artificially created from time to time unduly large profits or losses. It may be that one day ENI will be in a position to reassert some independence of policy by reason of its public status in Italy. But for the present, its heavy reliance on external sources of oil supply and its need to pay more than its vertically integrated competitors for the oil that it acquires make it a relatively cooperative partner of the international oil companies.

Germany, as noted earlier, was the most consistent follower of the free-enterprise system until the time of the oil crisis. Nevertheless, the evolution of the oil market after 1970 produced a change of approach inside the German government as well. Many doubts arose over the possibility that German interests might not be protected in an open market during a period of vast change. Accordingly, the German government introduced a stricter surveillance on the international companies' activities. Moreover, the German government decided to merge the two firms in which there was public ownership, Veba and Gelsenberg. Accordingly, Germany came to acquire a relatively large government-owned oil firm with a significant share of the German market. Germany also had broken old patterns by negotiating bilateral agreements with oil-producer countries, particularly with Iran. In Germany, therefore, the crisis has generated a system now much closer to the French model.

Britain, on the other hand, has always been influenced by the presence in London of two big oil companies, British Petroleum and Royal Dutch/Shell, long-term partners with the United States firms in the development of Middle East oil. Although British Petroleum's principal stockholder is the British government, the British companies have generally had looser links with their government than those on the Continent and have never discriminated in favor of their country.

The Days of the Crisis

The petroleum crisis beginning in October, 1973, had two phases: the first, a reduction in oil production; the second, an increase in prices. Consequently, the impact of the crisis on Europe also occurred in two different periods. The first phase, between October, 1973, and February, 1974, was associated mainly with scarcity, and the second, beginning in March, 1974, reflected the continuing problem of rising prices.

Between October and November, 1973, Arab countries comprising the Organization of Arab Petroleum Exporting Countries decided to ban the exportation of petroleum to the United States and the Netherlands and, in the aggregate, to reduce the production of petroleum by 25 per cent.[10] The prime political target of the offensive was the United States. However, it quickly became evident that Europe and Japan were the most vulnerable, for the embargo on the United States could not possibly generate

Table 6A
PETROLEUM CONSUMPTION, BY ECONOMIC SECTOR, 1972
(as per cent of total)

	Europe	Japan	United States
Industry	35	39	21
Transport	28	20	52
Private-Commercial	26	14	17
Electricity	11	27	10

Source: OECD, *Energy Prospects to 1985*, I (Paris, 1974).

an energy deficit on the American market of more than 1.5 per cent. By contrast, the reduced production plus the embargo on the Netherlands implied a possible reduction of 12 per cent of the energy supply of Europe. In addition, any shortage in the United States would primarily affect transportation, while for Europe and Japan, the entire industrial sector was threatened.

The impact on these areas was of two different kinds. One was the realization of the enormous vulnerability of their energy supplies; the other was the realization that Europe was powerless to help shape the outcome of the Middle East problem. Since the withdrawal of the last British forces from the Persian Gulf at the beginning of 1972, there had been no European presence in that area. Not that such forces had been capable in the past of preventing or resolving crisis situations, but their withdrawal was a striking affirmation of the decline of Europe's influence in that part of the world.

A further element of weakness for Europe derived from the distinctions the Arab countries made between individual European states. The Arabs applied a total embargo on the Netherlands, froze exports to "friendly" countries (France, Spain, Great Britain) at the September, 1973, level, and redistributed what remained proportionately among other countries. The embargo against the Netherlands, which was applied in retaliation for the country's favorable attitude toward Israel, assumed a special importance due to the strategic role of Rotterdam in the European petroleum system: some two million barrels of petroleum and petroleum products were funneled daily through Rotterdam to neighboring countries.[11] As a result, Europe found itself in a rather precarious situation.

The Arab embargo did not, however, precipitate any immediate crisis in actual supplies. As a general rule, European countries had about 90 days of oil reserves in storage and about 30 days more in shipment.[12] Coal stocks equivalent to about 80 days' consumption were also on hand. Moreover, the Arab embargo was neither well coordinated nor well enforced.[13] Despite these reassuring facts, the behavior of European governments in the crisis dealt a hard blow to the concept of European unity.[14]

The Dutch request for support from its European partners clashed head on with the desire of France and Great Britain to avoid a showdown with the Arab countries, resulting in a "vulnerable state of indecision" among European governments, combined with a tacit preference for "delegating" to the international oil companies the responsibility of allocating scarce oil supplies among the various countries. The position of these companies in the European market allowed them to assume that task. Notwithstanding the entry or the strengthening of various independent companies in the European market during the nineteen-fifties and -sixties, the seven great multinational companies directly controlled about 65 per cent of Europe's supply of petroleum in 1973.[15] The ineffectualness of both the producing and the consuming nations in the crisis ensured that the actual handling of the emergency would be assumed by those companies.

Not surprisingly, the reallocation of available resources was influenced first of all by the commercial objectives of the companies themselves. Needless to say, these did not necessarily coincide with an equitable distribution of the supply among countries.[16] One criterion for reallocation was to provide their own subsidiaries with the supplies they needed. Another was to favor the countries with the highest domestic prices for petroleum products. Although the search for maximum profit was not the only factor that determined the commercial policy of the companies, their market power was a decisive element in inducing importing governments to raise prices of the various products when they requested it. In fact, Germany, which adopted a substantially free-market policy, encountered a much less critical supply situation than those countries that did not immediately raise their internal prices at the request of the companies. As may be seen in Table 6B, Germany maintained a higher level of domestic prices for the major petroleum products than any other European country.

Still more significant is a comparison of the margins of average profit from the sale of oil products. In December, 1973, sales in the Netherlands showed a margin of about $24 per ton as compared to $20 for Italy, $14 for France, and a negative margin of $1.40 in the United Kingdom. In Rotterdam, where prices were 66 per cent to 100 per cent higher than other markets in Europe, a theoretical average margin of about $94 per ton of oil could be calculated.[17]

In one important respect, the October, 1973, crisis differed from the crisis periods of 1956 and 1967; cooperation between companies and governments did not develop. From the beginning of the crisis, mistrust and doubt prevailed regarding the capacity and will of the companies to satisfy national interests. "The dilemma," as Edith Penrose so well stated it at the time, "is a classic example of the power of multinational corporations which escape the control of any single government by being under the control of them all."[18] Another factor in the deterioration of the relationship between the European governments and the international companies was the conviction of the governments (not always supported by the facts) that the companies were in some way influenced by the interests and the policies of the United States government. There was also widespread suspicion regarding the role that the companies played in the crisis itself.

Table 6B
PRICES OF SOME PETROLEUM PRODUCTS IN EUROPEAN COUNTRIES
(DECEMBER, 1973-JANUARY, 1974)

	Belgium	France	Germany	Italy	Netherlands	United Kingdom
Hi-test gas (lire/liter)	55	59–90	67–99	63	68–65	42–46
Gasoil (Cars) (lire/liter)	40	46–59	59–92	51	61	40–43
Medium oil (lire/kg.)	32	22–34	40	25	15–18	30–33
Heavy oil (lire/kg.)	25	18–29	21–41	19	17	30–32

N.B. The first figure in each column refers to December, 1973, and the second to January, 1974. A single figure indicates that prices remained unchanged during the period.

Source: Our reworking of data from *International Crude Oil and Product Prices*, October 15, 1974 (Ramos and Parra), and from *Petroleum Times, passim.*

Although never verified, both the governments and the public suspected that the multinational enterprises had been preparing themselves to handle reductions in oil exports for some months,[19] and this suspicion tended to increase the general discontent.

The fact was, however, that the European governments never perceived exactly what was going on. Even while the European ports were experiencing difficulties in handling all the arriving tankers, and while Europe was running out of space for the storage of surplus oil, the governments of all European countries were announcing the need to adopt austerity measures.[20] Limits were commonly placed on the consumption of some products, while controls were also placed on their exportation. Belgium, the Netherlands, Italy, Spain, and Great Britain were all major exporters of refined products; all imposed such export controls.

The effect of these national restrictions was gravely to distort trade among countries that were traditionally net importers of refined products. Germany, for example, found itself cut off from its usual sources, because refineries in other European countries were prevented from exporting. The attempt to manipulate the exports of refined products to accommodate the domestic market was futile so long as the importation of crude oil was not also under control. The multinational enterprises, it was clear, simply rerouted their crude oil toward refining countries that had not imposed controls, thus nullifying any impact these controls might otherwise have had once inventories had been consumed.

Measures limiting consumption differed from country to country, depending on the relative importance of petroleum in the nation's energy supplies, on the balance of payments, and on political relations with the producing countries.[21] In general, the objective of the governments was to reduce the consumption of oil from previously predicted levels by 10 to 15 per cent. Since no reliable data existed regarding the price elasticity of demand for oil, governments were obliged to work largely in the dark. The policy that emerged in practically all countries was to try to spare the industrial sector by placing restraints on private consumption. Accordingly, gasoline consumption was widely curtailed, even though gasoline represented only 12 per cent of total consumption and the restrictions adopted (such as a ban on Sunday driving and the imposition

Table 7
PRINCIPAL MEASURES ADOPTED BY FOUR EUROPEAN COUNTRIES
DURING THE ENERGY CRISIS OF 1973-74

Restriction	France	Germany	Italy	United Kingdom
Speed limit	X	X	X	X
Ban on driving on stated days		X	XXX	
Rationing of gasoline	X	XX	XXX	X
Restrictions on heating and lighting	X	X	XX	XX
Control on exports	X	X	X	XX

N.B. The crosses are meant to reflect the degree of hardship caused by the measure in each country.

of speed limits) could only have negligible results. Restrictions were also applied to heating and lighting, by means of the rationing of fuel deliveries, the early closing of businesses and public offices, the reduction of street lighting, and the curtailment of hours of television transmission. The most critical product in terms of availability, however, was fuel oil, which represented more than one-third of total consumption and constituted the principal product used by industry. Specific measures to reduce deliveries were adopted in various countries, especially in Italy and Great Britain.

It is difficult to ascertain to what extent governmental policies helped to limit the growth of demand. A certain slowing down in oil consumption had already been apparent in the latter part of 1973. The winter of 1973-74 was also unusually mild in Western Europe. The decrease in consumption may have stemmed partly from the exceptional increase in the price of petroleum products, a result of the increased costs of crude oil and the rise in the level of excise taxes. In any event, *there was at no time a real shortage of petroleum on the European market.* Consumption simply responded to the increase in prices, thus, incidentally, casting some doubt on the presumed "rigidity" of petroleum demand. Between October, 1973, and April, 1974, the reserves of oil products in the countries of the European Community never descended below the 80-day equivalent of consumption; and in Italy the reserves in fact increased by 23 per cent.

Between December, 1973, and March, 1974, the availability of petroleum in the five principal European countries—France, Germany, Great Britain, Italy, and the Netherlands—was only about 5 per cent lower than it had been in the same period a year earlier. Availability was reduced by 16 per cent in the Netherlands, 12 per cent in Germany, 7 per cent in France, not quite 1 per cent in Great Britain, and actually increased by about 4 per cent in Italy.

The decline in Europe's oil consumption during the crisis is to be attributed, then, not to a lack of crude oil, but rather to a contraction of demand caused in turn partly by changes in prices, partly by the replacement of oil with other sources of energy, and

Table 8
Oil Reserves in EEC Countries: Selected Dates 1973-74
(in number of days of consumption)

Country	1/4/73	1/10/73	1/1/74	1/4/74
Belgium	98	114	88	66
Denmark	64	91	75	74
France	94	111	101	88
Germany	69	80	72	70
Great Britain	84	93	80	77
Ireland	96	64	62	62
Italy	80	79	97	98
Luxembourg	21	17	17	17
The Netherlands	64	97	91	90
EEC	80	90	80	81

Source: EEC, *Importations de pétrole brut dans la Communauté*, règlement 1055.

Table 9

AVAILABILITY OF PETROLEUM, COAL, NATURAL GAS IN 5 EEC COUNTRIES: DECEMBER 1972-
MARCH 1973 COMPARED WITH DECEMBER 1973-MARCH 1974

(in million of tons oil equivalent)

	Great Britain	Germany	France	Italy	The Netherlands	TOTAL
	December, 1972-March, 1973					
Oil	39.0	46.1	42.4	32.9	12.2	172.6
Coal	35.8	21.9	10.2	2.6	1.1	71.6
Natural gas	9.9	9.8	4.1	5.4	11.8	41.0
TOTAL	84.7	77.8	56.7	40.9	25.1	285.2
	December, 1973-March, 1974					
Oil	38.7	40.8	39.3	34.2	10.3	163.3
Coal	20.0	23.6	10.0	3.1	1.2	57.9
Natural Gas	12.5	11.9	5.5	6.3	12.5	48.7
TOTAL	71.2	76.3	54.8	43.6	24.0	269.9
	Percentage change between 1972-1973 period and 1973-1974 period					
Oil	−0.8	−11.5	−7.3	+4.0	−15.6	−5.4
Coal	−44.0	+7.8	−0.2	+19.0	+9.0	−19.1
Natural gas	+26.3	+21.4	+34.1	+16.7	+5.9	+18.8
TOTAL	−15.9	−1.9	−3.3	+6.1	−4.4	−5.4

N.B. In the case of oil, availability is calculated as the balance between importations of crude petroleum and petroleum products and exportations of petroleum products; in the case of coal and natural gas, in addition to the balance, account was taken of domestic production.

Source: Official statistics of individual countries.

partly by redundant governmental restrictions. In Germany, for example, by dipping heavily into its reserves, coal consumption was increased by 7.8 per cent, and natural-gas consumption by 21 per cent. As a result, the total consumption of energy in Germany declined by only 2 per cent. By increasing the consumption of natural gas, the Netherlands and France similarly managed to limit the decline in their energy supply to only 3 or 4 per cent.

In the case of Great Britain, the greatest difficulty resulted from the reduced availability of coal as a result of a miners' strike.[22] In the first quarter of 1974, the production of pitcoal in Great Britain was reduced by about 60 per cent, dropping from 38 million tons to 16 million tons. This reduction was only partly compensated by the increase in the consumption of natural gas, which went up 26 per cent. The three principal sources of energy—petroleum, coal, and natural gas—accordingly dropped by 16 per cent in terms of thermal equivalents.

Italy, meanwhile, paradoxically managed to register a 6 per cent *increase* in the consumption of energy during the height of the crisis. The explanation is to be found either in the total lack of elasticity in Italy's energy consumption patterns, or in the lack of government interaction, or in the distortion of various phases of the distributive cycle, which greatly encouraged stockpiling.

From the material discussed so far, a picture of the petroleum crisis emerges that is not so dramatic as governments and petroleum companies have labored, for various reasons, to depict. It seems evident that, so far as Europe was concerned, the oil crisis did not bring on an overall shortage in either oil or energy. The difficulties of the crisis were mainly the result of disturbances in the internal distribution of oil products, which were triggered by the resistance of governments to the price increases imposed by the oil companies, and which led to delays in delivery, discrimination against independent companies, and the speculative hoarding of stocks.

After the Crisis

On March 18, 1974, the Arab nations meeting at Vienna decided, over the objections of Libya and Syria, to ease their restrictions. Italy and Germany were added to the list of friendly countries, and the embargo against the United States was lifted. Paradoxically, the embargo on the Netherlands stayed in effect as did the special limitation on exports to Denmark. Saudi Arabia's intransigence regarding Denmark was attributed by *Le Monde* (April 4, 1974) to "the American wish to hinder the European-Arab dialogue," which was supposed to take place shortly thereafter. "Riyadh," continued the French newspaper, "has agreed to become Washington's partner in this anti-European maneuver."

After March, 1974, the remaining Arab restrictions had no real impact: The availability of oil on international markets easily exceeded the level of demand.[23] European refineries were working at 60 to 70 per cent of their capacity. The governments of the producing countries were finding it increasingly difficult to dispose of even the reduced amount of crude oil they were producing and impossible to sell free oil at their target prices. They were obliged, therefore, to turn back to the international oil companies both to distribute their output and to keep a tight hold on the structure of market prices.[24] The supply phase of the crisis had passed; the key question for public policy now became that of prices.

As early as December, 1973, the heads of member states of the EEC in Copenhagen had charged the Commission with the problem of sorting out the differences in prices and profit margins for petroleum products in the individual European countries. In mid-January, the EEC Commission proposed to get rid of the worst price distortions in the Common Market,[25] suggesting both the harmonization of internal price structures and the freezing of the prices themselves.

By March, 1974, the crisis mood had dissipated, and resistance on the part of governments to the rising prices of the oil companies began to increase, discouraging the OPEC countries from raising prices further. As the leading multinational enterprises began to reveal their exceptional profits for early 1974, particularly on operations outside the United States, resistance to higher prices increased still further.[26]

In the first half of 1974, the five largest American companies—Exxon, Mobil, Socal, Gulf, and Texaco—registered an increase in net profits of 68.3 per cent over the previous year. These profits, as it turned out, were partly illusory, a result of the higher prices placed on inventories. Less illusory, however, were the costs that European countries had to bear. Governments had come to distrust the oil companies with regard both to scarcities and rising prices. Even Germany, the principal proponent of the free market in Europe, in April, 1974, ordered the leading oil companies operating in its domestic market to reduce prices of diesel fuel and gasoline.[27] "It is inadmissible," declared German Minister of Commerce Hans Frederichs, "that we should be kept in the dark about sales, policies, prices, and profits of the international petroleum companies, which are operating in our territory and which are behaving like a state within a state."[28] From the German case, it clearly emerges that the only interlocutors of governments on the problem of prices could be the parent firms and not the local subsidiaries, inasmuch as only the former were able to provide precise data on the real cost of supplying petroleum in the various countries.

The objective of illuminating the operations of the international oil market and of the companies that participate in it has been one of the principal themes of German spokesmen in international gatherings. In the opinion of Chancellor Schmidt, the principal consumer countries will also have to develop a coordinated fiscal strategy for taxing the multinational firms, given their ability to escape the control of any individual state. Meanwhile, the EEC Commission has taken a similar position favoring a broad international approach. EEC anti-trust specialist Albert Borschette said before a meeting of the European Parliament in May, 1974, ". . . to solve all the problems posed by multinational corporations it is necessary to go beyond the limits of the Community in search of a wider structure such as the OECD or the United Nations."[29]

Complaints against the oil companies were to be heard in practically all the major consumer nations: in Germany, Belgium, Italy, and France, among the Europeans; and in Japan, as well, where the national petroleum association was accused of having concluded a cartel agreement from the second half of 1972 to the first half of 1973;[30] and, finally, even in the United States.[31]

The obstacles to reaching international agreement on the control of multinational enterprises have led individual countries to adopt their own national measures. In Great Britain, the government widened the powers of its price commission; in Belgium, where the dispute between the government and the oil companies had helped precipitate a government crisis in March, 1974, a mixed labor-industry Committee for Control of the Petroleum Industry was formed, with the task of backing the government in the formulation of a petroleum policy;[32] in Italy, in April, 1974, the International Committee of Economic Planning (CIPE) approved a "petroleum plan" directed at a complete review of petroleum regulations with the objective (although thereafter pursued in only desultory fashion) of strengthening public powers in this sector.

The Role of the European Economic Community

So far, the focus has been on the positions and policies of the individual nations of Europe, but what of the European Community itself? The European institutions concerned with energy, it will be remembered, include the European Coal and Steel Community, the European Economic Community, and the European Atomic Energy

Community. After its formation, Euratom languished and could be counted nearly a total failure. This failure was due chiefly to rivalry between member states, which blocked the effective coordination of national projects that had any direct industrial interest. Though many bilateral accords were concluded by member states, the European Community was effectively excluded.

The policies of the ECSC as they related to the energy field were indicated by its protocol of agreement on energy policy, and, as we have already seen, concentrated on making the price of coal more competitive. This was to be achieved not through an increase in the price of imported petroleum but through subsidies in the sale of coal.[33] On this score, ECSC was adapting itself to prevailing national policies, even if this went against the very rules which had been built into the ECSC treaty itself. The same lack of initiative could be seen in various memoranda presented by the Commission of the European Community to its Council of Ministers.[34]

The contents of a key 1972 memorandum on the subject are illustrative. The stated objectives are contradictory: low prices, security of supplies, and competition between the different energy sources. The only concrete measure for the achievement of these goals is an obligation on the part of the Community countries to stockpile 65 days' consumption of petroleum products, a period that was prolonged to 90 days by a decision taken in 1973. The inconsistencies between the low-price objective and the security objective were passed over; the Community was unwilling to confront the question of the price for obtaining greater security of supply.

Had the Community confronted the issue, it would have realized that the price for increased security would be the increased domestic production of energy—an expensive policy. The United States was already paying this price, but the European coal lobby was too weak to impose it and the interests of the various individual countries appeared to outweigh the objective of a common European policy. For the same reason, an attempt to introduce Community-sponsored legislation to create a so-called Community Petroleum Corporation also failed. A number of European oil companies asked the authorities in Brussels to sponsor legislation that would offer them the same tax advantages that American firms had long enjoyed. The discussion revolved principally around the "depletion allowance," by which oil exploration was strongly encouraged, and around other tax provisions (especially the so-called foreign tax credit), which American companies also enjoyed. These incentives had in the past been the mechanisms for the expansion of American companies abroad, and their absence in Europe constituted a serious barrier to the international development of European companies. The project contemplated that the European subsidiaries of United States firms would not be eligible for any such liberalized provisions under European law. To extend these benefits to such subsidiaries, as they requested, would have perpetuated the relative disadvantages of European firms. Faced with the problem, the Community found itself unable to act, and contented itself with producing a long series of memoranda on the subject.

This indecision was reinforced by disagreements among the member countries themselves. Proposed by ENI from the time the Common Market was established, the European Corporation was finally supported by France in 1963. By 1967, a loose agreement existed between Italian, French, and German companies.[35] But the hostility of American firms, strongly fortified by the mistrust of the Commission and the indifference of national governments, led to the jettisoning of the idea.

From the documents of the Community, it may appear as though nothing had happened in the world energy market until the outbreak of the 1973 crisis. By 1974, however, a document was drawn up suggesting the priorities that a common energy policy should address.[36] These were defined as relationships with other importing countries, relationships with exporting countries, and regulation of the market. But that was as far as the Community could go; the crisis had so aggravated the differences among countries that nothing further was agreed upon. Nevertheless, the Commission did identify some objectives of European policy. Equilibrium in the long-term balance of payments was identified as one objective; special attention to the protection of the environment and the quality of life was another.

In accordance with these objectives, plans were made to reduce energy consumption for 1985 by 10 per cent from earlier projections, through a more efficient use of energy. There were also proposals to augment the share of electricity in total energy consumption from 25 per cent to 35 per cent, and to reduce the share contributed by petroleum from a projected 60 per cent to 40 per cent of the total consumption of energy. This would be achieved by reversing the cutback in coal production, increasing the importation of coal, and increasing the domestic production of oil, natural gas, and nuclear energy. None of these goals was associated with a tangible program of action, however; they were rather predictions of what would happen to European energy, following the enormous price increases in oil, regardless of any political decision.

Though the institutions of the European Community found themselves unable to make much progress on a common policy in the energy crisis, they were not prevented from arranging and encouraging a considerable amount of interaction among representatives of the member states. On November 6, 1973, the nine foreign ministers at Brussels tried to reach a common position on the conflict in the Middle East. The resolution that emerged emphasized the obligation of Israel to give up the territories occupied in 1967, and endorsed United Nations Resolution 242 as a point of departure in the search for a peaceful solution. The resolution was favorably received by the Arabs; on November 18 they suspended their earlier decision to cut back supplies to Europe another 5 per cent.[37]

The Brussels meeting did not, however, resolve any of the most urgent problems that divided the countries of Europe. In mid-December, government heads met again at Copenhagen to search for a definition of a common course of action in the emergency. The fundamental cleavage was between the Benelux countries, who requested action on the Communal level, and the French and the English, who were eager to maintain their favored positions on the Arab preference lists. As a result, all that emerged at Copenhagen was an agreement voluntarily to provide certain statistics on the energy situation to Community organizations. The fact that, after about a month, no data had yet reached the Community was in keeping with the general spirit of the Copenhagen effort.[38]

With action for the short run having been blocked, the Community organizations turned to problems of strategy over the long term, leading to the document described earlier. Even these innocuous proposals by the Commission, however, got nowhere. Britain expressed strong opposition, claiming to be dissatisfied with the general character of the text;[39] its real worry, however, was to avoid the least diminution in national control over the resources of the North Sea. Great Britain, aided from time to time by the Netherlands, blocked every attempt to define a common energy policy, fearing that this might eventually force it to share its North Sea reserves with other Common Market countries.

As international oil supplies grew easier, however, the British opposition also relaxed.[40] On September 17, 1974, the foreign ministers of the Common Market countries were able to reach an agreement on the "long-term project" under the same conditions that had been rejected two months earlier. The only emendation imposed by the British was a provision that the European Community in Brussels might deal directly with developing countries. Even though the agreement was only a declaration of intention, it assumed importance both because it was the first substantive agreement of any sort reached since the beginning of the crisis, and because it responded to a widespread conviction that agreement within the Community was a prerequisite to solving the energy crisis for Europe.

Building on the modest success of this meeting, the energy question was inserted into the agenda of the "summit" held at Brussels on December 10, 1974. In addition to energy, the agenda included a discussion of economic and monetary cooperation in the fight against inflation, the creation of a European fund for regional development, and the problems of renegotiating Britain's position in the Community. After twenty years of nearly complete oblivion, energy once again was one of the principal problems on the Community agenda.[41]

At this stage, however, still another restraint on European policy-making became apparent. France was advocating a direct encounter between the producing countries and Europe without coordinating with the United States; Great Britain and Germany, on the other hand, were eager not to interfere with the United-States-led effort in the International Energy Agency.

Once again, in the new Brussels Conference, the member nations attempted to obtain the adherence of France to the IEA as an act of good will in dealing with the United States; in exchange, Bonn showed itself willing to join the trilateral conference (producing countries, industrialized consumer countries, and underdeveloped consumer countries) proposed by the French president at the end of October. On the problem of joining the international agency, France assumed a more flexible attitude, accepting the fact that the Community might participate as observer, but maintaining a negative position on its own participation so long as the producing countries continued to regard the IEA as hostile. In this situation, it is no wonder that the Common Market proved incapable of undertaking any important unified political initiative. It was not even capable of constructing a common strategy for the energy conference convened in Washington from February 11 to 13, 1974, with the participation of the nine Common Market countries and the United States, Canada, Japan, and Norway.

Prior to the Washington conference, numerous efforts were devoted to devising a common European policy with special attention to mediating the differences between France and other countries. The most important points of this common European platform were the refusal to transform the Washington conference into a permanent organization (so that the problems of energy could be brought back into the sphere of existing international organizations), and the desire of the Community to maintain its freedom in direct relationships with the producer nations. All the European countries wanted to avoid tensions with the Arab world.

These slender preliminary agreements were not sufficient to avoid a dangerous rift between the Europeans at the negotiating table: the Washington conference thus ended by being "une pénible journée pour la Communauté."[42] So strong were the tensions that France refused to subscribe to the most important and meaningful clauses of the final communiqué.

Once again, unanimity could be reached only on long-term problems about which it was really very hard to disagree. These included the necessity to increase efforts to rationalize energy consumption, prepare an allocation schedule for periods of crisis, develop alternative sources of energy, and promote programs of research and development in the energy field.

These results were minimal, especially when one considers the French refusal to participate in the Energy Coordinating Group and the obvious tensions between France and Germany. The Washington conference represents a low point in the already somewhat less than glorious history of the Common Market: when the participants were not in outright disagreement, they were forced by their lack of unity to talk in completely general terms about present difficulties and future dangers. The American position, which was directed at confronting energy problems through general cooperation among all countries, also generated profound skepticism among the Europeans, precisely because these proposals were formulated at a time when it was becoming dramatically clear how difficult it was going to be to conciliate opposing interests among producing and consuming countries, and even among the consuming countries themselves.[43]

Convergences and Divergences in the European Course of Action

The agreements reached through European organizations during the crisis were painfully few, because each government insisted on maintaining the widest possible room for maneuver in the crisis. That goal persisted even as the dimensions of the energy crisis began to be apparent and began to entail profound modifications in the relationships among the producing countries, oil companies, and consuming countries. Perhaps it was the very scale of these modifications, unlike the modest shifts in the crises of 1956 and 1967, that led to the national emphasis on freedom of maneuver. In the 1973 crisis, unlike the earlier ones, there was full-scale competition between the large consuming areas—Europe, the United States, and Japan. The market offered no escape route from the blackmail imposed by the OPEC countries.

The sense of frustration and demoralization in Europe over its inability to develop an effective role in the crisis was increased by the fact that practically all European countries at the time were caught up in internal political difficulties not directly related to energy. Between November, 1973, and March, 1974, governments fell in Great Britain, France, Italy, Germany, and Belgium. These temporary weaknesses prevented bold initiatives in response to the actions of oil-exporting countries. The failure to respond immediately and effectively contributed to the image of Europe's growing vulnerability, and to the political and economic price of the blackmail that would have to be paid.

Another important aspect of the problem in which the states displayed their impotence lay in their relations with the multinational corporations. The image of compatibility of interests, present until then, received a profound blow when the crisis came. The control that the major international companies exerted on the foreign trade in oil led the states to impose various internal regulations in their respective markets, aimed at restricting the autonomy of the petroleum companies. This was a reaction common to all European countries.

A second common reaction in Europe, closely connected with the first, was to strengthen state-owned corporations in the European oil market. This was supposed to increase the bargaining power of the states in their dealings both with the oil companies and the oil-exporting countries. The companies were no longer seen as a buffer force between the importing and exporting countries; instead, the conviction grew that the international oil companies had linked their long-term interests with the stronger side of the market, that is, with the OPEC countries. That conviction was reinforced by the persistence of high international prices for crude oil in spite of the prevailing situation of surplus.[44]

The third principal line of response in Europe to the oil crisis concerned the relationship between the various countries and the oil-exporting countries. Herein lies the nub of the differences between the United States and Europe, and sometimes even among the European countries themselves. The terms in which the debate has been conducted may be expressed in the slogan "American multilateralism against European bilateralism." The "bilateral" choice of the European countries in dealings with the producing countries originated simply from the fear, during the early part of the crisis, that they would run out of crude oil, and from the necessity of alleviating an alarming deficit in the balance of payments; it was short-run pragmatism, not long-term philosophy, that produced the emphasis. American bilateralism also existed, but it manifested itself through the big private American companies, rather than through state-owned corporations or the government. Europe, however, was responding to a basic change in world conditions, not creating it. The old structures that once had regulated world markets no longer served, and there was not much sense in defending them. Europe was also charged with adding to price increases, even above the levels set by OPEC, as a result of this bilateralism. The highest prices paid for oil, however, were those arising out of the purchases of Japanese trading companies and by independent United States refineries. The charges levied against the Europeans on this score therefore appear to be unfounded.

Europe sees multilateralism as an appropriate long-term strategy, but accepts bilateralism in time of crisis as a second-best, short-term solution, although it cannot say so openly because any such declaration would be regarded by the United States as a politically hostile move. All oil-importing countries continue, therefore, to pursue various bilateral possibilities, even though all but France profess a distinct preference for multilateral action. The fact that the EEC countries—again with the perennial exception of France—have joined the IEA does not in itself represent the defeat of the bilateral strategy. The pursuit of a collective IEA program, entailing stockpiling and sharing, is not incompatible with the pursuit of bilateral accords; it simply represents another aspect of a process that will reduce the power of the oil exporters' cartel. Bilateral contracts and agreements between European countries and producing countries are gradually multiplying with the passing of months. Although there have also been direct contacts between the European Community and the Arab countries, the long-term foundation for relations between Europe and the Middle East will continue to be based on the reconstitution of the petroleum market.

The expectation that more normal market conditions in oil will reassert themselves rests in part on the astonishing increase in petroleum discoveries during 1974 and 1975,[45] the beginning of an oil flow from the North Sea fields, and the gradual reduction of oil imports by the United States. These trends will restore the dependence of the oil-producing countries in the Middle East on the markets of continental Europe and Japan. Since the intermediary role of the international corporations has already been weakened, the relationship between Europe and the Middle East will become still more direct.

Accordingly, even if the European countries are not capable of developing a common policy on oil, they are nonetheless pursuing patterns of behavior that run parallel and that could lead to a common solution. The risks of failure and of disintegration are still enormous, as they always have been. But today's market forces in oil are working in Europe's favor: the problem is to let them work freely.

REFERENCES

[1]One of these is the fact that the costs of using different fuels may vary, especially the costs of investment and of plant operation; cf. L. Gouni, *Aspects économiques de la concurrence entre combustibles* (VI Conférence mondiale de l'énergie, 1962).

[2]Organization for European Economic Cooperation (OEEC), *Europe's Growing Needs of Energy, How Can They Be Met?* (Paris, 1956).

[3]M. A. Adelman, *Security of Eastern Hemisphere Fuel Supply* (1963).

[4]The reductions by country are: United Kingdom, from 560,000 to 285,000; Germany, from 510,000 to 190,000; Belgium, from 119,000 to 34,000.

[5]O. Thur, "Le funzioni economiche dei prezzi dell'energia," *Economia Internazionale delle Fonti di Energia (EIFE)*, January-February, 1960.

[6]M. Ippolito, *Contribution à l'étude du problème énergetique communautaire* (Paris, 1969), p. 195.

[7]*Ibid.*, p. 321.

[8]In 1960, the average yield of refineries located in Western Europe was 37 per cent of fuel oil and 21 per cent of gasoline, as against an average yield in the United States of, respectively, 11 per cent and 48 per cent.

[9]*Le Monde*, November 8-9, 1974; report of the Parliamentary Commission of Enquiry on the commercial, financial and fiscal conditions under which the petroleum companies operate in France (by M. Julien Schvartz), November 6, 1974.

[10]All told, these restrictions implied a contraction of 13 per cent in the supply of crude oil on the international market.

[11]*Petroleum Intelligence Weekly (PIW)*, October 29, 1973.

[12]*Middle East Economic Survey (MEES)*, October 19, 1973.

[13]In particular, it was not clear whether the embargo on the United States was also aimed at the quantities destined for countries that export refined products to the United States, including especially the Caribbean countries, Canada, Italy, and Greece.

[14]Louis Turner, "The European Community: Factors of Disintegration Politics of the Energy Crisis," *Foreign Affairs*, July, 1974.

[15]EEC, *Importations de pétrole brut dans la Communauté*, règlement 1055.

[16]Edith Penrose, "Multinationals: Dilemma for Individual Countries," *Times* (London), February 22, 1974.

[17]*PIW*, December 3, 1973, pp. 5-6.

[18]Penrose, *op.cit.*

[19]The excellent account of the crisis by Robert Stobaugh in this issue gives us the opportunity to verify how true this impression was.

[20]A sensation was caused by an article in *The Economist* on December 15, 1973, entitled "How Scarce Is Oil?"

[21]*Financial Times*, December 11, 1973.

[22]Beginning on November 12, 1973, the strike affected only overtime hours, but from February 10 to March 11, 1974, it was total.

[23]The surplus in supply calculated in the first third amounted to 1 million barrels a day; *The Petroleum Situation*, May 31, 1974 (The Chase Manhattan Bank).

[24]See the case of Abu Dhabi in *Platt's Oilgram News Service (PONS)*, March 11, 1974.

[25]*Comunicazione della Commissione al Consiglio sulle misure da prendere per far fronte all' attuale crisi energetica nella Comunità*, COM (74), 40 def. Strasbourg, January 16, 1974, p. 6.

[26]*The Economist*, April 27, 1974. According to one calculation reported in *PIW* on April 29, 1974, p. 6, the average margin of the petroleum industry on the European markets was multiplied 2.7 times between 1972 and October-December, 1973, growing from 35 cents per barrel to 95 cents per barrel.

[27]*PONS*, April 2, 4, 6, 1974.

[28]*Enterprise*, May 24, 1974, p. 45, *International Herald Tribune*, May 3, 1974.

[29]*PIW*, May 27, 1974, p. 2.

[30]*International Herald Tribune*, May 29, 1974.

[31]*New York Times*, May 29, 1974.

[32]*PONS*, April 16, 1974.

[33]D. Swann, *The Economics of the Common Market* (London, 1974), pp. 149-65.

[34]Among these, cf. EEC, *Previsioni e Orientamenti a medio termine per il settore petrolifero nella Comunità* (Brussels, September 25, 1972).

[35]Cf. *Mesures tendant à établir en faveur des entreprises petrolières communautaires des conditions équitables d'activité (Note des Societés pétroliers communautaires à la Commission de Bruxelles)*, February 15, 1967.

[36]Cf. *Towards a New Energy Policy Strategy for the European Community* (Communication and Proposals from the Commission to the Council), Com (74) 550 final, Brussels, May 29, 1974.

[37]This reduction remains valid for the other countries: in substance, in place of the predicted reduction for December of 750,000 barrels per day, there was a reduction of 375,000 barrels per day, after subtracting the needs of the Common Market.

[38]*PONS*, January 7, 1974.

[39]*Le Monde*, July 25, 1974.

[40]*The Times* (London), September 18, 1974.

[41]If we examine the agenda of the Council of Ministers of the Common Market, it becomes clear that the problem of energy enjoyed primary importance only in exceptional situations. For example, in 1967, the energy problem was treated in 8 sessions (out of a total of 39) but, after 1967, the discussions on energy diminished by an average of three a year, with the same importance assigned to the problem of transportion and considerably less than that assigned to the problem of agriculture and to financial questions. Cf. H. Wallace, *National Governments and the European Communities* (London, 1973), p. 70.

[42]*Le Monde*, February 14, 1974.

[43]"The Divergent Views at Washington Energy Summit," *PIW*, February 18, 1974.

[44]In the course of 1974, consumption of oil diminished by 4 per cent in the United States and by 6-7 per cent in Europe (excluding the Communist countries), while the production of crude oil was reduced by only 1.6 per cent. See also M. A. Adelman and Soren Friis, "Changing Monopolies and European Oil Supplies: The Shifting Balance of Economic and Political Power on the World Oil Scene," *Energy Policy Planning in the European Community: The Case of Electric Energy*, Eindhoven, May 29-June 1, 1974, published in *EIFE*, January-April, 1974. Adelman analyzes the effects of the vertical integration of the international corporations after the crisis.

[45]In 1974, world petroleum reserves increased by at least 12 billion tons (about 14 per cent), shifting the time span for the exhaustion of these reserves from 31 to 35 years.

YOSHI TSURUMI

Japan

Introduction

ON JANUARY 24, 1975, Prime Minister Miki of Japan told the Japanese Diet that he regarded the question of Arab oil as being indivisible from the Arab-Israeli conflict; a solution to the oil problem must be preceded by a solution to the problem of Jerusalem. Miki thus confirmed the diplomatic rift that had been developing between Japan and the United States ever since the crisis of the fall of 1973 over the disputes in the Middle East and oil. Miki's declarations positioned Japan closer to France and her European neighbors than to the United States, which was seeking to separate oil from other Middle Eastern problems.

Why was oil so important to Japan that it chose to chart its own diplomatic course in order to continue procuring oil from the Middle East? By 1973, the gross national product of Japan had grown to be the third largest in the world, outdistanced only by the United States and the Soviet Union. Two decades earlier, in 1945, the Japanese economy had been left in shambles by the war. The rapid industrial growth of postwar Japan had included rapid expansion of such heavy industries as steel, shipbuilding, petrochemicals, and machinery. These industries, in turn, relied heavily upon imported oil, whose price had been declining in Japan throughout the nineteen-sixties relative to other forms of energy, including coal and hydroelectricity.

After abandoning coal and exhausting the possibilities of hydroelectric-power development by the mid-sixties, Japan came to rely on oil for over 70 per cent of its total energy needs. Unlike the United States, whose oil consumption for private uses exceeded 40 per cent of its total oil consumption, three-quarters of the oil consumption in Japan went to industrial uses. Lacking domestic sources, by 1973 Japan had come to import over 99 per cent of the crude oil it consumed. On the eve of the oil crisis, Japan's imports totaled about 4.5 million barrels a day. About 40 per cent of these came from the exporters that held dual membership in both the Organization of Arab Petroleum Exporting Countries (OAPEC) and the Organization of Petroleum Exporting Countries (OPEC).

When OAPEC instituted in October, 1973, a 25 per cent across-the-board reduction of crude-oil exports, and thus withheld over 5 million barrels of crude oil per day from the export market (an amount greater than the total imports of Japan for 1973), it classified Japan as an unfriendly state. The result was panic over the possibility that a prolonged shortage of crude oil would destroy Japanese industry. On the eve of the oil crisis, the Japanese economy was already suffering from double-digit inflation. The oil shortage created feelings of uncertainty and helplessness among the Japanese public.

The Ministry of International Trade and Industry (MITI), to which both consumers and industries turned for decisive leadership in the crisis, reacted with apparent hesitation and bewilderment. By the end of the crisis period, however, MITI was found to have renewed some of its regulatory power over Japanese industry. It also attempted, though unsuccessfully, to further weaken the investigative authority of the Fair Trade Commission (FTC)—Japan's anti-trust body and a long-dormant phoenix of the Japanese bureaucracy, whose power had also been increased by the crisis. When the first stage of the oil crisis—namely, the actual shortage of supply—subsided in January, 1974, both MITI and the FTC emerged as renewed power centers in Japanese domestic politics. One way of examining Japan's behavior during this period is to consider the factors that produced this curious outcome.

MITI's Policy Toward the Oil Industry

Before World War II, the industrial energy sources of Japan were coal and hydroelectric power, which were adequately available both at home and from Japan's newly acquired colony, Manchuria. The private and industrial markets for automobile gasoline, bunker fuel, kerosene, and other oil derivatives were growing during the nineteenthirties, but the largest users of petroleum products were the Imperial Forces of Japan. A number of Japanese firms thrived as procurement agents and importers of petroleum products for the military as well as for the small yet growing number of industrial users. As Mira Wilkins' article in this volume reveals in detail, Japan imported petroleum products from the United States and the Dutch East Indies through Standard Vacuum—jointly owned by Socony-Vacuum (now Mobil) and Jersey Standard (now Exxon)—and through Royal Dutch/Shell.

During the immediate post-war years, occupied Japan was not permitted to reconstruct the oil-refining facilities that had been destroyed by Allied bombings, a policy widely attributed in the oil industry of Japan to the fact that the oil bureau of General MacArthur's headquarters was heavily staffed with American personnel on temporary leave from Jersey Standard and Mobil.[1] According to the Japanese individuals who were involved in negotiations with the headquarters oil group, it first resisted the establishment of oil-refining operations in Japan and only changed its attitude when the international oil majors began to prefer refineries closer to a major consuming market. As the Cold War frosted relations between the United States and the Soviet bloc, as the Southeast Asian countries one after the other claimed political independence and often subsequently fell into political instability, and as consumer and industrial demand in Japan for all petroleum products showed signs of marked increases, the international oil majors appeared to have found it both safer and more profitable to locate their oil-refining facilities in Japan. In particular, as Mao's Communist Party seemed certain of winning the civil war in China in 1949, Japan became increasingly attractive as a location for oil-refining bases for the international oil majors. Moreover, with ever-larger tankers available, it became obvious that great savings could be made by transporting crude oil rather than processed petroleum products. Thus, in July, 1949, General Headquarters permitted the Japanese government to begin the reconstruction of oil-refining facilities. With some differences in the specific format of the international oil firms' entry into the Japanese market, Exxon (Esso's parent company), Mobil, Shell, and Getty positioned themselves as *de facto* integrated oil firms in Japan, whose refining and marketing interests were tied to their crude-oil interests held outside Japan. Under the Allied occupation, the Japanese government was powerless to block such business links.

By 1952, when Japan regained its political independence, MITI had emerged as the agency entrusted with the industrial development of post-war Japan. The task of allocating capital, technologies, imported materials, and production facilities among Japanese industries had formerly been carried out principally by a few leading industrial groups called *zaibatsu*, each consisting of interrelated manufacturing firms and financial institutions. Such *zaibatsu* groups as Mitsubishi, Mitsui, and Sumitomo had dominated the Japanese economy. With the dissolution of the *zaibatsu* groups after the war, the task of industrial planning and resource allocation was taken over by MITI.

During the fifties, MITI promoted Japan's push toward heavy and chemical industrial development. Heavy energy- and oil-consuming industries, such as steel, non-ferrous metals, petrochemicals, and synthetic fibers, came to lead the industrial growth of Japan, especially after the mid-fifties. Thus, the question of procuring sufficient quantities of oil at as low a price as possible for Japanese industrial users confronted MITI officials. In order to reduce the dependency of Japanese industrial users on oil products from Japanese subsidiaries of foreign oil firms, MITI came to favor Japanese oil refiners and marketers in Japan that were not in such a dependent relationship. MITI suspected that foreign oil subsidiaries were not to be relied upon for cooperation in achieving the industrial goals of Japan. MITI's control over foreign exchange allotments for the import of vital natural resources, including oil, provided the government with an effective tool for regulating Japanese refining and marketing activities. A "rule of thumb" used by MITI was to structure the oil industry in such a way that independent Japanese oil interests should constitute at least 30 per cent of the market share in both the refining and wholesale stages of the oil industry.

Accordingly, between 1952 and 1954, MITI permitted four Japanese oil companies to enter the refining stage, which had been until then totally dominated by subsidiaries of foreign oil firms. By 1962, the number of independent Japanese oil refineries had increased to seven, and their combined refining capacity approached 45 per cent of the total capacity in Japan. During most of the nineteen-fifties, MITI supported the initiation of refining activities by such fully Japanese-owned marketers of oil as Idemitsu, Maruzen, Daikyo, and Nihon Kogyo. The rationale of MITI appeared to be that a limited number of large Japanese oil wholesalers with their own refining operations would best be able to compete with subsidiaries of foreign oil majors. MITI saw to it that the prices of crude oil imported by foreign-owned subsidiaries did not greatly differ from the prices paid by independent Japanese oil refiners to their foreign sources of supply. It welcomed, for example, Idemitsu's successful purchase of Iranian oil in

1953 and its subsequent purchase of Russian oil in 1955. However, as the world oil market became increasingly glutted in the late fifties and early sixties, MITI's concern over procurement of crude oil was dissipated. The only goal remaining was to keep the prices of refined petroleum products in Japan as low as possible for the benefit of the heavy industries of Japan.

For this purpose, MITI needed a new controlling tool. Its import licensing power was losing its regulatory effect over Japanese and foreign oil firms as various import restrictions were being eased in 1962 under mounting pressures from foreign countries, notably the United States. At this juncture, MITI promoted new regulatory legislation, the Oil Act (the *Seikyu Gyoho*), which would empower it to ration permits for the refining and sales operations of domestic oil firms. MITI argued that without the Oil Act it would not be able to help Japanese oil firms to compete with foreign subsidiaries. The ruling party, the Liberal Democratic Party (LDP), and business circles in Japan were easily persuaded to endorse the Oil Act, partly because they feared the consequences of relaxing import restrictions and the specter of foreign domination.

Eventually, MITI acclaimed the operations under the Oil Act as an "exemplary success" in guiding Japanese industry—an observation that was more self-serving than objective. The major thrust of its oil policy during the nineteen-sixties was to let a limited number of large steel and chemical firms play one oil firm off against another to obtain the lowest possible prices. For that purpose, MITI kept the structure of the Japanese oil industry fragmented by various means, including permitting more and more new entries into both the refining and wholesale stages of the oil business and deliberately keeping the refining capacities out of balance with wholesale abilities in order to impede the vertical integration of refining and wholesale operations. In short, MITI counted on a greatly fragmented oil industry inside Japan to engage in price competition in the sale of petroleum products, which would then be shielded from import competition by high tariffs on such products from the outside.

In practice, competing refiners and wholesalers repeatedly resorted to price-cutting tactics in order to expand their market share. Then each firm would exhibit its increased market share to MITI in order to demand a greater allotment of the permits required for new refining and distributing facilities. The scrambling for market shares was in turn exploited by such large industrial oil users as steel firms and public utilities to drive hard bargains with the oil industry in its fragmented state. MITI expected that price competition of this sort would keep refiners in Japan on the lookout for low-priced crude oil in the world market.

MITI's success in inducing price competition in the oil industry can be gleaned from various data on the oil firms' profit performance. Between 1960 and 1972, the average profit before taxes per kiloliter of petroleum products for twenty-nine oil firms declined from 1,300 yen ($3.61) to about 230 yen ($0.82), with a corresponding decline in the firms' net revenue per kiloliter of petroleum products sold.[2] Total profits were also relatively low. Although the combined sales of the nineteen firms specializing in refining in 1972 totalled 4,360 billion yen, about 10 per cent more than the sales of either the steel industry or the automobile industry for the same year, their combined profits were only 43 per cent of those of the steel industry, and 25 per cent of those of the automobile industry—both dominated by a few large oligopolistic firms.

As a semblance of a competitive market was created for petroleum products inside Japan, MITI's belief in keeping the oil industry fragmented became ingrained in its oil policy. Thus, when the steel, public utilities, and trading firms of Japan joined in financing Japan's first overseas venture in crude-oil exploration off the coasts of Kuwait and Saudi Arabia in 1958, MITI gave only lukewarm support.[3] When the exploration group struck oil and formed the Arabian Oil Company with 10 per cent equity participation by each of the governments, Kuwait and Saudi Arabia, MITI did not allow the Arabian Oil Company to integrate its operations into refining and marketing in Japan. Instead, MITI required all the refining firms in Japan to purchase crude oil from Arabian Oil on a pro-rata basis, according to existing refining capacities.

Until 1965, however, MITI did not take the price competition in the Japanese oil industry for granted. In order to monitor the oil industry from within, MITI attempted to create its own "protegé" oil firm, which was to be integrated vertically from the refining to the wholesale stages. This protegé was to be created by the merger of smaller oil firms under MITI's direction, and was expected to act as a model of "desired" behavior for other oil firms. However, smaller refiners like Toa and Fuji did not wish to be absorbed into the orbit of Nihon Kogyo, a large mining firm in Japan which MITI had chosen as the nucleus for its new protegé firm. Nor was Nihon Kogyo willing to share with the others the profits of its new oil refining business, upon which it was basing a strategy of diversification from its main line of copper smelting. A compromise was reached in 1965 by which a new oil marketing firm named the Kyoseki Company was formed jointly by Nihon Kogyo, Toa, and Asia, to be operated separately from the refining businesses of the participants. Later other, smaller refiners such as Asia Kyoseki, Fuji, and Kashima joined the Kyoseki Company. But each participant was left free to sell refined products to any customer, thus frustrating MITI's attempt to create a protegé firm that would be completely integrated from the refining to the marketing stage.

As the oil glut continued into the second half of the nineteen-sixties, MITI's initial interest in keeping down the number of refiners in Japan disappeared. Crude oil was readily available, and MITI increasingly succumbed to political pressure from petrochemical firms, trading companies, public utilities, and other business interests that wanted to enter the oil industry without worrying about their ability to procure crude oil. Various individual politicians and factions within the ruling party, the LDP, were quick to take advantage of MITI's retreat, and they began to negotiate with MITI on behalf of aspiring entrants over the "political" distribution of coveted allotments of new refining capacities. Rather than enter the wholesale stages—the success of which required uncertain investments in the building of distribution networks and the creation of brand "images" to withstand the fierce competition with the existing thirteen firms—new entrants swarmed to refining operations, for which plants could be purchased on a "turn-key" basis. The manufacturing expertise and technologies of oil-refining operations were readily procured from independent foreign engineering firms. Foreign major oil firms, seeking captive customers for their crude oil as well as supply sources for oil products for their sales subsidiaries in Japan, gladly extended technological aid to new oil refiners in Japan.

The channel of communication between MITI and new candidates for entry into the refining business was to be supplemented on another level through dealings between MITI and the central lobby for big business, the *Keidanren* (Economic Confederation), which was re-formed in 1946 on the base of a war-time structure for collaboration among Japanese industries. With the dissolution of the *zaibatsu*, the business community felt a need for a central body to coordinate its political activities. During the nineteen-fifties and sixties, the *Keidanren* built up its political influence mainly through the distribution of donations among the various political parties (excluding the Communist Party) and, more importantly, among diverse factions within the ruling LDP.

The political relationships that had long been evolving among the three groups—the bureaucracy, the LDP, and the *Keidanren*—were shaped partly by the idiosyncracies of Japanese society. One of MITI's persistent problems was to find post-retirement positions in industry for retiring high-level bureaucrats. A solution is of importance in a ministry employment system that can expect the retirement of all members of a "peer group" when one of its number reaches the position of vice-minister, the highest rank attainable by career officials. Historically the custom has evolved in order to encourage senior officials to relinquish their places to younger men, but it does leave many ambitious and vital men in their forties and fifties without jobs, unless places for them can be found elsewhere. MITI's control over the Japanese oil industry produced a tendency to regard that industry as a source of placement opportunities for its retiring officials; its fragmentation was thus useful for MITI's internal needs. The larger the number of firms in the industry, the greater the opportunities for placement. In the spring of 1974, about fifty former ranking officers of MITI, including five former vice-ministers and four chiefs of agencies, were found occupying various positions of top management in thirty oil firms in Japan.

All told, there were sixteen fully Japanese-owned and twelve jointly Japanese- and foreign-owned firms in the refining industry by 1973. The two groups divided the total crude-oil cracking capacity of Japan, which in 1973 reached 4.7 million barrels per day, into roughly equal parts. In 1973, twenty-eight separate refining operations, the largest of which had less than 12 per cent of the refining market, were competing with one another to sell their products to thirteen oil wholesalers, including six subsidiaries of foreign oil firms. The largest oil wholesaling firm was Nihon Sekiyu with about 17 per cent of the market share.

The proliferation of oil firms in Japan was nevertheless limited to "downstream" operations, i.e., the refining to retailing stages. None of Japan's refiners ventured abroad for crude-oil production. Financially too weak to undertake risky ventures of crude-oil explorations, they were content to count on the worldwide glut situation to ensure a ready supply of crude oil. Major foreign oil companies had come to supply over 80 per cent of Japan's needs for crude oil; the only Japanese overseas producer of crude oil, the Arabian Oil Company, supplies less than 9 per cent. The remaining 11 per cent was procured by trading firms from independent foreign oil firms.

MITI and FTC in the Crisis

When the Israeli-Arab war broke out in October, 1973, the Japanese oil firms,

MITI, and the Ministry of Foreign Affairs were predicting both publicly and privately that the war would be short-lived, that crude-oil tanker rates might go up, but that the Arabs would not resort to an oil embargo. When, on October 17, OAPEC followed up the OPEC announcement made the previous day of a 70 per cent increase in the posted price of crude oil with cutbacks in crude-oil production and export restrictions, MITI and the Japanese oil firms predicted that Japan's total importation of crude oil would be reduced by no more than 10 per cent. This prediction was based on the fact that Japan imported about 40 per cent of its crude-oil needs from the OAPEC countries as well as on the assumption that, at most, only one-quarter of this 40 per cent would be affected. This prediction did not take into consideration the possibility that the major international oil firms might shift their supplies of crude oil to embargoed customers, such as the United States and the Netherlands, even at the expense of non-embargoed nations.

As events unfolded throughout October, the political and economic impact of the oil shortage proved greater than anticipated. The prices of various products and services were immediately raised to make up for the "increased costs" of imported oil. The specter of food and other commodity shortages, still vivid from the memories of the immediate post-war years, rose up before the public. The oil shortage was reported in the mass media not by business reporters but by reporters ordinarily covering crime and political events. When a few oil tankers were found arriving at Japanese ports with less than a full load of crude oil, the story was immediately magnified into an incident in which the omnipotent foreign oil majors were redirecting oil to the United States at the expense of Japan. Neither MITI nor the oil industry discouraged such speculation. In retrospect, the seeming inaction of MITI and the oil industry toward allaying public fear of commodity shortages during the months from October through December, 1973, can be explained by the way they seized upon the public fear of an oil shortage to rebuild their own bureaucratic position.

As revealed in the articles by Penrose, Stobaugh, and Mikdashi in this volume, the world-wide market in crude oil began increasingly to show signs of shortage toward the end of the nineteen-sixties. But these signs went unnoticed by both Japanese oil firms and MITI prior to the Teheran Agreement of 1970, mainly because, since about 1968, increases in crude-oil prices had been absorbed by the international majors supplying crude oil to Japan, and they had not been passed on to Japanese oil consumers. Accordingly, the Japanese oil firms and the government bureaucracy were slow to notice the fundamental shift of the crude-oil market from glut to shortage until foreign crude-oil suppliers notified Japanese oil firms, after the Teheran Agreement, of forthcoming increases in crude-oil prices.

In 1970, the independent oil firms in Japan suddenly found themselves squeezed between the rising price for crude oil, on the one hand, and the fierce price competition in the market for petroleum products in Japan, on the other. Furthermore, Japanese independents learned that their competitors, the subsidiaries of the foreign-owned oil companies, were buying crude oil from their parent suppliers at about 33 cents to the dollar less per barrel than the prices the independents were being charged. With the Japanese economy now growing at a much slower rate than it had been during the previous decade, the independent oil firms began to fear that foreign oil firms might drive them completely out of business. This fear united the Japanese oil firms in an agreement to try to raise the prices of petroleum products.

In February, 1971, long before the oil crisis, a secret "study group" had been formed by the sales managers of five leading petroleum firms as an invisible "shadow" of the Marketing Committee of the Petroleum Industry Association. The five oil firms were Idemitsu, Nihon Sekiyu (Caltex), Daikyo, Maruzen (owned 20 per cent by Gulf) and Mitsubishi (owned 50 per cent by Getty). Their combined market shares were around 54 per cent—sufficiently large to give them price leadership in the fragmented oil industry. Kyoseki, the MITI protegé, was excluded from the study group because it was suspected of leaking confidential cost data to MITI.

This study group produced a plan suggesting a range of price increases product by product. The Marketing Committee of the Petroleum Industry Association, consisting of high-ranking marketing managers of the leading fourteen out of a total of thirty-one member firms, ostensibly "discussed" the plan. The plan did not suggest absolute prices for petroleum products, but spelled out ranges of recommended increases, leaving it to each firm to fix its definitive price within the range. The increases were tacitly cleared with MITI, although MITI had never openly ruled on the suggestion. Conforming to implicit and subtle modes of communication between MITI and the association, MITI's purposeful silence was accepted by the association as a sign of consent.

This about-face by MITI from its past policy of forcing the oil industry into price competition was prompted by its renewed concern over the weakening financial position of the Japanese oil firms. MITI's overriding policy lay in fostering Japanese oil firms, and it was this policy that was being threatened. Former officers of MITI, who had become the managers of the oil firms, in effect persuaded MITI to appear blind to the price fixing among them.

Unlike the Shell Oil Company, Esso (Exxon's subsidiary) and Mobil maintained the appearance of non-participation in the "discussion" of price increases. At the regular monthly meeting of the Marketing Committee, the Esso and Mobil managers made a point of leaving when the official meeting was adjourned and informal discussion of prices was started. Their expressed fear of provoking the Justice Department of the United States was accepted by others as sufficient to excuse them.[4] Naturally, Esso and Mobil followed the price increases when they were put into effect.

As the price of crude oil went up between 1971 and 1973, oil firms in Japan increased their prices at rates higher than the increases for crude oil. Between January and August, 1973, Japanese consumers of petroleum products saw their oil bills go up five consecutive times. The oil crisis in the fall of 1973 led to a new round of price increases. When, on November 1, Esso increased the wholesale price of automobile gasoline by 7,000 yen ($25) per kiloliter (equivalent to an increase of 12 cents per U.S. gallon), Japanese independents followed suit on the assumption that Esso must know something they did not about the world oil situation. Whatever speculation one might venture for this tantalizing shift in Esso's behavior from price follower to price leader in Japan, the move by Esso was interpreted by MITI and Japan to mean that the price levels of petroleum products in Japan were being determined by Exxon, Esso's foreign parent firm.

One might wonder, however, why such large oil users as steel, public utilities, and petrochemical firms did not resist the oil firms' price increases after 1970 and particularly during the fall of 1973. The oil users' acquiescence to the oil firms can be attributed mainly to the fact that, since about 1970, they had been seeking to increase their own prices. During the nineteen-sixties, various Japanese industries had waged fierce price battles in order to expand their sales and market shares in Japan. Their cost-cutting tactics had been facilitated by various process innovations that economized on labor and raw materials. However, by about 1970 the rate of technological innovation within large manufacturing firms had leveled off,[5] and the firms now faced increased labor costs. As a result, toward the end of the nineteen-sixties, leading firms in key manufacturing industries were abandoning their price wars and beginning to collude in efforts to increase the prices of their products. MITI condoned these mergers and other forms of cooperation among large Japanese firms on the premise that they would then be able not only to withstand anticipated foreign competition domestically, but eventually also to become multinational firms themselves. MITI's new policy further encouraged large firms in various industries to cooperate in fixing prices.

When one after the other of the oil firms announced their various price increases beginning in 1970, the manufacturing firms became concerned over the possibility of production cutbacks. The price of oil itself was not a major problem for them, since they could pass on these increases to their customers, often after having tacked on their own price increases. The possibility that oil could not be had at any price, however, was quite a different matter. As this fear grew in Japan throughout the months of October and November, 1973, the Japanese government did very little publicly to allay panic. It reiterated its optimism in assessing the impact of the oil crisis on the Japanese economy. Not until November 16 did the cabinet call for two therapeutic measures in answer to the mounting public outcry. They were 1) a 10 per cent voluntary reduction in use of oil and electric power by industries; and 2) the closing of gasoline service stations on holidays, after the pattern already set by the United States, the United Kingdom, and the Netherlands. But the cabinet took no further steps, leaving the prices of various goods and services to rise freely. Behind the scenes, however, a fierce power struggle was being waged between FTC and MITI.

FTC had originally been established in 1947 under the occupying forces to prevent the reemergence of the old Japanese industrial combines, the traditional *zaibatsu* groups. However, from the time Japan regained her political independence in 1952, the alliance of three groups—MITI, LDP, and *Keidanren*—had served to weaken FTC's power of investigation and action. In 1972, a new bill was introduced in the Diet concerning the investigative and regulatory mandates of FTC. The opposition parties were trying to strengthen FTC's power so that it could control big business in Japan for the benefit of consumers and smaller businesses. MITI and its allies within LDP were confronted with a dilemma: how to make FTC sufficiently strong to restrict the predatory tactics of foreign-owned subsidiaries in Japan without enabling it also to go after Japanese-owned big business. Staff investigators and committee members of FTC—the majority of whom, as we have seen, had come from the Ministry of Finance, MITI's bureaucratic rival—were trying to gain public support for increasing the investigative authority of FTC.

On November 27, 1973, FTC, encouraged by the rising public complaint over price increases, suddenly initiated an investigation of the Petroleum Association and thirteen petroleum sales firms on suspicion of being engaged in unauthorized price and production cartels. Two days later, on November 29, FTC enlarged its investigations to include the petrochemical firms and industry association, also charging them with forming an unauthorized price-fixing cartel. Next, it made the same moves against the aluminum, tire, and pulp industries. On the basis of the evidence collected from the oil firms, FTC prodded the Public Procurator's office, by using public pressure and against the covert objections of MITI, to investigate the oil firms still further. This investigation culminated, on May 28, 1974, in the indictment of twelve leading oil firms—including Shell's sales subsidiary in Japan—for alleged price fixing.

This sudden aggressiveness mainly reflected an attempt to resist MITI's efforts to weaken FTC's investigative mandates. Prior to the oil crisis, MITI had been increasingly frustrated by its loss of power over the Japanese industries. More and more, firms in Japan were expanding their direct manufacturing activities abroad and were thus escaping MITI's sphere, which was limited to domestic control.[6] In addition, the import restrictions on goods and technologies, upon which MITI has previously built its control over industrial activities in Japan, were also declining.

From around November 10, amidst the public outcry for direct governmental action, MITI began drafting its two oil bills. One was to permit MITI to tighten its control over the production, distribution, and pricing of petroleum products; the other was to empower it to regulate the prices, production, and distribution of all other manufactured goods. Into these two oil bills, MITI attempted to incorporate clauses authorizing various forms of cartel agreements among the firms. The drafts of the two bills, which the Japanese government made public on November 22, contained clauses that would further weaken FTC's investigative authority.

This series of agggressive moves by FTC against oil and other industries during the last week of November represented its bid for support from the Japanese public and the Diet. With the election of the Upper House scheduled for the summer of 1974, Prime Minister Tanaka, increasingly apprehensive about losing his Upper House majority to the combined opposition parties, knew that he could not ignore the public sympathy that FTC had thus generated. On November 30, MITI, the Prime Minister's Office, LDP, and FTC agreed upon a compromise; it deleted from the draft bills the clauses explicitly authorizing cartels. At the same time, a "private memorandum" between MITI and FTC acknowledged that certain cartel actions were to be excepted from litigation by FTC—a time-honored method of Japanese compromise, ostensibly conceding victory to one party, while permitting the other party to retain the substance of its demands.

Once revived, however, FTC continued during 1974 to maintain its public visibility by attacking unauthorized price fixing and unfair trade practices among trading companies and manufacturers of automotive tires, agricultural equipment, dairy products, laundry detergents, and toilet paper. The consumers, who came to resent having been manipulated during the oil crisis by the industries and MITI, applauded the "active" FTC. Thus, FTC had injected itself into the political processes of Japan amidst the aftermath of the oil crisis of 1973.

In December, 1973, MITI concentrated on having the two oil bills passed by the Diet. For this purpose, it encouraged the mass media to play further upon the themes of the oil shortage and the evils committed by Japanese and foreign oil firms. The vice-minister of MITI, Yamashita, labeled the oil industry of Japan "the root of all evil" and popularized the phrase through his press conferences. This publicity helped MITI to blame the oil industry for the seeming helplessness of the Japanese government and to convince the Diet and public of the need for the two oil bills it was sponsoring.

As it turned out, during the month of December, 1973, Japan actually received over 2.2 million more tons of oil than the amount MITI was publicly admitting as having been delivered to Japan, an amount corresponding to the capacity of about ten mammoth tankers. The customs office of the Ministry of Finance was weekly reporting to MITI the actual and scheduled arrivals of crude-oil tankers at Japanese ports, and the oil firms and trading firms in Japan were "ordered" by MITI to report not only actual deliveries of crude oil but also amounts of purchases abroad. It was therefore unlikely that MITI had not been informed about the amounts of crude oil scheduled to arrive at Japanese ports during the month of December, 1973, but it had reason not to dispel the public fear of an oil shortage while the Diet was debating MITI's bills. On December 21, 1973, the two oil bills were enacted.

In sum, at the first stage of the oil crisis, during which oil users in Japan feared real shortage, the main actors concentrated on harvesting as many political benefits as possible. The political leadership of LDP, which had long been dependent on the Japanese bureaucracy for policy guidance, remained immobilized, while the bureaucracy, oil industries, and industrial oil users were absorbed in exploiting the oil crisis to their own advantage.

The Oil Crisis and Japan's Diplomatic Stance

The oil crisis destroyed the premise upon which post-war Japan's external relations had rested. Perhaps partly because of the unpleasant results of Japan's defeat in World War II, the Japanese population had a mass abhorrence—amounting almost to a phobia—about considering questions of national security and defense. As a result, within the Japanese bureaucracy and political circles the notion had grown up that economic matters should be kept separate from political matters (a policy known as *Seikei Bunri*). Japan was to concentrate on the former, while the latter was to be left to the United States. The United States cultivated this Japanese dependency, and Japan naively assumed that the rest of the world would also accept its separation of economics from politics. Japan had long believed that the Arab countries would treat it as a friendly nation simply because it harbored no malice toward the Arabs.

Thus, in October, 1973, when OAPEC classified Japan as "unfriendly" to the Arab cause, it struck Japan as a "thunderbolt from the blue sky"—a phrase used both privately and publicly by various members of the Japanese bureaucracy, including the Ministry of Foreign Affairs. In a sense, the persistent decline in influence and prestige that the Ministry had suffered after 1945 within the Japanese bureaucracy was a logical consequence of Japan's neglect of external politics. Lacking any contacts with either financial or industrial communities, and abandoning its role of gathering and interpreting information for the other branches of the Japanese bureaucracy, the Ministry of Foreign Affairs was increasingly dependent upon the United States Department of State for information as well as for policy guidance.

During the 1973 oil crisis, the Ministry approached Secretary of State Kissinger for advice. His advice was not to capitulate to OAPEC's position in the Middle Eastern dispute. When the Ministry of Foreign Affairs relayed this advice to LDP, *Keidanren*, and the rest of the Japanese bureaucracy, it only served to confirm suspicions already prevalent among business and political circles of Japan of the existence of a Kissinger-Israel conspiracy. According to one version of the conspiracy, the United States was happy to see the foreign-exchange reserves of Japan and West Germany drained off to the United States via the OPEC nations. The Kissinger-conspiracy theory combined with the spreading suspicion that the American-based oil majors were diverting oil to the United States at the expense of Japan was sufficient to convince Japan that it should endorse OAPEC's position in the Middle Eastern dispute.

After November 18, when the EEC countries were exempted by OAPEC from restrictions on oil exports in exchange for making public statements sympathetic to the Arab countries' position, the leading personalities in Japanese business circles conveyed to Nakasone, the minister of MITI, their collective wish that Japan should endorse OAPEC's position. On November 22, only four days before the Arab summit meeting, the Tanaka cabinet took that step. This was the first open break with American foreign policy in post-war diplomatic history that Japan had dared to make.

Once it had become aware of the need for cultivating closer political ties with oil-exporting countries, Japan wasted no time in sending a series of special envoys, between December, 1973, and the spring of 1974, to the oil-producing countries in the Persian Gulf and to Syria, Egypt, Algeria, Sudan, Morocco, and Jordan. The purpose of these missions was to offer economic and technical assistance to these countries in the hope of improving Japan's relations with the OAPEC countries and with Iran. By mid-1974 the total amount of economic aid that Japan had promised to Iran, Syria, Egypt, Saudi Arabia, Algeria, Sudan, Jordan, and Morocco had reached about $563 million. The projects included plans to widen the Suez Canal ($140 million) and to develop petrochemical and other industrial complexes in Iran ($250 million). The seeds of Japan's involvement in the Persian Gulf and North Africa were thus sown, although the initial involvement was perceived by LDP and MITI mainly in terms of simple economic trade-offs between Japan's foreign aid and guaranteed access to oil.

Once the fear of oil cutoffs subsided, the oil-importing countries became increasingly concerned over two problems related to their balance-of-payment questions: the actual transfers of economic resources from oil importers to oil exporters, and the "recycling" of petro-dollars (financial claims) from importers to exporters. Japan attempted to encourage its industries to export goods and services to the OPEC countries, and at the same time it borrowed petro-dollars either directly from the governments of Saudi Arabia, Kuwait, and Abu Dhabi or indirectly from American banks. As a result, by September, 1974, Japan's balance of trade began to register surpluses. With recycled petro-dollars, Japan's overall balance of payments was materially improved. Superficial as it was, this apparent recovery from the "oil shock" led Japan to continue in the spring of 1975 to differentiate its position from the United States by expressing an increasingly lukewarm and cautious attitude toward the Kissinger-supported plan, the International Energy Agency (IEA).

Earlier, Japan had been stung by the vehemence with which the United States government and industries privately blamed the Japanese government for failing to prevent Japanese trading firms, utility firms, and oil refineries from bidding up the price of OPEC oil during the fall of 1973. Due to the fragmented structure of the Japanese oil industry, many bidders swarmed to compete for the limited amounts of oil offered by Algeria, Nigeria, and Iran. The Japanese government and industry, still panic-stricken over the prospect of oil cutoffs, did nothing to hold the Japanese bidders in check. However, European and American bidders representing power utilities firms and other oil users were also actively participating in these international auctions. Indeed, Americans turned out to be the highest bidders in the Nigerian and Iranian sales. Since the Japanese trading firms and oil firms had agreed among themselves not to go beyond a thirteen-dollar limit per barrel, the Japanese government and industry felt that the criticism coming from the United States was unwarranted. More importantly, Japan felt that it could hold its own firms in check, but that the governments of other industrialized nations would not or could not do the same. On these grounds, Japan expected that Kissinger's demand on all the oil-importing countries for oil conservation by 10 per cent ought to be modified to encourage a large user like the United States to conserve more than smaller users. Besides, in view of its own diplomatic fence-mending attempts with the Middle Eastern countries and the other OPEC nations, Japan had grown increasingly uneasy over a confrontation between the IEA and OPEC. If Japan were to underwrite, as it would be expected to do, the financial risks of the IEA, along with Germany and the United States, the Japanese government and industry were privately saying in the spring of 1975 that their peculiar situation, including their difficulties with energy conservation, ought also to be taken into consideration by the United States.

Earlier, in July, 1974, the Oil Group of the Overall Energy Investigation Council of Japan, whose members had been selected by MITI, published an interim report suggesting the general directions that the Japanese government might follow in the immediate future in order to assure Japan of procuring her increasing needs for oil abroad. Unlike earlier reports, this one acknowledged the power exerted by the major international oil companies in allocating crude oil among various consuming countries, and it recommended that Japan solicit their assistance in obtaining crude oil for the country. Gone was the thought that Japan could reduce her reliance on the international oil majors for her supply of crude oil.

But more importantly the report recommended involvement of the Japanese government in concluding direct deals with the oil-producing countries for the purchase of oil and for exploration concessions. It also strongly suggested a need for Japan to integrate her oil firms backward to the crude-oil production stage abroad. MITI did not articulate its rationale for the policy beyond its own intuitive hunch that vertical integration of the Japanese oil firms via the regrouping of existing but fragmented oil firms might enhance Japan's chances for making direct deals with oil-producing countries. It had not yet occurred to MITI that Japan's direct access to crude-oil explorations abroad might also improve Japan's bargaining position vis-à-vis foreign oil majors. Accordingly, MITI's plan for reorganizing the oil industry in Japan was to link the major foreign oil firms' interests with those of Japanese oil firms throughout the vertical chain. Three Japanese oil firms, which had been scrambling for such tie-ups in competition with Japanese trading firms, had already undertaken joint-exploration projects for crude oil and natural gas with foreign oil firms. In the spring of 1975, Mitsubishi-Shell, Idemitsu-Amco, and Teikoku-Exxon had begun exploration of the continental shelves around Japan for oil and natural gas.

Outside Japan, the Oil Development Public Corporation of Japan (a brainchild of MITI) had, for some time prior to the fall of 1973, subsidized Japanese banks, trading firms, and steel firms to explore for crude oil. From 1974 to 1975, it also stepped up its financial and technical assistance to Japanese private-firm groups for overseas exploration for oil and natural gas. By the spring of 1975, about fifty Japanese projects for crude oil and natural gas were operating abroad. Since the trading firms and banks that were active in these overseas projects already owned their own refining subsidiaries in Japan, an oil strike by any of them would form a nucleus around which MITI might regroup the fragmented oil industry.

Meanwhile, suggestions were repeatedly made that Japan approach mainland China for oil. Some observers went so far as to say that China would become a major supplier of crude oil and coal to Japan in exchange for Japan's manufacturing technologies. However, this development appears to be the least plausible of all the alternatives. For one thing, the ideological foundation of China's industrial development would not permit it to become a major supplier of raw material to capitalist Japan. China's sales of crude oil to Japan might well be increased from about 12 million barrels (0.6 per cent of Japan's needs) in 1973 to, say, 3 to 6 per cent of Japan's needs (a ten-fold increase from China's present exports to Japan) by 1980, but more seems unlikely. Moreover, from the Japanese point of view, there was no guarantee that China would turn out to be any more stable a supplier than Saudi Arabia; and from China's point of view, it would be more advantageous to retain as much additional crude oil as possible for its own "oil diplomacy" vis-à-vis the oil-starved developing nations in Asia and Africa. A limited "arm's length relation" appears to be the more likely eventuality.

In short, the oil crisis of 1973 and its aftermath jolted Japan into assuming an independent diplomatic posture. The politics and economics of oil began to be mixed internally and externally by the Japanese government. Painfully aware of its dependence not only upon the OPEC countries but also upon the rest of the world, it would not be surprising to find the Japanese government taking the lead in the near future as a promoter of free international trade and investment. And the day might also be near when Japan will demand equal partnership with the United States in defining common political goals and diplomatic policies in the Pacific area.*

*The field research for this article was financed in part by the Japan Institute of Harvard University, with the endorsement of the Center for International Affairs, Harvard University, and in part by the Multinational Enterprise Project of the Harvard Business School, which is supported by the Ford Foundation and coordinated by Professor Raymond Vernon.

REFERENCES
[1]Based on interviews by the author with the Japanese reporters and oil personnel who had dealt with GHQ during the years 1946-1949.

[2]Computed from *Yukashoken Hokoku Sho* (Financial Reports filed with the Ministry of Finance) of the respective firms, 1960-1972.

[3]The Arabian Oil Company was initiated by Taro Yamashita, who, during the 1930s and 1940s, made his reputation as a maverick industrial promoter in Manchuria. While no Japanese oil firms showed any interest in participating in Yamashita's "outlandish" project, steel firms, public utilities, and trading firms, which wanted their own captive supply source of crude oil, financed the project. Once the company struck oil, however, the steel firms and utilities that used crude oil either as a fuel or as raw material found it unnecessary to have their own refining capacities in Japan. Yamashita himself was also content to stay away from the refining and marketing activities of oil in Japan—a sphere of influence long conceded to MITI and the oil firms. Thus both Yamashita and MITI disappointed the governments of Saudi Arabia and Kuwait. According to the knowledgeable sources of the Japanese oil industry, Saudi Arabia and Kuwait indicated at that time their desire to have their minority-owned independent oil firm, Arabian Oil, integrate its operation with refining and marketing stages in Japan.

[4]These conclusions were drawn from the indictment of the petroleum firms by the Procurator's office, as well as from the author's interviews with newspaper reporters covering the antitrust case.

[5]Hiroki Tsurumi (Rutgers University), "A Bayesian Test of the Product Cycle Hypothesis Applied to Japanese Crude Steel Production," a paper given at the meeting of the Econometric Society, San Francisco, December 27-30, 1974.

[6]Yoshi Tsurumi, "Japanese Multinational Firms," *Journal of World Trade Law* (February, 1973).

MARSHALL I. GOLDMAN

The Soviet Union

IF THERE IS ANY SINGLE MESSAGE that emerges from the Soviet handling of the oil crisis, it is that no economic opportunity offered by the crisis was overlooked. While it is doubtful that the Soviet ministry for the petroleum industry drew up a scenario predicting the oil embargo of 1973 and suggesting how the Soviet Union could benefit from it, from the early nineteen-sixties, consciously or inadvertently, Soviet and East European petroleum officials were positioning themselves to take advantage of just such an opportunity. To understand how this came about, it is necessary to go back to a somewhat earlier period when petroleum was a glut on the market.[1]

The Buyer's Market

Exports of oil from the Soviet Union are not new. Prior to World War I and the development of the Middle Eastern oil fields, Russia was one of the world's largest exporters of petroleum, mainly to West European markets. Except for a temporary suspension of the flow immediately after the Revolution, oil exports continued from the Soviet Union to Western Europe until World War II much as they had before the Communist takeover. After World War II, as the Soviet Union began to recover from the destruction of the war, and as old oil fields were refitted and new ones discovered, it sought to reclaim its former markets. In the political and economic conditions of the early nineteen-fifties, however, this was a strategy hard to implement. In the period of the Cold War, the Western countries tried to ensure that they would not become dependent on vital commodities from the East, and Soviet sales efforts outside the COMECON countries (Council of Mutual Economic Assistance—originally most of the Communist countries of Eastern Europe and now Mongolia and Cuba as well) were bitterly resisted. As a result, the Soviet Union was hard-pressed to find independent oil buyers willing to risk the displeasure of the major oil producers.[2]

After considerable frustration, Soviet oil exporters began to direct their selling efforts to the developing countries, many of which were short of the hard currency needed to pay for their oil imports. Like most new competitors, the Soviet Union found it necessary to use the wedge of lower prices. They were also willing to sell their oil on a barter basis. Armed with such favorable terms, the Soviet Union sought to sell its oil in countries such as Cuba, India, Ghana, Ceylon, and Guinea. Reacting to Soviet competition, the Western oil companies consistently refused to process Soviet oil through their refineries and pumps and threatened various forms of opposition. But with both savings in hard currency and the political concessions being offered by the Soviet Union, the developing countries were attracted to Soviet oil. Eventually, after Cuba and other countries began to nationalize company properties, the Western oil companies reluctantly began processing and selling Soviet petroleum in their facilities around the world.

Breaking into the Western European market took considerably more time. Asserting that they were only trying to regain their historic market share, the Soviet Union again offered the inducement of low prices. Beginning in 1960, the Soviet Union offered Italy and West Germany crude oil at a price of about 8 rubles a ton (about $1.20 a barrel). Where there was no need for competitive pricing, however, the Soviet Union charged considerably more. At the same time as they were offering Western countries petroleum at 8 rubles a ton, their export price to countries such as East Germany (GDR) was almost 18 rubles a ton (about $2.70 a barrel).[3]

The big breakthrough for the Soviet Union in Western Europe came with the decision by Enrico Mattei of Italy's ENI to seek crude-oil sources outside those controlled by the Western oil companies. As he built up his own refineries and network of filling stations, he sought to protect his position from the extreme pressures of the Western companies. By 1958, with the first closing of the Suez Canal as a stimulus, Italy was importing considerable quantities of Soviet oil. In the period from 1959 to 1970, Italy became the Soviet Union's largest customer for oil products, buying more than any of the countries of Eastern Europe.[4] At one point, over 22 per cent of the oil Italy consumed came from the Soviet Union.[5]

Once Italy had demonstrated that it was possible to buy Soviet oil without being victimized by the Soviet Union or punished by the majors, other West European countries began to step up their purchases as well. This became easier once the Druzhba (Friendship) pipeline from the Volga oil fields in the Soviet Union to the West was opened in 1964. Oil could be shipped cheaply not only from Baku to Mediterranean destinations, but also from the pipeline terminal at Ventspils, Latvia, on the Baltic to Northern and Western Europe. The second closing of the Suez Canal in 1967 and the threat of further disruption of the usual oil flow pattern led others to turn to the Soviet Union. By 1971, when the United Kingdom authorized the direct importation of petroleum from the Soviet Union, most of the countries of Western Europe had become regular purchasers of Soviet petroleum. Except for Finland, which at times obtained as much as 80 per cent of its total petroleum from the Soviet Union, and Iceland, which obtained 65 per cent, no country was overly dependent on the Soviet Union. Nevertheless, as Table 1 shows, the Soviet Union's hard-currency earnings came to about 484 million rubles in 1972 (about $586 million).

Supplier's Market

The closing of the Suez Canal in 1967 had provided an early indication to the Soviets that price cutting might no longer be needed for the sale of Soviet petroleum in

Table 1

U.S.S.R. Production of Crude Oil and Export of Oil and Oil Products

	Million Barrels		Million Rubles**	
	1972	1973	1972	1973
Production of *Crude Oil*	2,888	3,086		
Exports				
Austria	7	10	14.9	28.6
Belgium	18	12	35.7	68.6
Denmark	6	4	10.2	32.5
France	23	39	43.3	91.0
Germany*	48	46	94.7	263.3
Greece	7	6	17.2	17.6
Iceland	3	4	8.6	12.9
Italy	62	64	109.0	153.1
Japan	7	15	15.6	41.2
Netherlands	18	23	40.2	135.6
Norway	3	4	6.7	11.4
Spain	6	4	11.5	9.1
Sweden	32	23	58.0	58.3
Switzerland	6	5	14.0	36.1
United Kingdom	2	6	3.6	17.3
Total Hard Currency *Countries*	248	265	483.2	949.6
Bulgaria	58	68	118.6	135.7
Cuba	51	54	92.2	114.2
Czechoslovakia	95	105	210.5	235.1
Germany	84	95	161.8	184.8
Hungary	40	46	93.8	112.9
Korea	3	4	13.8	16.8
Mongolia	2	2	10.9	12.1
Poland	82	90	182.2	214.0
Yugoslavia	25	29	51.3	99.3
Total COMECON	440	493	935.1	1,124.9
Finland	63	73	162.0	221.8
Egypt	10	3	26.2	10.0
Ghana	4	4	8.8	9.1
India	3	4	7.1	10.8
Morocco	7	6	13.4	13.5
Total Soft Currency *Countries*	87	90	217.5	265.2
Subtotal	775	848	1,635.8	2,339.7
REPORTED TOTAL	784	867	1,664.0	2,403.0

*Includes West Berlin
**1972 1 ruble = $1.21
  1973 1 ruble = $1.34
Source: Vneshniaia Torgovlia SSSR 1972 (Moscow, 1973); hereafter *VT SSSR*. Also *VT SSSR 1973*.

Western markets. The eagerness of the Soviet authorities to take full advantage of any pricing opportunities is illustrated by the contract they negotiated with Austria for the sale of Soviet natural gas. Because of its location, Austria does not have much to choose from in the way of natural-gas supplies. When the Soviet Union brought its natural gas by pipeline to the Austrian border, the Russians knew that there was no potential competition, at least for the near future. Consequently, in 1967 the Austrians were required to pay 12.5 rubles for 1,000 cubic meters of gas over a seven-year period, a price not much lower than the 13 to 14 rubles the Poles and the Czechs were made to pay. According to a report in the October, 1968, *Petroleum Press Service*, this was equivalent to about 4.33 pence a therm. By contrast, the Dutch at the time were charging the equivalent of 3.11 to 3.48 pence a therm for Groningen gas, while the British were charging 2.87 pence a therm for gas from the North Sea.[6] Illustrative of the Soviet responsiveness to competitive factors in price setting was the difference between their selling price in Austria and their buying price from other sources. Even as they were charging Austria 12.5 rubles, they were paying Afghanistan and Iran only about 5.5 rubles for gas imported from those areas. At that time neither Afghanistan nor Iran had an alternative outlet.

With the closing of the Suez Canal in 1967, the Soviet Union began to move aggressively to obtain new markets for its petroleum. Its sales to Western Europe increased rapidly. In some cases, it worked out swap arrangements with oil companies that at one time had insisted on avoiding all contact with Soviet oil. British Petroleum (BP) agreed to send its Abu Dhabi oil from the Persian Gulf to Soviet customers in Japan, in exchange for which the Soviet Union delivered from the Black Sea region to BP's customers in Western Europe. The Soviet Union set up its own petroleum marketing outlets in Finland, the United Kingdom, France, and Belgium. By 1971, the Soviet Union had even succeeded in breaking into Morocco's petroleum market at Algeria's expense and into Ghana's petroleum market at Libya's expense.

Domestic Needs

By the early nineteen-seventies, however, Soviet officials were beginning to find that, despite increased production, there were limits on the amounts available for export. This was due to a change in priorities at home. After several decades of heavy reliance on coal, Soviet planners had decided to shift the Soviet fuel balance from coal to oil and natural gas. Oil and gas provided more efficient and cleaner forms of energy. Besides, the Soviet automobile production was increasing—from 294,000 units in 1969 to 917,000 in 1973, a tripling of output in four years. Natural-gas pipelines had been extended into almost all large Soviet cities. The net result was that, between 1950 and 1970, Soviet energy generated by coal had fallen from 60 to 40 per cent.[7] For much the same reasons, a similar shift in the fuel balance was taking place in Eastern Europe, imposing additional pressures on the Soviet Union, as the area's prime supplier of liquid fuel.

The increasing demand at home coincided with growing fears among Soviet geologists over the adequacy of their sources of supply. Many of the older wells were drying up. While some rich deposits were being discovered in Siberia, these were often difficult to reach and to transport. Specialists both inside and outside the Soviet Union began to question whether the Soviet Union would have enough oil to satisfy its own and Eastern Europe's needs.[8] The Soviet Union does not issue official estimates of its reserves, but the *Oil and Gas Journal* in late 1972 estimated that Soviet oil reserves were about 80 billion barrels (almost 11 billion tons).[9] If that figure is correct, it would mean that Soviet reserves are about double the estimated reserves in the United States.[10] Yet, unless new reserves are found, the Soviet supply of petroleum will last only twenty to twenty-five years at present rates of consumption.

Similar fears were being expressed with regard to Soviet natural-gas reserves.[11] Reported reserves were estimated to be about 15-20 trillion cubic meters, a seventy-to-eighty-year supply. Most of the Soviet Union's natural-gas reserves are located in remote and hostile regions; if such deposits are to be exploited, more sophisticated equipment will be required. For that reason, the Soviet Union has devoted special efforts to importing American drilling and pipelaying equipment.[12] Whether the Soviets can generate enough exportable surplus to maintain the pace it desires is a disputed question.[13]

Building a Pipeline to the Middle East

The Soviet Union and Eastern Europe have not limited their efforts in seeking new sources of supply to Siberia; petroleum supplies from the Middle East have also become important. In the earlier years, when petroleum was a glut on the market, the Soviet Union and some of the East Europeans had taken quantities of oil as a special favor for the oil-producing nations in a complex package of trade and oil. As late as 1973, several Middle East countries were continuing to honor the old arrangements and the old price schedules, despite the evident change in the world's oil conditions. The unknown factor was how long the Middle East producers would be content with the old arrangement.

To understand the place of oil in these arrangements, it is necessary to recall that the Soviet Union and Eastern Europe have usually sold their manufactured equipment to the less-developed countries on the basis of long-term credits, eventually repayable in exported goods. On this basis, the Soviet Union extended considerable economic aid to oil-exporting countries. Algeria received $400 million in credit; Iraq, Iran, and Afghanistan received more than $500 million; and Egypt received over $1 billion.

For years, these developing countries repaid the Communist countries with products such as cotton, bananas, sugar, and jute. When some aid recipients offered to include petroleum and natural gas on the list of such exports, these too were accepted. So long as oil was in oversupply, it was not much more valuable to the Russians than the cotton and sugar which they also resold on occasion.[14] Indeed, until 1972 the Soviet Union placed limits on its willingness to accept oil in repayment for loans to Iraq.[15]

The first reported imports of Middle Eastern oil by the Soviet Union took place in 1967, as Table 2 shows. Though Soviet statistics on imports are incomplete, it is clear that Algeria was also one of the early suppliers. These early shipments, as well as those

Table 2

Crude Oil Imports of the U.S.S.R.
(in thousands of tons)

	1967	1968	1969	1970	1971	1972	1973
Algeria	447	909	3,628	3,621	5,490	4,178	NOR
Iraq	NOR	NOR	NOR	NOR	NOR	29,936	80,703
Libya	NOR	NOR	NOR	NOR	NOR	13,685	12,556
Saudi Arabia	NOR	NOR	147	NOR	NOR	NOR	NOR
Egypt	NOR	NOR	6,854	14,821	14,953	7,117	1,532
Syria	NOR	NOR	NOR	NOR	NOR	2,309	247
Subtotal	447	909	10,629	18,442	20,443	57,225	96,602
REPORTED TOTAL	733	6,597	18,325	25,655	37,382	57,174	96,756

NOR: None Officially Reported
Source: Country figures—*VT SSSR 1968*, p. 264; *1970*, p. 251, 262, 274; *1972*, p. 237, 276, 282, 286; *1973*, p. 237, 265, 274, 284.
   Reported total—*VT SSSR, 1971*, p. 141; *1972*, p. 41; *1973*, p. 41.
   VT SSSR, 1968, p. 49.
   VT SSSR, 1970, p. 49.

from the United Arab Republic, were intended as repayment for Soviet foreign aid, but, by the late nineteen-sixties, the Middle East oil exporters had come to use their Soviet sales to serve other purposes as well. By that time, Iraq and Libya had begun to demand increased concessions from the Western oil companies, holding out the Soviet Union and Eastern Europe as alternative purchasers.[16]

By the late nineteen-sixties, some Middle East oil was being used to repay loans from the Communist countries and some was exchanged in current barter deals. The exact path of the oil once it reached Soviet hands is hard to trace. Nevertheless, there is good reason to believe that the Soviet Union simply transshipped some of the petroleum to Western Europe, Japan, and the United States. As Table 2 indicates, Soviet data for 1971 fail to account for 17 million barrels of oil imports from the Middle East. Moreover, the Soviet-owned distribution companies in Western Europe were selling more oil in these markets than they were reported to be receiving from the Soviet Union itself. Nafta B, the Soviet-owned distributor in Belgium, sold 11 million barrels of oil in 1969, even though Belgium imported only 555,600 barrels of petroleum from the Soviet Union. In 1970, the gap was even greater; then the Soviet Union exported 8,796,000 barrels to Belgium, but Nafta B sales totaled 18,325,000 barrels.[17] Occasional reports of Soviet offers of Middle East oil in third countries came from other sources. One of the major sellers of Soviet oil in the United States, for instance, reported Soviet offers of Iraqi oil for American use.[18] As we shall see, such arrangements made it possible for the Soviet Union to profit handsomely in 1973.

Eastern Europe

The Soviet Union was not the only country to have worked out such agreements with the Middle Eastern oil producers. Almost all the other members of the Council of Mutual Economic Assistance negotiated similar deals. Rumania in particular took the lead in working out barter ties with petroleum producers in the Middle East. Rumania has had a long experience in exporting its own oil over many years, and it was in a position to feed imported petroleum into its existing export network.

Though Rumania has been alert to the possibility of dealing with the Middle East, the initiative for such deals seems to have come in part from the Middle Eastern countries. Some of these looked upon the East Europeans, and Rumania in particular, as they had upon the Soviet Union itself, as a useful bargaining counter in their dealings with the Western oil companies. As early as 1966, the National Iranian Oil Company (NIOC) was insisting that a consortium of Western oil companies operating in Iran give up some of its output to NIOC for barter in Eastern Europe. Iran obtained an agreement from the consortium that it would produce an extra twenty million tons of oil (140 million barrels) during the period 1967-71. Reportedly a deal was also signed between Iran and Rumania for 73 million barrels of oil for the period 1970-75.[19] This oil was then to be used by NIOC expressly for barter with Rumania and other countries in Eastern Europe.[20] Such transactions made it possible for Iran to increase its oil production and foreign revenues without necessarily increasing competitive pressures on existing oil supplies and sales in the West.

Rumania's inclination, as well as that of the other Eastern Europeans, to participate in such arrangements was increased by the hints that began to emanate from the Soviet Union in the mid-sixties that it did not wish to continue supplying Eastern Europe with raw materials at low prices. The East Europeans would be well advised, suggested Soviet economists and politicians, to look to the developing countries, particularly in the Middle East, for supplies. Pressures of this sort continue up to the present time.[21] As one Soviet authority put it:

The steadily growing demand of the COMECON member countries for oil and the desire of these countries to improve their consumption can be met not only by deliveries of oil via the Friendship Oil Pipeline and expansion of their own oil production, but also from the developing countries of the Middle East and the African continent. Increased oil deliveries from these countries can be carried out at the cost of rendering technical assistance to the developing countries, in establishing their national oil industries, and by engaging in other forms of cooperation.[22]

The East Europeans were often surprised by assertions such as this, since they saw themselves already paying higher than market prices for Soviet raw materials. Soviet officials were quick to respond to East European complaints with assertions that the apparent high prices for Soviet oil were offset by the fact that the East Europeans were selling their own goods at high prices to the Soviet Union.[23] In any event, since the late nineteen-sixties, the Rumanians have worked out barter deals with Libya, Iraq, and Algeria, in addition to their arrangements with Iran. At the same time, East Germany, Bulgaria, Czechoslovakia, Poland, and Hungary worked out exchanges for oil with Iran, Iraq, and Libya.[24] One report indicates that Bulgaria arranged for delivery of 37 million barrels of petroleum from Iraq and Syria in 1972, an unusually large amount for such a small country.[25]

Because of the closing of the Suez Canal in 1967, not all of the oil contracted for under these various barter arrangements was actually delivered. The closing of the canal made it particularly expensive for Iran to deliver oil to countries in Southeastern Europe. Thus, out of the 140 million barrels set aside for NIOC delivery to Eastern Europe in the period 1967-71, according to one source, only 31 million barrels were delivered, and most of them went to Rumania.[26] Shipments of oil from the Persian Gulf rose markedly in 1970, and again it was Rumania that received the major portion. By 1972, Rumania was importing 15 million barrels a year.

What is the explanation for Rumania's active role, particularly with Iran? With the canal closed, Iran had to find some other means to send its oil to Rumania, since the Aramco pipeline from the Persian Gulf to the Mediterranean was not available. Shipment around South Africa and back into the Mediterranean was very expensive. There was, however, another way to ship Iranian oil to Eastern Europe, namely by way of the Elat-Ashkelon pipeline in Israel, which had been opened in 1970. This meant, of course, that Rumania had to maintain diplomatic relations with Israel. By more than coincidence, Rumania is the only East European country to have that distinction. There were even rumors that oil destined for East Germany and the Soviet Union had, upon occasion, been picked up at Ashkelon, at least until the pipeline was temporarily cut off as a result of the blockade of Israel during the Yom Kippur War.[27]

Given the nature of the transactions between the Middle Eastern and East European countries, it is difficult to ascertain just how much petroleum was involved. Data from the Middle Eastern countries are incomplete, contradictory, or unavailable. Thus, Iranian oil officials in New York assert that Iran exported 37 million barrels to Eastern Europe in 1972 and 16 million barrels in 1973. At the same time, in response to a written request, NIOC executives in Teheran report a different set of figures. According to their data, petroleum exports to Eastern Europe were the following:

	millions of barrels
1968	2.4
1969	11.3
1970	18.0
1971	15.6
1972	15.4
1973	18.9
1974	10.8

The Teheran figures seem to be the more accurate ones, but there is no way to be sure. Nor do the data from the Eastern European countries fill in the blanks. Rumania was reported to be the largest purchaser of Iranian petroleum. Rumania's import statistics show that it imported a total of 21.3 million barrels of petroleum in both 1971 and 1972 from all countries and that its imports increased to 30 million barrels in 1973.[28] In addition to the imports from Iran, increased purchases of 10 million barrels by Rumania of Libyan petroleum in 1973 probably helped to fill part of the gap, but there are still unexplained discrepancies.

Despite the fragmentary nature of the data, there is evidence that the East Europeans are trying to become less dependent on the Soviet Union. Between 1969 and 1973, for instance, the East Germans reduced their dependence on Soviet oil from 94 per cent to 81 per cent.[29] As of 1973, Poland still depended heavily on the Soviet Union for its crude oil, but it had already diversified its inflow of refined oil products so that the Soviet Union was only supplying 43 per cent of them.[30] Even the Czechs, who were wholly dependent on the Soviet Union for their crude oil in 1969, had reduced their dependence to 92 per cent in 1973.[31] In the extreme case, Yugoslavia was importing only 28 per cent of its crude oil from the Soviet Union in 1973; however, Yugoslavia is generally more integrated into the Western European trade pattern than are the other countries of Eastern Europe.[32]

While searching for extra supplies, some of the East Europeans probably went further than the Soviet Union had anticipated; they turned not only to the Middle East, but also to the West. Rumania made a deal to obtain oil from Shell. Poland signed a contract with BP to build a refinery in Poland and supply it with 22 million barrels of crude oil a year; East Germany and Hungary also signed contracts with BP. In some cases, BP and Shell were even authorized to open up their own filling stations. It was anticipated that some of the oil that came through these channels, as well as that supplied directly by the state-owned enterprises of the Middle Eastern countries, would be brought through a pipeline that is reportedly to be built from the North Adriatic coast of Yugoslavia at Omisalj. One branch of the pipeline is to head north to Poland, passing through Hungary and Czechoslovakia, and another branch east to Rumania. The goal is to complete this pipeline by 1976.[33] The probability that the pipeline will be completed on time was increased when Kuwait agreed to provide part of the financing and presumably part of the petroleum for it.

The Yom Kippur War—An Opportunity

By the outbreak of the Yom Kippur War, the Soviet Union was the third largest producer of oil in the world after the United States and Saudi Arabia, and it had well-established ties with most of the Middle Eastern producers; it was, therefore, along with the Rumanians, well prepared for the embargo that followed. Solidifying its ties with the Middle East countries, however, was neither simple nor without its price. After the debacle of the Six-Day War in 1967, the Soviet Union had lost considerable prestige in the area. Although its subsequent willingness to reequip Iraq, Syria, and Egypt did much to restore its influence, it also contributed to an increase of tensions in the area. The vacillations in Soviet influence that occurred between 1970 and the outbreak of war in 1973 are well chronicled. Whatever hesitation some segments of the Soviet government might have had about the wisdom of another war, the Soviet Union quickly rallied to the Arab cause once it began.[34] Additional supplies of Soviet weapons were dispatched along with words of support and advice. One piece of Russian advice was for the various Arab states to embargo shipments of their oil to the Western world and Japan. These suggestions persisted into March, 1974, after the fighting had ceased.[35]

The oil embargo turned out to be a powerful weapon for the achievement of Arab goals, and it generated significant economic gains for the Soviet Union as well. Because the Communist countries had been such overt supporters of the Arab cause, they were exempted from the embargo so long as they promised not to reexport any Arab oil to the embargoed countries.[36] In these circumstances, the Soviet Union managed to increase its purchase of Arab oil in 1973, especially from Iraq, which had refused to cut back its production or restrict its sales. Iraq had ample quantities for sale, quantities that it could not send to its traditional Western customers.[37] One solution was to step up shipments to the Communist countries. For 1973, Iraq sent a record shipment of 80 million barrels to the Soviet Union, more than the total imports from the Arab countries combined in any previous year. The likelihood is that at least a portion was sent during the embargo.

It is not clear whether the Soviets, in their subsequent handling of this oil, adhered to their commitment not to ship to embargoed countries. The official position of the Soviet authorities was that "oil purchased in Arab countries" would be sent only to "other Socialist countries."[38] In any event, while the Soviet Union was taking possession of oil originating in the Middle East, it was simultaneously shipping oil to the West. The pragmatic approach to the oil trade once more seems to have dominated.

The heightened interest of the Soviet Union in exploiting the scarcity conditions of 1973 was reflected in the rise of its oil exports by 81 million barrels over the previous year. This increase in exports took place despite the fact that petroleum consumption within the Soviet Union seems to be increasing faster than its production.[39] Undoubtedly the 40-million-barrel increase in imports from the Middle East in 1973 made possible at least a portion of those additional exports. The result, in any event, was a windfall for the Soviet Union. As indicated in Table 1, the Soviet Union earned $700 million more in hard currency from petroleum exports in 1973 than in 1972. Since Soviet exports to the hard-currency countries increased by less than 15 million barrels over 1972, the added $700 million was due mainly to an increase in prices.

The pricing policy of the Soviet Union in this period reflected its persistent pragmatism where oil was concerned. In the last quarter of 1973, most sales from the Soviet Union were above $10 a barrel. In 1972, the average price per barrel of oil sold to Western Europe amounted to about $2.30. Denmark's purchases of Soviet oil jumped from about $2.20 per barrel in 1972 to almost $10 per barrel in 1973. The average price to Switzerland rose from $2.90 to $9.30, and to the Netherlands from $2.70 to $7.60.

The almost three-fold increase in price is a sure sign that at least a portion of the petroleum sold to the Netherlands by the Soviet Union was shipped after the start of the embargo and after prices began to rise. When a Swedish correspondent reported that the Soviet Union was selling petroleum to the Netherlands in December, 1973, the Soviet government expelled the correspondent.[40] An analysis of Soviet price statistics makes it possible to verify his reports.

The Soviet Union also used the occasion of the embargo to increase its shipments of petroleum to the United States. As of 1971, such sales totalled only $650,000, but, by 1973, they had climbed to $75 million. Some $40 million worth of shipments arrived during the last quarter of 1973, in the midst of the embargo.[41]

Natural Gas

Although Soviet policies for natural gas are not the main focus of this paper, the Soviet Union's policies with regard to its natural-gas sales during the embargo

Table 3
Natural Gas Imports and Exports of the U.S.S.R.
(in million cubic meters)

Imports	1965	1966	1967	1968	1969	1970	1971	1972	1973
Afghanistan	0	0	207	1500	2030	2591	2513	2849	2735
Iran	0	0	0	0	0	965	5623	8197	8680
TOTAL	0	0	207	1500	2030	3556	8136	11046	11415
Exports									
Austria	0	0	0	142	782	956	1428	1633	1622
Czechoslovakia	0	0	265	587	889	1342	1639	1937	2363
Poland	392	828	1025	1000	994	1002	1488	1500	1710
TOTAL	392	828	1290	1729	2665	3300	4555	5070	6800*
Net Imports (+) or Exports (−)	−392	−828	−1083	−229	−635	+256	+3581	+5976	+4615

*Includes 1105 million cubic meters not identified as to destination. Exports to West Germany begin in October, 1973, but are not specifically attributed to West Germany in Soviet sources.

Source: Various issues of *VT SSSR.*

strengthen the picture of economic pragmatism that typifies the Soviet handling of its oil sales. The Soviet Union in 1973 actually increased its exports to non-COMECON countries by 1.1 billion cubic meters. Soviet statistics as indicated in Table 3 do not identify the recipient of this natural gas nor when it was shipped, but since the West German terminal of the natural-gas pipeline from the Soviet Union was officially opened on October 1, 1973, the likelihood is that the extra gas was shipped to West Germany during the last quarter of the year. Throughout the embargo, the Soviet Union maintained this flow of natural gas to West Germany. It also continued its work on a pipeline to Finland, which was opened in January, 1974, and yet a third pipeline to Italy, which was opened the following May.[42]

Eastern Europe and the Soviet Union During the Crisis

While the Soviet Union charged what the market would bear in the hard-currency countries during the embargo, it was more restrained in its approach to its COM-ECON partners. Limited by five-year contracts (1970-75) for the delivery of oil to other COMECON countries, the Soviet Union did not raise prices in 1973 and 1974. As a result, it found itself in a situation that was unique in COMECON experience. Traditionally, the Soviet Union had charged its East European allies more for petroleum than its Western customers. In 1961, for instance, the price of Soviet oil to East Germany was about 2.30 rubles a barrel, and the price to West Germany about 1.10 rubles a barrel; but by 1973, the East Germans were paying 1.90 rubles, while the West Germans were paying an average price of 5.18 rubles. Had the Soviet Union insisted on charging the East Europeans the hard-currency equivalent of $10 a barrel in 1973, it would have brought the Soviets an additional $6.7 billion in revenue.

Calculations of this sort led the Soviet Union to break its long-term contracts with the East Europeans and to raise the price of its petroleum in January, 1975, a year ahead of time. In the fall of 1974, it had already been pressing for such a change, but, at that time, it seemed unlikely that the Soviets would move quite so quickly.[43] When they did act, however, the prices were only doubled, leaving them still about one-third below world price levels. There is a good chance, however, that Soviet prices will be increased again in January, 1976, when the five-year contracts officially expire. Of course, as in any barter arrangement, doubling the prices of oil could be partially offset by allowing the East European countries to raise the prices of the goods they sell to the Soviet Union. To some extent, this is likely to happen; but the East Europeans none-theless appear to feel that the new Soviet price for oil will have a serious impact on their relatively fragile economies.[44]

It remains to be seen, however, if the East Europeans can free themselves from their heavy dependence on Soviet petroleum. Although we saw earlier that determined efforts were being made to diversify their sources of supply, most East European countries still depend on the Soviet Union for more than half of their crude oil and petroleum products; the difference is lower only for Rumania and Yugoslavia. With the opening of the Adria pipeline through Yugoslavia in the late nineteen-seventies, the East Europeans may conceivably have an opportunity to buy oil from the Middle Eastern nations. The latter are not likely to see much economic advantage in the barter deals that the East Europeans typically have to offer. Nonetheless, Kuwait's willingness to finance part of the Yugoslav pipeline suggests that economics is not the only determinant. The usefulness that some Middle East exporters may see in garnering political support from Eastern Europe in exchange for their petroleum may yet determine whether the new supply link is forged.

How Long Can It Continue?

Although the broad lines of Soviet policy toward the oil crisis, it can be assumed, were shaped primarily by considerations of political and military strategy, the Soviet Union nevertheless managed to use the occasion to garner some considerable economic advantages. Future policy will also no doubt be shaped by a similar mixture of broad strategic policy and narrow economic considerations.

One critical factor in the economic calculation is the country's own need for imported oil. As indicated earlier, there is considerable doubt over the Soviet Union's potential for increasing its oil production. To some extent, future Soviet output and delivery of petroleum are dependent on the Soviet Union's obtaining the Western equipment and technology that will allow it to exploit more geographically remote deposits. There is also uncertainty about how much restraint can be placed on this growth in demand. In any event, even if the Soviet Union is forced to cut back on exports, it is likely to continue to accumulate higher hard-currency earnings because the price of petroleum has risen so much. Thus the Soviets could reduce petroleum exports by one-half in 1975 and still earn twice as much as they earned from petroleum exports in 1972.

Failure to curb their growing demand would also reduce their ability to reexport a part of their Middle East oil. The Soviets may not be in a position, therefore, to pass on a part of their Middle Eastern oil for very much longer. Their opportunities to act as a middleman will also be reduced or eliminated if Arab suppliers begin to insist on payment in hard currency and at world prices. Apparently the Iraqis began to do just that in May, 1974.[45]

If trade relations between the Middle East and the Soviet Union continue to rest on barter, the Soviet advantage might also be diminished by a change in the prices assigned to the Middle East oil exports. Of course, the Soviet Union could respond by raising the price of its exports, as it has been known to do in the past, [46] but whether it will be able to match increases in oil prices by the Arabs depends on its bargaining power. The Soviet Union may be able to secure improved terms for the sale of Soviet arms, but it is less likely to be successful in raising the prices of its industrial goods.

Whatever the outcome, the Soviet Union may still see advantages in acting as a middleman. In the past, they have been happy to take bartered oil and resell it for hard currency to the countries of Western Europe. While it is true that they will have to give equipment and arms to the Arabs in exchange for oil, Soviet officials seem not to object to that arrangement since it provides them with access to the convertible currency that the middleman role in petroleum generates.

The East Europeans, on the other hand, are in a vulnerable position. If the oil exporters raise their prices or insist that payments be made in convertible currency, the effects may be serious. These countries are dependent on foreign oil and are hard pressed for convertible currency. The effect of increased oil prices for them, therefore, will be to drive them back into a position of dependency.

Conclusion

The Soviet Union is the first or second largest producer of petroleum in the world and the fifth largest exporter. Soviet behavior during the oil crisis suggests strongly that its huge economic stake in the oil trade can affect the policies that it is likely to adopt in this area. Its future role in the petroleum markets of the Western world depends on several factors. First, it will have to develop new reserves. Second, its domestic demands for energy will have to be satisfied by holding in bounds the amount of petroleum it supplies to Eastern Europe. Finally, the Arabs will have to continue to supply the Soviet Union with added quantities of petroleum either in exchange for bartered goods or at concessionary prices in hard currency.

Based on the available evidence, it is plausible to assume that certainly one or two of these conditions will be met, at least in the years just ahead. So long as the oil exporters have a stake in keeping their ties with the Soviet Union alive, they are likely to provide petroleum on terms that satisfy the last condition. And so long as that is the case, the Soviet Union can be expected to exploit the economic advantages of its relationship. This may take some delicate balancing of foreign policy on the part of the Soviet Union, but, on the basis of the evidence so far, the Soviet Union seems fully equal to the task.

REFERENCES

[1]For a fuller discussion of Soviet petroleum and natural-gas policy, see Marshall I. Goldman, *Detente and Dollars, Doing Business with the Soviets* (New York, 1975), Chapter 4.

[2]For a sample of the type of pressure applied at the time, see United States Senate, *Hearings Before the Subcommittee to Investigate the Administration of the Internal Security Act and other Internal Security Laws of the Committee of Judiciary, Export of Strategic Material to the U.S.S.R. and Other Soviet Bloc Countries; Problems Raised by Soviet Oil Developments*, 87th Congress, second session, 1963, Part 3, pp. 386, 406. See also P. H. Frankel, *Mattei: Oil and Power Politics* (New York, 1966), p. 140.

[3]Converting rubles into dollars at the official exchange rate may sometimes be misleading because the Soviet government does not freely allow such conversion except by tourists and for a few other special purposes. However, Soviet imports and exports are almost always measured in world prices and reported at the official rate of exchange. Because the Russians generally insist on convertible currency when they sell petroleum to the developed countries, it is not too misleading to convert these ruble figures into dollars when petroleum is sold to the hard-currency countries.

[4]*Vneshniaia Torgovlia (VT) 1918-66*, (Moscow), pp. 126-27, various annual issues of *VT*.

[5]*Hearings, op. cit.*, Part 3, p. 382.

[6]*Petroleum Press Service*, October, 1968, p. 376.

[7]The United Nations Statistical Office, *World Energy Supplies*, Series J, No. 17 (New York, 1974), p. 27.

[8]Leslie Dienes, "Soviet Electric Power: Problems and Trends in Resource Use," paper presented at the Conference on Soviet Resource Management and the Environment, University of Washington, June 6-7, 1974, pp. 34-40.

[9]*Oil and Gas Journal*, December 31, 1973, p. 87.

[10]*Statistical Review of the World Oil Industry, 1973* (London, The British Petroleum Company Ltd., 1974), p. 4.

[11]Dienes, *op. cit.*, p. 40; *Oil and Gas Journal*, December 31, 1973, p. 87; *Petroleum Economist*, March, 1974, p. 100; *Trud*, August 12, 1967, p. 2.

[12]See Appendix II, Goldman, *op. cit.*, and Ray Vernon and Marshall I. Goldman, "U.S. Policies and the Sale of Technology to the Soviet Union" (mimeographed, October, 1974), p. 3.

[13]United States Congress, *Subcommittee on Priorities and Economy in Government of the Joint Economic Committee, Allocation of Resources in the Soviet Union and China*, April 12, 1974, p. 24; see Goldman, *op. cit.*, Chapter 5.

[14]Marshall I. Goldman, *Soviet Foreign Aid* (New York, 1967), pp. 110-11.

[15]*Petroleum Press Service*, October, 1972, p. 390.

[16]See Edith Penrose's article in the issue.

[17]*Petroleum Press Service*, January, 1971, p. 29.

[18]Personal letter, April 17, 1973.

[19]*Petroleum Press Service*, March, 1968, p. 99.

[20]*Petroleum Press Service*, March, 1971, p. 109.

[21]G. Prokhorov, "Mirovaia sistema sotsializma i osvobodivshiesia strany," *Voprosy ekonomiki*, November, 1965, pp. 84-85. O Bogomolov, "Khoziaistvennye reformy i ekonomicheskoe sotrudnichestvo sotsialisticheskikh stran," *ibid.*, February, 1966, pp. 85-86. M. Sladkovskii, "XXII s'ezd KPSS i problemy ekonomicheskogo sotrudnichestva sotsialisticheskikh stran," *ibid.*, April, 1966, p. 96. N. Volkov, "Struktura vziamnoi torgovli stran SEV," *Vneshniaia torgovlia*, December, 1966, pp. 10-12.

[22]*Foreign Trade* [U.S.S.R.] June, 1973, p. 14.

[23]See Franklyn Holzman, *Foreign Trade Under Central Planning* (Cambridge, Mass., 1974), pp. 269-314.

[24]*Petroleum Press Service*, March, 1971, p. 109; *Moscow Narodny Bank*, September 13, 1972, p. 3; January 9, 1973, p. 3.

[25]*Petroleum Economist*, June, 1974, p. 205.

[26]*Petroleum Press Service*, March, 1971, p. 109.

[27]Personal correspondence. See also *Petroleum Economist*, January, 1974, p. 13.

[28]*Anuarul Statistic Republicii al Socialiste Romania, 1974* (Directa Centrala de Statistica, Bucharest, 1974), p. 333.

[29]*Statistisches Jahrbuch der Deutschen Demokratischen Republik, 1974* (Staatsverlag D.D.R., Berlin, 1974), pp. 291, 295, 297, 307.

[30]*Rocznik Statystyczny Handlu Zagranicznego*, 1974 (Glowny Urzad Statystyczny Polskiej Rzecny-pospolitej Ludowej, Warsaw, 1974), p. 113.

[31]Chamber of Commerce of Czechoslovakia, *Facts on Czechoslovak Foreign Trade, 1974* (Prague, 1974), p. 45.

[32]*Statisticki Godisnjak Jugoslavije, 1974* (Socijalisticka Federativna Republika Jugoslavija, Savezni Zavod za Statistikii, Belgrade, 1974), p. 250.

[33]*Petroleum Economist*, June, 1974, p. 205; *Petroleum Press Service*, February, 1973, p. 273; *Wall Street Journal*, August 27, 1973, p. 13; *New York Times*, January 10, 1975, p. 51.

[34]Observers differ over whether or not the Soviet Union tried to head off the war. See Galia Golan, "Soviet War Aims and the Middle East," *Survival*, May-June, 1974. I am indebted to Linda Miller for this citation.

[35]*London Times*, November 20, 1973, p. 18; *New Times*, Nos. 45-46, November, 1973, p. 8; No. 48, November, 1973, p. 20; *New York Times*, March 13, 1974, p. 24; Jean Riollot, "Moscow and the Oil Crisis," *Radio Liberty Dispatch*, January 10, 1974; Radio Moscow, September 29, 1973; Radio Moscow in Arabic, September 25 and November 5, 1973.

[36]See the articles in this issue by Edith Penrose and George Lenczowski.

[37]See the article in this issue by Robert Stobaugh.

[38]*Soviet News*, December 4, 1973, p. 508; *New Times*, November, 1973, No. 48, p. 22.

[39]Dienes, *op. cit.*

[40]*New York Times*, December 14, 1973, p. 11.

[41]Bureau of East-West Trade, *Export Administration Report, East-West Trade*, U.S. Department of Commerce, Third Quarter (Washington, D.C., 1973), p. 84; First Quarter (1974), p. 60.

[42]*Soviet News*, January 22, 1974, p. 28; *Izvestiia*, June 11, 1974, p. 4.

[43]*New York Times*, October 1, 1974, p. 5.

[44]*Financial Times*, February 24, 1975; *New York Times*, February 4, 1975, p. 6.

[45]*Washington Post*, May 28, 1974, p. A-18.

[46]*Wall Street Journal*, July 3, 1974, p. 6.

NORMAN GIRVAN

Economic Nationalism

THE OPEC OFFENSIVE OF 1970-73 was long in the making; ultimately its roots lay neither in the formation of the Organization of Petroleum Exporting Countries in 1960, nor even in the Arab Petroleum Congresses or the oil policy discussions within the Arab League beginning in the late nineteen-forties. It was rather a direct response to the structure of unequal power relations that existed—and continues to exist—in the contemporary world capitalist system. It formed the most concrete and dramatic manifestation of a more general phenomenon in contemporary international relations, that of Third World economic nationalism. It represents the vanguard of a movement by which states in the Third World—or more precisely, the classes that control these states—are seeking to redefine their role in the world economic system and, in particular, to gain a more nearly equal role in their relations with the centers of power. The OPEC offensive should therefore not be analyzed exclusively within the framework of the international oil industry, but rather within the context of the domination-dependency relationships that characterize the contemporary world system and the tensions, reactions, and conflicts generated by them.

What was unusual about the OPEC action? Why did it generate such profound tensions, and why has the issue of OPEC behavior become so important an item on the international agenda? Surely not because the OPEC actions were in themselves unique or unprecedented. Cartels, cartel-like behavior, and administered pricing are certainly not new to modern capitalism. Nor is the use of the trade embargo an unusual political weapon in modern international relations—as every Cuban knows. To be sure, oil is a vital commodity, and OPEC quadrupled the price; but grain is an even more vital commodity to the world's population, and its price also quadrupled after the poor harvest of 1972-73, contributing probably more to the world inflationary process than the increase in the price of oil.

The OPEC action was unusual and unprecedented because it was taken by a group of Third World, primary-product-exporting countries. For the first time in modern history, some non-industrialized countries had succeeded in securing market power in world trade in their export commodity, and they had used this market power to impose a substantial improvement in their terms of trade with the industrialized West. This rudely disrupted the established pattern of center-periphery power relations that had governed the economic relationships between the developed countries and the

145

Third World. It also served as an inspiration, a lever, and possibly a financial base for similar attempts by other Third World exporting countries to secure market power in other commodities. One journal paraphrases Che Guevara as speaking of creating "one, two, many OPECs";[1] a prominent American economist warns of "the threat from the Third World," and points to attempts to form producer cartels in bauxite, phosphate, copper, tin, coffee, bananas, iron ore, and mercury, and to the potential for similar attempts in timber, rubber, nickel, tungsten, cobalt, tantalum, pepper, and quinine.[2]

Nor is Third World economic nationalism seen as stopping at attempts to control prices for the primary products upon which the export incomes and the economic livelihood of these countries depend. Market power is seen as only one component in a general strategy for securing control over marketing and ultimately over production of the natural resources that sustain the Third World economies. State participation in ownership—whether the percentage of equity held is twenty, fifty, or a hundred—has in the course of the past few years become routine for these industries in the Third World. National economic sovereignty, in short, is seen as inseparable from the objective of more nearly approaching equality in international economic power relations.

An outgrowth of this trend toward sovereignty is the development of joint ventures in basic industries among governments of the Third World countries, in which the participating states provide capital, infrastructure, and raw materials. Thus economic cooperation among Third World countries accompanies state involvement in resource industries and attempts to secure market power in the international trade of primary products. Formally, the policies and objectives of Third World economic nationalism were crystallized in the resolutions adopted at the Algiers summit meeting of the nonaligned countries in September, 1973, and the declaration of principles for a New International Economic Order adopted at the United Nations Special Session on Raw Materials in May, 1974. These events revealed a unanimity of stance among Third World governments that formed an important part of the political foundation upon which the OPEC offensive was built.

Center-Periphery Relations and the World System

At the same time as political decolonization was taking place in the aftermath of the Second World War, an international order was being established in the non-Communist world that was marked by profound—and growing—inequalities. At one end of the system stood the United States, closely followed by Western Europe and Japan, the centers of advanced industrial capitalism, characterized by sustained economic growth and technological progress, almost full employment, structurally transformed and flexible economic systems, and a considerable amount of international economic, financial, and military power. At the other end lay the countries, variously called "underdeveloped," "developing," or "Third World," that included the majority of nations and peoples in the world. These lay at the periphery of the international capitalist system and were characterized by underdeveloped and dependent economic systems and by a low standard of living for the overwhelming majority of their populations. The two ends of the system were—and continue to be—linked by complex relationships, marked by profound inequalities in power, in the international division of labor, and in the international distribution of income.[3]

While it is true that the contemporary systems of development-underdevelopment, domination-dependence, and center-periphery became established in the post-war years, it is also true that these forms represent merely the most recent manifestation of a system that had been evolving over the better part of the last five centuries. In a sense, the seeds of the present world system were sown toward the end of the fifteenth century when, in Western Europe, the embryo of a commercial capitalism began to emerge from out of the old feudal order, and was further encouraged by the European maritime nations' "discoveries" of America and of a sea route to the East. The centuries that followed were disastrous for the vast majority of non-European peoples, who were visited by massacre, pillage, forced migrations, slavery, indentured labor, and finally coerced into a reorientation of their socio-economic systems around the production of goods that would satisfy the requirements of the external market. In short, these countries had been led from undevelopment to underdevelopment, from autonomy to dependence.[4]

These consequences of European economic and political expansionism assumed different forms and had different effects as its methods progressed from primitive plunder to more sophisticated commercial and industrial exploitation. In the sixteenth century, the bustling and creative civilizations of highland America were destroyed, their autonomy lost, and a process of socio-economic reorganization initiated that subjected the indigenous population to exploitation by a small, landed, colonial ruling class. From the seventeenth to the nineteenth centuries, the tropical lowland areas of the Americas were incorporated into the plantation system based on chattel slavery, which provided the labor for the large-scale production of agricultural staples for European markets. This system generated vast profits for the European planters and even vaster ones for the merchant class that financed production and trade and organized the market. In the eighteenth and nineteenth centuries, the Indian subcontinent was similarly exploited, its surpluses systematically drained to help finance the British industrial revolution, and its traditional textile industry destroyed to facilitate the growth of British industry. As the industrial revolution spread from Britain to Western Europe and then to the United States and Japan, these countries used their rapidly growing economic and military power to carve up the rest of the world among themselves and to assign to the colonial areas what was to become their characteristic role in the emerging international order: that of providing cheap raw materials for the factories of the industrial center countries and cheap food for the factory workers. By the end of the nineteenth century, the development-underdevelopment dichotomy of the economic system had been firmly established, and the countries in what had by then become the underdeveloped periphery of this system had in most cases completed their adaptation to the needs of the industrialized countries.

It is hardly surprising that the development of such a system was continually marked by tensions and conflict. Wars by the peoples of the emerging centers waged against the peoples of the submerging periphery, as well as other forms of conflict, reaction, and rivalry marked the relations between the colonial and imperial powers until the Second World War. In the post-war period, however, it seemed that a relative stability had been achieved, based on the recognition of Soviet, East European, and Chinese socialism, political independence for most of the colonial world, and American dominance in the international capitalist order. But this apparent stability—sometimes called the *Pax Americana*—only served to conceal certain fundamental unresolved conflicts in the twentieth century, just as the *Pax Britannica* had in the nineteenth.

One source of conflict was found in the challenge of the socialist states to the hegemony of the capitalist powers and in the resulting geopolitical tensions generated by socialist revolutions in such otherwise peripheral countries as Cuba and Vietnam. Another, which surfaced in the sixties, arose from the challenge by Western Europe and Japan to American economic and financial preeminence in the capitalist bloc. The conflict expressed in the OPEC offensive stemmed from still another source: the rebellion by Third World countries against the inequalities inherent in the participation forced upon them in the international capitalist order, which the process of political decolonization had done little, if anything, to correct.

The fundamental inequality lay in the degree of development of the productive forces in the center and in the periphery. The bulk of high-productivity, high-income activities, especially manufacturing, are located in the center countries, while the bulk of low-productivity, low-income activities, especially primary production, are located in the periphery. This unequal division of labor does not, however, amount to specialization in manufacturing by the center countries and primary production by the periphery, for the developed countries in fact also account for a considerable share of world output in primary products, while manufacturing has spread to many countries in the periphery. What is notable is that primary production in the center countries yields relatively high incomes to those engaged in it, while such manufacturing activity as takes place in the periphery yields relatively low incomes to its workers. This phenomenon has given rise to the theory of *unequal exchange*, which purports to demonstrate and to explain lower rates of remuneration to workers in the periphery, as compared to the center, for equal levels of labor productivity.[5]

In addition to the unequal international division of labor and unequal exchange, unequal development is also found in qualitative differences in the degree of autonomy and the orientation of the economic systems characteristic of the two groups. In the center countries, the process of economic growth and capital accumulation is self-sustaining, that is, it is concentrated largely in the production of goods consumed by the population of the country, and it is characterized by strong links between consumer-goods production and capital-goods production. The typical peripheral economy, on the other hand, is concentrated in the production of export commodities for consumption in the center and import-substitutes for consumption by the elite groups in the domestic society. There are in most cases virtually no capital-goods production and few links between the export industry and manufacturing production. As a result, the main commodity circuits are linked only on the international level and through integration with the center countries.[6] Hence the countries of the periphery tend to be economically dependent, structurally underdeveloped, and poor; while self-sustaining growth, structural interdependence, diversification, and steadily rising living standards for the mass of the population tend to be characteristic of the central economies.

The answer to the problem of the underdeveloped countries that was formulated by the developed countries after World War II was a set of attitudes, policies, and measures that might be called "international developmentalism." The underlying assumption was that underdevelopment results from deficiencies in capital supply, skilled manpower, technology, and the economic infrastructure, and from the backwardness of the socio-economic institutions concerned. Development could therefore be initiated by increasing the supply of critical but scarce resources and by institutional "modernization." Policies and measures were accordingly adopted to promote the flow of capital to the underdeveloped countries in the form of economic "aid" and private investment, in order to train manpower and to develop transport, communications, educational systems, and so on. At first, international developmentalism was enthusiastically accepted by governments in the Third World, partly because they genuinely believed in it, partly because their weak position in the international system left them little choice but to accept it, partly because the elites who controlled these governments were interested in expanding their interests and in adopting the Western way of life.

By the end of the sixties, however, it had become increasingly clear that the combination of political decolonization and international developmentalism was not working. The gap in incomes between the rich and the poor countries continued to widen (in the 1950-68 period, the per capita gross domestic product grew at an annual average of 2.5 per cent in the underdeveloped countries compared to 3.3 per cent in the developed world).[7] According to one estimate, the 70 per cent of the world's population found in the underdeveloped countries enjoyed only 17 per cent of world income in 1963; this declined to around 15 per cent by the end of the decade, and all indications pointed to a probable continuation of the downward trend.[8]

The failure of international developmentalism also found expression in the endemic crisis in the external economic relations of the underdeveloped countries. By any one of a number of indicators in this area, they continued to be a very poor third in the world. Their share of world trade declined from 21.3 per cent in 1960 to 17.6 per cent in 1970,[9] a decline matched by a drop in their share of direct investment.[10] International economic relations both among the capitalist center countries and between them and the socialist countries were growing much faster than those between these two blocs and the underdeveloped countries. The Third World was, in a sense, becoming further "marginalized" so far as the developed countries were concerned. This was not matched by a corresponding fall in the dependency of the Third World upon the central nations, however, nor with a growth in self-reliance within or between these countries. On the contrary, their trade with the developed countries was growing faster than their trade with one another, and external dependency was being intensified rather than transcended. In effect, a new circle of dependency was being established. First, the strategies of international developmentalism placed great emphasis on the importation of products from the developed world, which in turn meant a high priority assigned to the generating of foreign exchange and therefore to export expansion. But, as the underdeveloped countries sought to increase exports, they encountered a progressive de-

cline in their terms of trade with the developed countries. According to UNCTAD estimates, the net barter terms of trade of the underdeveloped countries declined by 12 per
cent between 1954 and 1970,[11] and the losses resulting from such declines amounted to
$10 billion for these countries in the period 1960-72.[12]

Thus the growing demand for imports and the increased payments for technology
combined with a slow growth of exports and a decline in relative export prices to produce a persistent and growing balance-of-payments deficit. Such deficits were financed
by increasing external indebtedness—it quadrupled during the sixties, and reached the
startling figure of $80 billion by the end of 1973.[13] The external debt solution was ob-
viously only exacerbating the problem, for the need to pay interest and principal on
the rapidly growing debt soon became the grounds for new borrowing, thereby completing the vicious circle. By 1969, interest and profit payments on the external public
and private debt of the Third World were alone absorbing some 15 per cent of their
total export receipts and over one-half of the new capital coming into these countries.[14]

By 1960, the crisis in the post-war Third World was becoming increasingly evident. Governments were encountering severe and growing difficulties in meeting their
external payment obligations and their targets for investment and economic growth.
Their currencies were depreciating, inflation had for many become endemic, and
unemployment was growing. In human terms, the situation was appalling; the non-
aligned countries declared at their September, 1973, meeting in Algiers that:

Of the 2,600 million inhabitants of the developing world, 800 million are illiterate, almost
1,000 million are suffering from malnutrition or hunger, and 900 million have a daily income
of less than 30 U.S. cents.[15]

Populations, fed for a generation on the expectations of international developmentalism
and promises of political independence, were becoming increasingly discontented. Governments were coming under increasing pressure. Like it or not, most of them now had
little choice but to question the assumptions on which the post-war international economic order had been based and the attitudes and actions that went along with them.

Economic Nationalism

Any analysis of Third World economic nationalism that is cast only in economic
terms is necessarily incomplete and ultimately misleading. The actions of Third World
governments can only be understood in terms of the classes and social groups that control these states, their relations with other classes and groups within their own countries, their relations with the governments of the developed countries and with the
managers of the large multinational corporations that do business in the Third World.
While any generalization of necessity does some violence to reality and has limited applicability in any specific case, it can nonetheless be asserted as a general principle that
in the same way as the impact of the central countries on the Third World generated
an economic structure characterized by dependent underdevelopment, it also resulted
in a characteristic pattern of class formation associated with that structure.[16] Typically,
the majority of the population was comprised of low-income rural or urban workers

(including "own-account" workers, such as peasants and artisans). There was a considerable amount of underemployment and unemployment—a condition that ensured a supply of cheap and abundant labor for the export sector and, later on, for the sector producing import-substitutes for the local elite. The elite or ruling class was composed of three distinct groups, at least initially: a landowning class, exploiting the cheap labor of the rural workers for the production of cash crops for export to the center; an urban commercial middle class, organizing the export-import trade with the center; and a bureaucratic class associated with the government. Later—mainly in the post-war period—the bureaucratic class assumed growing importance and was joined in some countries by an industrial middle class which in turn allied with the multinational corporations producing import-substitutes for the local elite.

Until just after the Second World War, most of Africa and Asia were under direct or *de facto* colonial rule. Their governments were thus in the direct service of central capitalism and were in no position to assert an economic nationalism that would be against the interests of the developed countries. Except where national liberation was the result of a popular revolutionary process, as in China, the period just after political decolonization was ordinarily marked by the quiescence of the new regimes to the international economic order and the terms of international developmentalism. In many of these regimes, the landowning classes and the urban commercial middle class were important members of the dominant elite, and both of these groups perceived their interests to lie with the expansion of the external market and accommodation to international capitalism and the center countries. The bureaucratic middle class and the populist leadership were in most cases too weak or too inexperienced to confront the center countries or international capitalism, especially when this might also have meant defying other segments of their own elite ruling class at home.

Nevertheless, recent events show that the acquiescence of the ruling class to the terms imposed by the center is not necessarily stable or durable. On the contrary, where the local ruling class perceives a configuration of international and domestic circumstances that allows it to secure greater independence and increased financial resources from the center, there is every reason for it to take advantage of that pattern. The increased financial flows provide the material resources for the local ruling class to expand and to develop its material base, especially through industrialization. In addition, the assertion of economic nationalism is a useful ideological instrument for the political mobilization of the mass of the population in a "struggle against imperialism." The circumstances for a successful assertion of economic nationalism are provided when: 1) the nation-state possesses powerful sources of leverage over the center, such as critical raw materials in short supply, and 2) the local ruling class, especially technical people and professionals employed in the government, has developed the numbers, sophistication, and skills necessary to perceive and take advantage of the opportunities.

Although the OPEC offensive represents perhaps the clearest and most successful expression of the development of economic nationalism so far, attempts in this direction are by no means limited to the oil-exporting countries or to the early nineteen-seventies. Conflicts between Third World states and the foreign companies involved in their natural-resource industries were evident as long ago as 1937-38, when expropriations of American-owned oil companies took place in Bolivia and Mexico. Company-country disputes were also frequent in the copper and bauxite industries in the Middle East, Latin America, and Africa throughout the fifties and sixties. Conflicts and disputes tended to crystallize around a number of fairly well defined issues common to all the countries and the industries, although the relative importance of each issue naturally varied from case to case.[17]

The pricing of output and the taxation of company profits have for obvious reasons been among the most contentious, since they affect the level and rate of growth of government revenues from the industry and, through that, the size and rate of growth of the state bureaucracy, of the government's capital expenditures, and of the state's contribution to the private sector and politically favored groups in the population. Ultimately, pricing and taxation issues center on the division between the state and the multinational corporation of the economic revenues generated by the industry. The revenue question also forms part of another area of contention: that of the net foreign-exchange contribution to the national economy of the companies in the industry. Because of the importance of the foreign-exchange constraint in the development of most underdeveloped countries, the foreign-exchange issue can be as important as the revenue issue in some cases. Since both the rate of reinvestment by the companies and the rate of expansion within the country that they decide to undertake affect the rate of growth of revenue and foreign-exchange receipts from the industry, these factors often also become points of country-company dispute.

Yet another issue arises out of the strategic importance of petroleum and mineral raw materials as industrial input—for example, oil as a source of energy, or petrochemical feedstock, and copper, bauxite, and iron ore as materials for heavy industry. When governments seek to use such resources as the basis for a domestic industrialization effort, they may come into conflict with foreign-company strategies for using them as the basis for vertically-integrated processing facilities elsewhere in the world, especially in those center countries where the companies are based. Similar conflicts may arise in connection with the purchasing policies of the companies. The governments may wish to use the demand generated by the industry as the basis for backward linkages within the industrial sector.

Finally, country-company tensions might arise over the industry's existence as a foreign enclave, physically, socially, and economically isolated from the national society, engaging in discriminatory or culturally alien practices, and seeking to place itself beyond the reach of national jurisdiction and to defy or ignore the policies of the government. These issues help to explain why country-company conflicts often become identified in the eyes of the government with the issue of national sovereignty, and why policies of nationalization of foreign industries have become so widespread in the Third World.

The fact that an explosion of economic nationalism did not occur earlier was not due so much to the absence of fundamental conflicts of interest between countries and companies in the past as to the absence of conditions which allowed these conflicts to be openly expressed and successfully resolved. Until the early sixties, a large part of the Third World was still under colonial rule; and even among the formally independent states, such as those of Latin America, the balance of power was tilted heavily in favor of the multinational corporations and the center countries. A state that expropriated a multinational firm, for example, could at the very least expect an economic blockade, such as those which took place after the Mexican and Iranian oil nationalizations in 1938 and 1952 respectively, and quite possibly a military intervention as well, as happened in Guatemala in 1954, Suez in 1956, and Cuba in 1961. Production technology and marketing outlets in these industries were also so tightly controlled by the companies as to make it difficult and costly for a state to attempt to defy them and operate a resource-based industry successfully. Finally, a sizable and sophisticated bureaucratic class had not yet developed in these countries, and the disillusionment with international developmentalism was not yet complete.

By the end of the sixties, however, the growing importance of the socialist countries and the rise of Third World influence in the United Nations helped to shift the international climate to a direction more favorable for the assertion of economic nationalism, particularly the idea of sovereignty over natural resources. This shift occurred just as the failures of international developmentalism were becoming evident. Additional leverage was being made available to Third World resource-exporting nations, as Western Europe and Japan emerged as economic powers in the capitalist world; both had become heavily dependent on imported oil and mineral raw materials from the Third World. The growth of the Soviet and Chinese economies, though based to a far greater extent on self-sufficiency in raw materials, also represented additional potential competition for the resources of the Third World. These developments tended to loosen the tight monopolistic control over production technology and market access exercised by the multinational corporations, especially· the American companies. In addition, growing fears of long-term resource shortages, the rise of the conservationist ethic, and actual shortages of certain critical commodities (such as oil in the early nineteen-seventies) tended to shift the advantage in favor of suppliers.

Another factor facilitating the apparent success of economic nationalism since 1960 has been that the multinational corporations themselves have begun to adapt to it and even turn it to their advantage, thus diluting their resistance. Country cartels in raw materials can be used by the companies as a justification for raising downstream prices and profit margins: there is some evidence—or at least suspicion—that this took place in 1974 in the cases of petroleum, bauxite, and bananas.[18] Government participation in the ownership of these industries can be made profitable to the companies through the negotiation of favorable valuation and payment provisions, reduced taxation and foreign exchange burdens, and remunerative contracts for the supply of managerial, marketing, and technical services. Through such devices the multinational corporation can retain control over the partially or fully nationalized enterprise and secure the same or even an increased cash flow from it, while the Third World government undertakes the job of policing the labor force, providing much of the capital, and assuming the market risks. By defusing the ownership question and incorporating the Third World state into the industry, the supply of raw materials to the companies and to the center countries can actually be made more stable. In addition, many of the companies concerned are engaged in substantial diversification efforts—the oil companies now have major interests in alternative energy sources, the copper companies in aluminum and coal, and so on. Thus their profitability and growth are rapidly becoming less dependent on one particular resource material or on their operations in Third World countries.

Most, if not all, of these factors were in operation between the petroleum industry and OPEC in the 1970-73 period. The political backing of the non-aligned countries, the heavy import dependence of Western Europe and Japan, the fears of long-term inadequacy of energy resources, the actual shortages of oil, and the weak resistance of the major multinational oil companies to the OPEC claims were all factors that contributed to OPEC's success. To be sure, a great deal was also unique to the case of oil and of OPEC: the levels of output, income, and employment in the developed countries are far more sensitive to oil supplies than to supplies of any other industrial material imported in large quantities from the Third World. The OPEC countries were able to play off the independent oil companies against the majors. Libya was willing to act as a "leader" in taking apparently enormous risks with the companies; by succeeding,[19] it showed the other countries what was possible. The 1973 Arab-Israeli war also played a special role in provoking production cutbacks and embargoes and making it even easier to impose price increases. Indeed, were the case of OPEC and the petroleum industry not in some way unique, it could hardly have resulted in the enormous success for Third World economic nationalism that it became. But to stress the uniqueness of the factors leading to the OPEC offensive to the exclusion of others—the disillusionment of the Third World with the post-war international economic order and their rebellion against it—is to commit a serious error of understanding which can only lead to mistaken prognostications about the future.

Toward a New International Economic Order?

Is a new international economic order really in the making? Are we in fact witnessing a structural shift of international economic power to the OPEC countries and possibly to other Third World raw-material-producing nations at the expense of the multinational corporations and the developed capitalist countries? What of the other underdeveloped countries—should we now speak of a "Fourth World," the countries without resources of oil or other valuable mineral commodities, whose position has allegedly been worsened as a result of the OPEC offensive? These are some of the questions which it seems relevant to ask, though at the inevitable risk of committing such errors of interpretation as arise out of proximity to the events as they unfold.

To begin with, the durability of OPEC's achievements in securing effective market power in the international crude-oil trade is by no means certain. The special circumstances prevailing in the 1970-73 period, especially the oil shortages of 1973, had to a large extent disappeared by early 1975, when surpluses of supplies and capacity began to reappear. Indeed, the OPEC countries have probably been regarding themselves as on the defensive in the pricing question since the increases of December, 1973. Undoubtedly the decision of the January, 1975, OPEC meeting to freeze prices for the time being implies the formal recognition of a declining real price for the producer's oil as a result of continuing inflation in the West. Moreover, the surpluses and excess capacity that began to appear in early 1975 are a reminder that OPEC has yet to face the crucial test of an effective cartel—the ability to secure unanimous agreement among the members on production cutbacks in order to maintain agreed prices. Ultimately the durability of OPEC's market power will depend on whether the industrialized West will be able to cut its rate of consumption of OPEC oil below the minimum rate of production that the OPEC countries can tolerate, given their development objectives. In the short run, the advantage appears to be clearly on OPEC's side; but in the longer run, the industrialized West may well have more room to maneuver than OPEC, because of the flexibility of its economic systems and its advanced technology, implying the ability to economize on energy consumption and to develop alternative sources of energy.

If the durability of OPEC's achievement is in doubt, so much more so are the possibilities of other raw-material-exporting nations' duplicating OPEC's successes. Attempts to form effective cartels in other strategic commodities, such as copper, bauxite, and iron ore, have not been particularly successful, and the producers have yet to demonstrate that they can exercise market power. These efforts have failed in part because import demands by the industrialized West are more elastic for other commodities than they are for oil as a result of the availability of substitutes or of domestic production; partly because the producers have not been able to secure effective agreement on production restraints during times of excess supply; and partly because of the importance of such developed countries as Australia and Canada in the production of some of these commodities.

On the other hand, we should not by any means underestimate the determination of Third World nations to use whatever leverage they may be able to exercise in the markets for their export commodities in order to improve their terms of trade, secure increased bargaining power in the international economic order, and mobilize financial surpluses for development. Whatever our doubts may be about the long-term durability of OPEC's market power, there is no question that in the short to medium term the oil-exporting nations constitute a force to be reckoned with in international economic relations. OPEC's levers are the price it determines for crude oil, the rate at which it allows production to grow, and the use it makes of the huge financial surpluses which member countries are accumulating.

These levers are particularly significant to Western Europe and Japan, both of which are much more dependent on OPEC oil as an energy source than is the United States, and both of which lack the reverse leverage that the United States possesses in the form of its huge economic and military presence overseas. Hence, both are assiduously attempting to come to terms with OPEC through large technology-for-oil bilateral transactions, greater political support for the Arab cause, and a conciliatory attitude on the international oil question. This development is naturally viewed with suspicion and concern by the United States. For, although the OPEC offensive produced a temporary boost to the financial preeminence of the United States through its relatively greater impact on the economies of Western Europe and Japan, its long-term implications are more worrisome to the United States because of the potential for a Western Europe-Japanese-OPEC economic and financial alliance, which would inevitably reinforce the decline of American hegemony in the Western world.

OPEC's relations with the rest of the Third World, meanwhile, are necessarily ambivalent. It is obvious that Third World states suffering from chronic deficits in their balance of payments or in government spending, such as those in the Indian sub-continent and most of Africa and Latin America, are finding it increasingly difficult to class themselves in the same category as the oil-exporting states, which earned surpluses amounting to some $60 billion in 1974, and whose reserves might rise to $635 billion by 1980.[20] They are finding it particularly difficult with respect to those states having the bulk of these reserves, which are at the same time the OPEC members with the smallest populations: Kuwait, Libya, Saudi Arabia, Qatar, and the United Arab Emirates; these countries have a combined population of eleven million, compared with 278 million for OPEC as a whole. To the non-oil-exporting Third World states, OPEC's victory for economic nationalism has in many ways been a costly one, faced as they are with large increases in fuel-import bills, with contractions in non-oil imports, reduced economic growth, and internal price inflation. There are many in these countries who are also bitter that a large majority of OPEC surplus revenues end up being "recycled" to the West.

Still, it would be overhasty to conclude from this that it will be easy to breach the tactical alliance between the rest of the Third World and the OPEC countries where a confrontation with the center countries in demanding a new international order is concerned. The non-oil-producing Third World countries are willing to continue to support the OPEC cause for at least two reasons. One is that they are willing to work under the assumption that the OPEC countries are likely to be more generous and sympathetic with financial assistance and attach fewer strings to their "aid" than the center countries. OPEC pledges to other underdeveloped countries, for example, were reported to amount to some $10 billion in 1974, an amount representing nearly seven times the flow of OECD aid in proportion to gross national product (2 per cent for the OPEC countries compared to 0.3 per cent for the OECD countries).[21] Another example is provided by Venezuela's plan to lend back to Central American countries roughly 50 per cent of their payments for oil imports from Venezuela to be used for development projects.[22]

Another reason that OPEC can expect continuing support from other Third World countries is that oil can be used to exert more general pressure for improving the terms of trade for *all* primary products exported by the Third World. Some of the more progressive OPEC members, led by Algeria, have already utilized this lever and are pushing for OPEC to do so and thus to create a new international economic order more reflective of the interests of the Third World. The Declaration of Dakar, adopted by the non-aligned nations in February, 1975, approved the idea of using OPEC revenues to finance buffer stocks of other primary products; and at the Paris conference of April, 1975, the Third World bloc held firm in demanding that the proposed world energy conference be broadened to include the discussion of other primary products. For, whatever the differences in affluence that now separate OPEC from non-OPEC Third World states, the fact remains that their *structural* position in relation to the international economic system remains essentially the same: their economies are still fundamentally underdeveloped and dependent on primary products exported to the center countries.

It is this aspect of the center-periphery system that the OPEC offensive and Third World economic nationalism will not by themselves change. Current development strategies of OPEC countries involve the purchase of advanced technology from the developed countries and their large multinational corporations: nuclear power plants, gas liquefication facilities, petrochemical plants, large steel mills, and so on. Most of these projects are to be carried out with the state as either whole or part owner in joint ventures with the multinational firms. Oil revenues are, of course, also being used for the accumulation of armaments and for the modernization of the social and economic infrastructure. These developments imply a spread of heavy industrialization into many of these countries and a resulting modification of the traditional international division of labor. They also involve the extension of the state as an economic agent and the growth of a large bureaucratic class. In themselves, however, these developments do not necessarily guarantee self-sustaining growth and the superseding of the center-periphery system, for they are still dependent upon the import of technology from the center countries and upon the export to them of much of the output of the new industries. Furthermore, this pattern of development is usually highly capital-intensive technologically and highly concentrated geographically, and it therefore does not normally involve profound reorganization of the agricultural economy. There is no evidence that it can solve the problems of unemployment and of rural poverty, which are *the* fundamental conditions for the alleviation of mass poverty in these countries. The growth of a state bureaucracy and of the other elements of the middle class allied to it together with the continuation of mass poverty represent a potential for the emergence of new contradictions within OPEC countries that might replace in intensity those which have so far divided OPEC from the center countries.

A new pattern of relationships is being established in the world economic system, both internationally and intranationally. The OPEC offensive and its associated Third World economic nationalism, together with other developments in the last decade, such as the Indochinese movement toward national liberation and the economic resurgence of Western Europe and Japan, have signalled the end of the post-war international order. It is not yet clear what will replace it—if indeed it will be replaced by any order at all. If any lesson can be learned from the oil crisis, it is that in the times in which we live any system or set of arrangements that rests upon domination and dependency is inherently unstable. One hundred years ago, perhaps, the West could afford the arrogance of believing that it could order the world in its own image. But not today. For if it is true that for the majority of people in the world the nineteenth century was the century of imperialism, it is also true that for them the twentieth century is unquestionably the century of liberation.

REFERENCES

[1]"Matières Premières et Développement," *Les Dossiers Jeune Afrique et Economica*, July-December, 1974, p. 5.

[2]C. Fred Bergston, "The Threat from the Third World," *Foreign Policy*, 11 (Summer, 1973); and "The New Era in World Commodity Markets," *Challenge*, September-October, 1974.

[3]For an analytical and descriptive treatment of the center-periphery system, see articles in Norman Girvan, ed., *Dependence and Underdevelopment in the New World and the Old (Social and Economic Studies*, 22: 1 [March, 1973]); and Samir Amin, *Accumulation on a World Scale* (New York, 1974).

[4]See references in note 3 above. Also Walter Rodney, *How Europe Underdeveloped Africa* (London, 1972); A. Gunder Frank, *Capitalism and Underdevelopment in Latin America* (New York, 1967); Paul Baran, *The Political Economy of Growth* (New York, 1957).

[5]Arrighi Emmanuel, *Unequal Exchange* (New York, 1972); Samir Amin, *L'Exchange Inégal et la Loi de la Valeur* (Paris, 1974).

[6]Samir Amin, "Accumulation and Development: A Theoretical Model," *Review of African Political Economy*, 1 (August-November, 1974).

[7]Aninal Pinto, "The Center-Periphery System Twenty Years After," in *Dependence and Underdevelopment, op. cit.*, p. 41.

[8]*Ibid.*, pp. 40, 41.

[9]Fourth Conference of Heads of State or Governments of Non-Aligned Countries, Algiers, September 5-9, 1973, *Fundamental Texts: Declarations, Resolutions, Action Programme for Economic Cooperation*, p. 65.

[10]Pinto, *op. cit.*, pp. 51-52.

[11]"Long-Term Changes in the Terms of Trade 1954-71, Report by the UNCTAD Secretariat," in *Proceedings of the United Nations Conference on Trade and Development, Third Session, Santiago, Chile, IV* (April 13-May 21, 1972), p. 73.

[12]From the Report by the Secretary-General of UNCTAD to the United Nations Sixth Special Session on Raw Materials, New York, May, 1974; cited by the Senegalese Delegate to the UNCTAD Conference on Raw Materials, Dakar, February, 1975.

[13]*Fundamental Texts, op. cit.*, p. 65.

[14]Pinto, *op. cit.*, p. 57.

[15]*Fundamental Texts, op. cit.*, p. 63.

[16]For a partial discussion of this question, see Samir Amin, *Contemporary Peripheral Social Formations, Unequal Development and the Problems of Transition* (Dakar, 1973).

[17]See, for example, Raymond Mikesell, ed., *Foreign Investment in the Petroleum and Mineral Industries* (Baltimore, 1971); Norman Girvan, *Corporate Imperialism, Conflict, and Expropriation: Essays on Multinational Corporations and Economic Nationalism in the Third World* (forthcoming).

[18]Norman Girvan, "Economic Nationalists vs. Multinational Corporations," in *Corporate Imperialism, op. cit.*

[19]See the essay by Edith Penrose in this issue.

[20]"World Bank predicts huge rise in OPEC reserves," *Financial Times*, January 8, 1975, p. 4.

[21]Jonathan Power, "OPEC as an Aid-Giver," *International Herald Tribune*, March 5, 1975, p. 6. The same source points out that ". . . if one adds the substantial purchases of World Bank bonds and contributions to the International Monetary Fund oil facility, the total is even higher."

[22]Roberta Salper, "Venezuela; Steps Towards Hemispheric Unity," *The Elements*, 5 (February, 1975), pp. 4-5.

Part 4: THE MULTINATIONAL OIL COMPANIES

MIRA WILKINS

The Oil Companies in Perspective

OF ALL THE FACETS of the oil crisis, the least understood, perhaps, is the role that was played by the large multinational enterprises. Each of these giant companies used its own strategies, which reflected its individual corporate dynamics and occasionally came into conflict with the interests of the others. Since it was generally assumed that the firms would automatically act on behalf of their home countries, that they were in fact "flag" companies, there was widespread dismay when they obeyed the embargo rules laid down by the Arab states. To act thus as instruments of foreign states seemed unpatriotic, if not positively immoral. Another surprise was the rise in oil prices. According to the popular theories, either the American government allowed prices to rise because it was pressured by the corporations (whose interests were in any case identical to those of the government), or, alternatively, the corporations acquiesced in the oil exporting countries' wish to raise prices (which they in any case favored). To test the validity of these alternative explanations for corporate strategies and to understand the relations of the major corporations with their own countries, the historical background of the interactions between the multinationals and their home and host governments must be investigated.

I

Over the years, the international oil industry has come to involve many participants. In 1900, one mighty multinational oil enterprise, Standard Oil of New Jersey (now Exxon), dominated the industry. In the early years of the century, it attempted, in vain, to block other entries into the world oil business. By 1914, however, having failed in an effort to purchase both Royal Dutch and Shell, Standard had a second giant with which to compete. Royal Dutch had merged with Shell. These two companies—Standard of New Jersey and Royal Dutch/Shell—were then, and are still, by far the largest in the world oil industry, yet their strategies—governed by different leaders and histories—have on occasion differed. Standard Oil of New Jersey (Jersey Standard) had begun in refining; its initial foreign stakes were therefore in marketing oil products. Over the years, it made sufficient investments in crude-oil production to balance its supply needs and its market requirements. By contrast, Royal Dutch/Shell was frequently "crude short"; it made up for the deficiency by continuously seeking new crude-oil sources and by negotiating long-term oil purchase contracts. Jersey Standard, sensitive since the 1911 antitrust decision to possible antitrust action by the United States government, tended to be cautious about arrangements vulnerable to antitrust action. Shell, headquartered in Europe, naturally had different perceptions regarding cooperation and competition.[1]

World War I demonstrated the growing dependency of the industrialized nations on oil, and, in its aftermath, both Americans and Europeans became concerned that United States oil reserves (the United States was at that time still an exporter of oil) and possibly even world oil resources would soon be exhausted. In response to this concern, Jersey Standard, Royal Dutch/Shell, and the many other companies by now in the field sought to develop new crude-oil production. The resulting exploitation of both new and known oil reserves produced a glut in world oil supplies by 1928. By that time a third huge oil enterprise, Anglo-Persian, the predecessor of British Petroleum, which had been founded before World War I, had grown to importance. Unlike the other two giants, which had vast marketing networks, Anglo-Persian's strength lay in its crude-oil holdings. Its initial investments before the war had been in Iran, where it had developed oil production. Although it subsequently remained "long on crude oil," its strategy also included integration; consequently, like its rivals, it invested in a marketing organization to sell its oil output.[2]

The twenties somewhat resembled the sixties in that new corporations were entering the international oil business and the demand for oil was rising rapidly. However, supplies of oil were increasing even more rapidly than demand, as new sources came into production. In 1928, the leading three enterprises tried to stabilize the market by joining with other firms to develop new oil finds in Iraq. The group decided to act in concert within a "red line" area that included almost all the former Ottoman Empire. This so-called "red line agreement" is frequently seen as the first corporate attempt to control oil supplies.[3] Far more important, however, was the "as is" or Achnacarry Accord of the same year. This comprehensive agreement brought together the three leaders, Jersey Standard, Royal Dutch/Shell, and Anglo-Persian, in an effort to stabilize world markets, to minimize competition, and to organize the market. In deference to Jersey Standard's concern over the antitrust laws, the American market was explicitly excluded from these arrangements.[4]

Powerful as they were, Standard Oil of New Jersey, Royal Dutch/Shell, and Anglo-Persian found they could not regulate the world's oil industry[5] because of economic and political factors outside their control. Whether it was the oil discoveries in East Texas in 1930-31 (which sent oil prices plummeting), the American oil tariff of 1932 (which meant that the vast new Venezuelan oil found in the twenties had to be redirected from the United States to Europe), Anglo-Persian's new oil agreement in Iran in 1933 (which showed host nations could "tear up" oil agreements), the entry of Gulf, Standard Oil of California, and the Texas Company into Middle Eastern concessions, the government expropriations in Bolivia (1937) and Mexico (1938), or host-government measures in developed and less-developed countries from import restraints to government-run cartels, the existing multinational oil companies found themselves totally unable to enforce their "as is" provisions. The major companies did, it is true, renew their efforts in the nineteen-thirties to stabilize and to raise world prices and also to divide world markets, partly by bringing additional companies into the agreements. The results were minimal. Economic uncertainty in the thirties discouraged the oil companies from taking risks; joint ventures between and among firms, while far from unknown before, now proliferated.[6]

By 1939 the international oil industry bore little resemblance to what it had been in 1928, when the big three companies had attempted to enforce the "as is" rules. Standard Oil of California, the Texas Company, Socony-Vacuum (now Mobil), and Gulf had joined Jersey Standard, Royal Dutch/Shell, and the predecessor of British Petroleum as significant powers in international oil. There were now seven giants instead of three.

The new American multinationals included two old Standard Oil companies— separated from Jersey Standard in the 1911 antitrust suit. Standard Oil of California (Socal), after years of futile searching for oil, found it in Bahrain in 1931, and this discovery in turn encouraged interest in the nearby mainland, Saudi Arabia, where Socal obtained a concession in 1933. Three years later, it joined with the Texas Company in a number of joint ventures to produce and refine oil in Saudi Arabia, Bahrein, and the Dutch East Indies, and to market the oil products in Africa and east of Suez. Socal's principal strength in the international oil industry would remain based in its stake in Saudi Arabia.

Socony-Vacuum was another old Standard Oil company. Standard Oil of New York and Vacuum Oil had both long had foreign investments. Each had expanded its marketing network in the twenties, the former in Asia, Australia, Africa, and Eastern Europe; the latter, in Europe, Australia, and Africa. They had merged in 1931 to form Socony-Vacuum. In 1933, Socony-Vacuum tied its international marketing east of Suez with Jersey Standard's existing producing properties in the Dutch East Indies to form Stanvac, and thus it became another major multinational. In 1939, Socony-Vacuum's sole Middle Eastern production stake was an 11.875 per cent interest in the joint-venture company in Iraq; both in the thirties and later, it would be "crude short."

The Texas Company (Texaco), established in 1902, had created over the years an international marketing organization as well as other foreign interests. Not until it combined with Socal could it be ranked as a major multinational. Gulf—whose predecessors had started in the Texas oil fields in 1901—made its first foreign entry in Mexican crude-oil production (1912). Subsequent new investments in overseas producing properties made Gulf a key multinational enterprise with abundant crude-oil supplies.

While Shell and Anglo-Iranian (formerly Anglo-Persian) had long had joint ventures in international marketing, while Socal and Texaco were linked in certain markets through Caltex, while Socony-Vacuum and Jersey Standard were associated east of Suez in Stanvac, Gulf was the "odd company." Although it did participate in joint ventures in production—for example, sharing a concession in Kuwait with Anglo-Iranian—unlike the other majors, it appears to have been short of international marketing outlets. In time, it would look for long-term contracts to dispose of its crude. In a sense, all these corporate interrelationships represented efforts by the major companies to create order in world markets and to balance supply and demand; they were also an effective means of meeting world oil requirements.[7]

World War II and the immediate post-war period did not result in increased competition in the oil industry, but did produce some other changes. Both during and after the war, just as after World War I, a concern over the inadequacy of available oil supplies emerged, especially in the early fifties, when the United States shifted from net exporter to net importer of petroleum and petroleum products. The immediate post-war years saw new corporate attention paid to investing in the vast Middle Eastern oil fields. In 1947, Standard Oil of New Jersey and Socony-Vacuum made arrangements to join with Socal and Texaco in the latters' important concession in Saudi Arabia. The concession was held by Arabian American Oil Company (Aramco), which was now to be 30 per cent owned by Socal, Texaco, and Jersey Standard, while Socony-Vacuum—now Mobil—held the remaining 10 per cent. Mobil, with a smaller international marketing organization and fearing the investment risk, opted to take the 10 per cent interest rather than a larger share.[8]

In the immediate post-war years, a few independents entered the world oil industry (for example, Aminoil and Getty Oil in 1948-1949 in the Kuwait-Saudi Arabia Neutral Zone). But because it takes time to find oil, by 1952 the seven major companies were still producing 90 per cent of the crude oil and marketing 75 per cent of the oil products outside North America and the Communist countries. As in earlier years, however, the corporate giants failed to maintain their preeminence. British Petroleum's monopoly position in Iran was shattered in 1954 when a large number of companies (both majors and independents) joined in the Iranian consortium. Neil Jacoby has tabulated a list of 350 firms that invested in the foreign oil business (outside North America and the Communist countries) between 1953 and 1972. Although some of the companies on his list were not strictly speaking new in 1953, having had earlier operations abroad, many enterprises either entered anew or enlarged already existing but relatively small foreign stakes. The companies included large and medium-sized American firms that earlier had had few or no foreign investments, European and Japanese private firms that were developing multinational businesses, and a number of government firms—both consumer and host government enterprises.

As a result of these many new entries into the field, the control of the major seven diminished. By 1968, the majors were producing a little more than 75 per cent of the crude oil outside North America and the Communist countries, compared with their 90 per cent in 1952. This was still a high percentage, but it was a declining one. Likewise, the international giants marketed slightly more than 50 per cent of the oil products sold outside North America and the Communist countries, compared with 75 per cent in 1952. In every part of the oil industry—concession areas, proven reserves, production, refining capacity, tanker capacity, and product marketing—the percentage held by the top seven companies declined in the fifties and sixties.[9]

Oil resources were developed in the nineteen-sixties primarily by the majors in the Persian Gulf states and Nigeria *and* by the majors, independents, and consumer government enterprises in Libya and Algeria. Despite the oil companies' constant complaints about "overproduction" and excess capacity, the companies made substantial new investments in producing properties. These included investments by the majors to match the new stakes of independents, as well as investments by all the enterprises to obtain more diverse and thus more secure sources of supply and to take advantage of lower cost and more accessible oil discoveries. The companies made even larger corporate investments in consuming countries, in both refining and marketing, to obtain further outlets for the abundant petroleum.

The development of more integrated corporate operations failed to hold up the price of oil. The growth of the new oil supplies resulted in a drop in market price. Because of its cheapness, new dependence on oil developed in Western Europe and Japan, and because of its desirability compared to coal, reliance on it also rose in the United States.[10] The major multinationals could not control the price decline in Europe and Japan, nor were they very happy about it.

In 1970-73, however, the trend was reversed. The market price of oil rose again, as gaps in supply, often caused by political events, occurred with increasing frequency. Corporate strategies in the seventies continued to center on balancing supply and demand and maintaining corporate profits. As in times past, the investment decisions and marketing plans in the oil industry were made by the corporate managers of each individual enterprise, although now they found themselves more often on the defensive. While the multinationals persisted as the principal actors in the international oil industry, their powers were becoming ever more constrained.

II

That each major multinational corporation had its own history, its own management, and its own international logistical system did not mean that the home governments of the companies had no influence on their strategies and perceptions. Yet home governments and companies did not necessarily agree. Of today's seven giant multinationals—Exxon, Mobil, Texaco, Gulf, Shell, Socal, and British Petroleum—all began as private enterprises. Only one, the predecessor of British Petroleum in 1914, had attracted a government equity interest. And, while BP has since then been controlled by the British government, all indications suggest that, at least since 1921, it has in fact functioned much as the other multinational corporations, and not as an instrument of its home state.[11] The home-government equity holding in British Petroleum has on occasion acted to the firm's detriment, however; host governments are wary of dealing with a foreign "state" enterprise.[12]

In their international business, the American companies have been limited by government antitrust and trading-with-the-enemy legislation, but they have also been aided by government rules on double taxation and, at times, by Washington's general commitment to support legitimate private enterprise abroad. Over the years, American multinational managers have asserted that the British government has been more supportive of its corporations than their own,[13] but in fact the amount of assistance given by the British to BP and Shell has varied both in time and also according to the interpretation of particular analysts.[14] As for the Dutch government, it seems to have been less knowledgeable, less powerful, and less prepared to act on oil matters than the

British state.[15] Indeed, except in war, rarely did any of these home governments exert its prerogative to force the multinationals to perform in a particular way, although each company often informally reviewed its problems with home-government representatives.

The French and Italians established publicly owned oil companies in the nineteen-twenties. Apparently these governments have maintained greater control over their oil corporations than have the American, British, and Dutch, although these corporations have also developed strategies and styles over the years which have enabled them to remain aloof from direct state intervention.[16]

In any case, the seven major multinationals are not and, with the exception of British Petroleum, have never been "flag" companies; they have, however, never been averse to their governments' appreciating their efforts. They believed that "nationals operating in oil abroad need to have the political risks reduced by intelligent and sympathetic diplomatic support by the government," although, on the other hand, "this does not mean government participation, which should be rigorously avoided in any phase of their operations." These two quotations, from the 1944 recommendation of the United States National Oil Policy Committee, reflected the major oil corporations' views both before and after the war. The companies thought of themselves—and in fact were—the principal actors in the international oil industry.[17]

The following examples may perhaps illustrate the complex government-company relationships in the post-war years. While many foreign-policy issues affected oil industry operations, there is rarely any evidence that the oil companies contributed to these decisions. The Truman Doctrine, for instance, committed the United States to defend Greece and Turkey against Communism, and in the process created security for corporate Middle Eastern oil investments; yet, Texaco's chairman of the board testified that the promulgation of the doctrine caught him by surprise.[18] Neither his company nor, it seems, other oil companies had played any direct role in its formulation. On the other hand, when the United States government supported the formation of the state of Israel in 1948, this made operating conditions awkward for oil investors in Arab lands. Over the years, representatives of American companies and State Department officials certainly discussed the implications of American foreign policy, but there is no evidence that the corporations influenced American policy toward Israel; in fact, there is evidence that no such influence existed. Instead, the multinationals responded by adapting to it: In 1948, for example, the Iraq Petroleum Company rerouted an existing pipeline from the Kirkuk fields in Iraq so it would bypass Israel, and the partners in Aramco revised the route for a new pipeline under construction from Saudi Arabia to the eastern Mediterranean to achieve the same objective.[19]

During the Korean War, the home governments recognized that adequate oil had to be sent to the Far East in the face of shortages from the aftermath of the 1951 Iranian oil expropriation. When, in 1951, the United States government agreed not to invoke antitrust legislation, together the companies arranged the logistics of supply.[20]

The companies had eventually defeated the expropriation itself with an effective boycott of Iranian crude oil. Washington helped by "requesting" that American independents not buy Iranian oil.[21] It did so because the State Department was concerned that Russia might move into Iran, and it took the initiative to settle the dispute before this could happen. A Jersey Standard director, Howard Page, told the present author that his company participated in the resulting Iranian consortium on the urging of the United States government; his company also recognized the dangers that the Soviet Union might move in and dump Iranian oil on world markets. Mobil, which was crude short, may have had different interests in obtaining a stake in Iran, although there is evidence that it too had worries about the risks in investing. All of the majors wanted to obtain an "orderly re-entry" of Iranian crude oil into the world market.[22]

After an August, 1953, National Security Council decision that enforcement of United States antitrust laws was secondary to America's national security interests, the American majors met with the European-headquartered giants to formulate a common strategy in the Iranian negotiations. The group prepared the detailed plans on the proposed consortium. The United States Departments of State and Justice wanted American independent firms to become involved, and on March 30, 1954, the United States Ambassador to London, Winthrop W. Aldrich, cabled Secretary of State John Foster Dulles that, during discussions in London among the companies, an agreement had been reached to invite the independents to join. Aldrich added "we [are] very much relieved this matter apparently settled in principle at company level without necessity [to] apply government pressure." Apparently, the Italian government wanted its state enterprise (ENI) to participate also, but the British government and the major companies did not. ENI was excluded. By contrast, the French government company, Compagnie Française des Pétroles, which had from the beginning been involved in the Iraq Petroleum Company, was included in the Iranian consortium.[23]

In July, 1956, the Egyptians nationalized the Suez Canal Company. British Prime Minister Anthony Eden feared that Egypt would place Middle Eastern oil "under the control of a united Arabia led by Egypt under Russian influence." In October, when the second Arab-Israeli war began, British and French forces intervened to keep the canal open and to support Israel. The United States opposed the British and French action. The canal was closed. In the "transport crisis" that followed, the United States joined with European governments to plan for international oil allocations with the multinational oil corporations participating in the planning. The canal was reopened on April 1, 1957, and the crisis passed. Governmental involvement notwithstanding, it was the oil companies that coped with the allocation problems in 1956-57.[24]

The international oil world in 1959-60 seemed fraught with more than the usual amount of political uncertainty. Castro had expropriated the major oil refineries in Cuba. The multinational oil companies worried over their huge assets in nearby Venezuela, and in 1960 they withdrew funds from that country.[25] Records are not available, but there is evidence that oil company officials conferred with Washington to discuss the general state of affairs. The insecure situation in Venezuela made the Middle East and North Africa far more significant as oil sources. The Russians were exporting substantial amounts of oil to Western Europe and to the less-developed countries, leading to fears that the Soviet Union would start using oil as a political weapon.[26] Moreover, to add to the uncertainty, in September, 1960, the major oil exporters joined in forming the Organization of Petroleum Exporting Countries (OPEC), which subsequently added other oil-exporting nations to its roster. No one knew how much power OPEC might eventually wield.

John J. McCloy, a lawyer for the seven major multinational companies, discussed the Middle East with President Kennedy with regard to protecting the free world from becoming too dependent on Russian oil. McCloy indicated that the companies might need to act in unity "to offset the pressures" of the OPEC nations. At this time, little Middle Eastern oil was entering the United States, which had imposed oil-import quotas in 1959. But America's European allies were still dependent on Middle Eastern oil, and American-headquartered enterprises had large investments in it. McCloy discussed with Attorney General Robert Kennedy the possibilities of collective action by the companies, if needed, and later, on an informal basis, with each subsequent Attorney General, keeping the Justice Department informed in case formal clearances were required. The issue of antitrust continued to shadow the American multinationals' relations with their government. In 1953, the Justice Department had inaugurated a suit against the companies, and had obtained limited consent decrees from Jersey Standard and Gulf in 1960 and from Texaco in 1963.[27]

In the nineteen-sixties, there were numerous oil crises that many oil company officials apparently had forgotten by 1974. The home governments of the companies were consulted, for example, on the events in 1961, when Iraq ordered the Iraq Petroleum Company to relinquish 99 per cent of its concession, in the mid-nineteen-sixties, when the Indian government demanded American oil companies to refine Russian oil, and, in 1967, when the Six-Day War broke out and the Suez Canal was again closed. During the Six-Day War, with the Justice Department's permission, the companies coped effectively with reallocation of supplies, and no nation was short of oil. The State Department knew of these oil matters through the American-based companies, although it was sometimes informed imperfectly and usually informally. John J. McCloy reports that oil company executives "came down to the State Department constantly to talk . . . about the [political] situation in the respective countries. . . . It was simply a matter of exchange of information." At the same time, the State Department had apparently developed no expertise in oil and no unified policy. The logistics of international oil—delivering oil where and when it was needed—as well as all investment decisions stayed in corporate hands. A similar situation seems to have prevailed in Britain and the Netherlands. There were many informal discussions, but the international oil business remained in the hands of the oil companies.[28]

In 1969, Colonel Houari Boumedienne of Algeria took measures against foreign investors in oil, just as the American-headquartered El Paso Natural Gas Company (not an oil company) was negotiating with Algeria on natural-gas imports for the United States. Apparently the American government did not want to upset these delicate deliberations, and it thus made no vigorous representations on behalf of the American oil companies. The representations from other home governments proved an inadequate deterrent to Boumedienne's expropriations.[29]

Libya acquired a new government in September, 1969. Aware of Algeria's successful nationalizations, it pressured its nation's oil exporters for higher oil revenues. Occidental, an American independent with substantial stakes in Libya, was the most vulnerable target. Threatened with the loss of its properties, Occidental capitulated to Libya's demands in September, 1970; the other foreign oil companies in Libya reluctantly agreed to similar terms in the month following. George Piercy of Exxon later explained that Libya "was willing to cut its revenue to get . . . its . . . price objective." Libya had bargaining power in its large currency reserves and a very small population, so it could accept "a reduced income."[30]

The home governments were informed by the companies of Libya's moves, but they apparently did not intervene in the negotiations. However, when other oil-exporting countries set out to follow Libya's example, and there seemed to be a "ratchet" effect from one oil-exporting nation to another, and when OPEC, meeting in Caracas in December, 1970, insisted on added payments to all the producing countries, then the majors, independents, and consumer government firms felt concerted action to be essential. Shell called the companies together.[31] Wary of the antitrust consequences, the American firms turned to the Justice Department for a formal clearance, which they received. The State Department was informed of the negotiating plans, as were the home governments of the European companies.[32]

Thirteen companies originally signed the message to OPEC that was written by the companies in New York; eleven more subscribed later. American, British/Dutch, French, Belgian, German, Spanish, and Japanese companies participated.[33] The independents in Libya, aware of their vulnerability, asked for assurance that, if they took a strong stand, the majors would back them. The United States Justice Department gave qualified approval on January 15, 1971, to a Libyan Producers Agreement that guaranteed the major multinationals would "take care of" the independents, should the latter be required by the Libyan government to reduce production or be expropriated. This meant that the majors would—out of their diversified oil supplies—provide the independents with adequate crude oil to meet their commitments. The companies established a London Policy Group to prepare a unified negotiating stand. In New York, a group of company representatives gathered to "comment on policy decisions proposed in London" and to provide expertise. Home governments were kept informed.[34]

On January 16, 1971, the companies' message was delivered to Iran, to each of the other OPEC member countries, and to the OPEC office in Vienna. "We believe," the message read, "that it is in the long-term interest of both the producing countries and consuming countries alike, as well as that of the oil companies, that there should be stability in the financial arrangements with producing governments." The statement proposed "all-embracing negotiation" between the oil companies' representatives and OPEC.[35]

On the same day as this message was delivered, United States Undersecretary of State John N. Irwin began a Middle Eastern trip, ostensibly to provide the companies with support. In fact, however, he effectively undercut the oil companies' unified stand. He talked with the American ambassador, the Shah, and the Iranian minister of finance, and he left the impression with all of them that the United States government would accept *separate* negotiations for the Gulf states and Libya. Irwin, assured by the statements of the Iranians that the Gulf producing states would enter into a five-year agreement and hold to it, and feeling that the Iranian posture was moderate (at least compared to the extremism of Libya), recommended to the State Department that the oil companies would be well advised to open discussions with the Gulf producers and to conduct *separate*, parallel negotiations with Libya.[36] The State Department passed this suggestion on to the oil companies on January 18-19.

The companies, which were to start talks with the Gulf producers in Teheran on January 19, thus found the State Department proposing a last-minute reversal in strategy; they were dismayed at Irwin's position and the State Department advice. As late as January 30, a few were still arguing for "all-embracing" negotiations, but by then they no longer seemed feasible. The companies agreed to conduct separate negotiations.[37] Later, a representative of the independent Bunker Hunt Oil Company, one of the last holdouts, would see this as a disastrous turning point. More tactfully, George Piercy of Exxon would look back on Irwin's role and say that while he, Piercy, was "disappointed, . . . it may have had a moderating influence on some of the oil-producing countries."[38]

It has been suggested that the major oil companies did not bargain against raising prices during the negotiations because they wanted higher prices themselves. It has also been suggested that the American government was likewise not particularly concerned over oil prices rising. While it does seem clear that the major oil companies were not content with the low prices of the nineteen-sixties, it also seems obvious that if prices were to increase, they would want to be the ones to profit. They were not eager to have the host governments obtain a larger share of the revenues at their expense. What the major corporations most wanted was predictability, stability, and security of supply.[39] They feared that, should the market price for oil fall, the share of the profits going to the host governments would rise. Thus, the companies bargained vigorously *against* increases in posted price so long as they were attached to a formula permanently commiting them to pay large sums to the host governments. In establishing their negotiating posture, moreover, there was by no means unanimity among the majors; it has been reported that Exxon was more ready to yield to the Gulf producers' demands than was Shell, for example.[40]

On February 14, 1971, the Teheran agreement between the companies and the Gulf producers was signed. It gave the producing states an initially sizable rise in per-barrel revenue with annual modest upward adjustments for the subsequent five years. Presumably in exchange, it assured security of supply and stability in the industry for the same period.[41] The companies capitulated at Teheran in exchange for the promise of stability. Further negotiations were then carried on with individual countries. The Tripoli talks, which began in March and ended on April 2, gave Libya an additional premium for its valuable low-sulphur oil as well as for its nearness to European markets.[42] Agreements were concluded separately on oil carried from Iraq and from Saudi Arabia to the eastern Mediterranean by pipeline.

The Teheran-Tripoli agreements proved to be unstable, as subsequent events led to the price, embargo, and participation crises of 1973-74. When the United States devalued the dollar in August, 1971, and again in February, 1973, with no thoughts about oil, the producing states immediately demanded and obtained modifications of the Teheran-Tripoli accords.[43] The companies kept their home governments informed of negotiations on "the positions taken by the producing countries and the interests of the consuming governments at stake in the outcome of the negotiations."[44] Because the United States government had delayed approval of new offshore oil leases and of the Alaska pipeline, the growth of new oil supplies had been retarded, and this in turn negatively affected the companies' negotiating position. When Libya attempted, in 1972, to market oil from the BP concessions that it had expropriated the previous year, the United States and British governments (along with the major oil companies) tried without success to block the sale of this "stolen" oil. The Soviet Union, Bulgaria, Rumania, the Brazilian state oil company (Petrobras), and the Italian state oil company (ENI) bought from Libya. Moreover, when the United States government contacted private American companies to ask them not to purchase the "hot" Libyan oil, Washington received the reply, "We would be delighted not to take the oil, if you can show us where we can get some oil."[45] The State Department was equally ineffective in aiding the companies in their 1973 boycott of Libyan oil. Again, American domestic firms, which had contractual obligations and no alternative sources of supply, were prepared to buy the Libyan oil.[46] Neither in 1972 nor in 1973 did the American government order private American companies not to make the purchases.

As noted earlier, over the years the companies operating in the Arab states had recognized that American policies favoring Israel made their position difficult. On May 3 and May 23, 1973, Aramco executives paid courtesy calls on King Faisal of Saudi Arabia. On those occasions, the King insisted that the American government change its Middle Eastern foreign policy. At the second meeting, he warned that Aramco might "lose everything." Corporate officials communicated the substance of both meetings to Washington. On May 30, executives of Aramco's parent companies (Exxon, Mobil, Socal, and Texaco) met with White House and State and Defense Department representatives. The reaction in Washington, according to corporate correspondence, was "attentiveness," but the feeling there was that the King was "calling wolf." The corporate dispatch noted "there is little or nothing the U.S. Government can do or will do on an urgent basis to affect the Arab/Israeli issue."[47] Corporate needs and United States government policies were separate and distinct. Home governments were consulted, and they listened and occasionally intervened, but basically they left international oil problems to the companies.

III

Because the major oil corporations operate in many nations, their business is subject to the laws of numerous sovereignties. Historically, there seem to have been fewer significant company-host government contacts involving distribution than production. The relations in the distributing arena have tended to be more routine (tax paying and local government building permits for offices, warehouses, and gas stations, for example). There have been conflicts between companies and governments over prices, and there have been price controls. There have been rules of substance, such as orders to refine in the consuming country, significant import controls, regulations on size of inventories, consumer government-run domestic cartels, and so forth. Sometimes there have been expropriations of corporate marketing organizations, but never in a substantial consumer-nation market. In wartime, consuming states (in their roles as host governments) have mobilized the multinationals to their cause. In sum, there has been a very broad range of company-governmental relationships in individual consuming states.

By contrast, in most nations where a multinational corporation *produced* oil, it entered into a basic legal contract with the host state. It obtained a concession. The concessionaire received an exclusive right to explore and to produce oil within a specified area for a defined period of time—usually ranging from 40 to 75 years. The concession specified the royalties the company would pay the host government and sometimes other payments as well. Under it, certain corporate obligations were indicated—for example, to work the concession within a defined period, or to build a refinery after oil was found.[48] The signatories to the concession were the host government and the company (or companies) that obtained it. The home government of the companies was never a signatory, although sometimes, but by no means always, it might be helpful to the company in acquiring it.

Though concessions were formal legal documents, they were in no sense sacrosanct. Expropriations did occur; concessions were renegotiated; changes were made. From the expropriations of the Russian Revolution (1917), to those of the Mexican revolution (1938), to those of the Iranians (1951), the Bolivians (1937, 1969), the Algerians (1969-71), the Libyans (1971, 1973), and the Iraqis (1972, 1973), important expropriations were far from unknown to oil enterprises that produced abroad. Concessions were, on occasion, subject to renegotiation—for example, in Iran in 1933 and in Venezuela in 1943; through the years, existing concessions were often modified.

Over time, the major multinational corporations learned from experience that open, outright defiance of host sovereign states was always counterproductive. Host governments did have the power to expropriate. Companies found it in their best interests to negotiate and occasionally to accept the policies of host producing states even though they might not like them, rather than risk total loss. They had learned this very dramatically in the 1938 Mexican expropriation. In Mexico the companies had been intransigent, and had lost their properties.

In Venezuela, in 1943, it seemed possible that the same situation might recur, but this time companies renegotiated new concessions.[49] When, by 1948, Venezuela was making new demands, the companies accepted the Venezuelan legislation that established a mandatory 50-50 profit-sharing arrangement between the Venezuelan government and the foreign oil companies. After 1950, the 50-50 concept was adopted by the oil-producing countries in the Middle East. British Petroleum's predecessor Anglo-Iranian stood firm against it in Iran; its properties were expropriated in 1951. On the other hand, Jersey Standard had learned to live with it and had even discovered that the American double-taxation rulings could exempt the company from an extra American tax burden. Jersey Standard was helpful in working with its partners in Aramco and with the Saudi Arabian government on ways of meeting that government's desire for added revenues. Aramco was not expropriated. The 50-50 profit-sharing arrangement spread in the Middle East with a resulting rise in host-government revenues of some three- to four-fold.[50]

In the late nineteen-fifties, the market price of oil started to decline, and the companies sought to lower the posted price of oil (that is, the tax reference price). In response to this, in September, 1960, Venezuela, Iran, Iraq, Kuwait, and Saudi Arabia founded the Organization of Petroleum Exporting Countries. OPEC, which later came to include other petroleum-exporting nations, succeeded in maintaining the posted price of oil in the nineteen-sixties, while the market price fell. The result was that, during the sixties, host producing states received larger per-barrel revenues at corporate expense.[51]

While the market price of oil was dropping, producing states had only minimum bargaining power with respect to the multinational corporations. While they did make gains in the sixties, these were relatively small—at least compared with what followed in the early seventies. The companies were well aware, however, that the host producing states were determined to obtain higher revenues from their natural resources. The multinational corporations had also watched the creation during the fifties and sixties by host producing states of their own national oil companies,[52] and the corporations were aware of how significant these could become. In the nineteen-sixties, the issue of conservation was added to the picture when Venezuela started to talk about the conservation of oil resources and began to curb crude-oil output. Up to this point, production levels (except when there was nationalization) had *always* been determined by the companies, not the governments. Control over output obviously ultimately affects price levels.

In the nineteen-fifties and -sixties, the companies operating in Arab countries were aware that they were often regarded as representatives of "American" or "British" policy.[53] Anti-British feeling in the Middle East after the Suez crisis of 1956-1957 resulted in the sabotage of the pipelines of the Iraq Petroleum Company and of a well in Kuwait.[54] Anti-American sentiment was rampant during and after the Six-Day War of June, 1967. During both crises, Arab countries instituted short-term embargoes.[55] Yet, even with these political adversities, the companies continued to function. Indeed, in 1956-57 and again in 1967, the companies effectively reallocated world oil supplies. In 1967, because of the vast oil surpluses, the multinationals had far less difficulty than they had in 1956-57 in filling world demands.

In the early seventies, the market price of oil began to rise as increasing demand met with limited supplies. Supplies were curtailed owing to nationalizations (in Algeria, Iraq, and Libya), the spread of host-country production cutbacks (in Libya and Kuwait), the May, 1970, closing of the pipeline from the Saudi Arabian fields to the eastern Mediterranean as a result of a bulldozer severing the line, and the American government's inaction, which had halted the development of American off-shore resources and the Alaskan pipeline. The upward movement of prices gave new bargaining power to the host producing nations. Step by step, the companies conceded higher per-barrel revenues to these states. On January 1, 1970, the producer governments' revenue had been 91 cents a barrel for Saudi Arabian marker crude. It was up to $1.25 (February 15, 1971) after the Teheran negotiations, and to $1.69 (June 1, 1973) after modifications were made in the Teheran agreement to adjust to the devaluations of the American dollar.[56]

The achievement of higher per-barrel revenues gave host nations a great sense of power. With an eye to more gain, and outside the framework of the Teheran-Tripoli price negotiations, the Gulf producing states turned to the issue of *participation* in equity ownership of the existing oil-producing ventures in their nations. Participation amounted to achieving nationalization through negotiation rather than through expropriation. By March, 1972, the companies had little choice but to accept the principle of participation. In December of that year, the companies signed agreements with Saudi Arabia, Qatar, and Abu Dhabi under which the countries would acquire 25 per cent ownership of the oil-producing ventures in these nations, to be raised to 51 per cent by 1982; Kuwait signed a similar agreement, which was not ratified by its national assembly; Iran made separate arrangements with the consortium.[57]

By 1973, the companies were thoroughly on the defensive in their negotiations with the host producing states. During 1971 and 1972, they had made compromises, hoping that participation agreements would provide for the full development of oil from the more moderate Persian Gulf states. On February 28, 1973, the Iraq Petroleum Company reached a settlement with Iraq on the 1972 nationalization, opening the way for expansion of output there.[58] Greater oil production was essential; by the spring of 1973, oil shortages had begun to occur in the United States, and President Nixon in his April energy message had removed all oil import quotas.[59] The United States would therefore be buying more foreign oil.

In this context, Libya's strident 1973 demands were far from welcome. On June 11, 1973, Libya nationalized Bunker Hunt. Colonel al-Qadhafi said it was a "strong slap on the cool arrogant face" of the United States for its support of Israel.[60] Under the Libyan Producers Agreement set up in 1971, the major oil companies tried to stand by Bunker Hunt. Mobil and Shell were "crude short" and could not give assistance; neither could Gulf, which was subject to prorationing in Kuwait. Texaco, Socal, Exxon, and BP did help, at least until September.[61] In August, Libya took over 51 per cent of the Occidental and Oasis group, again aiming at the more vulnerable independents. The majors boosted output in Saudi Arabia and Iran to compensate for the cutbacks in Libya and Kuwait.[62] Saudi Arabia and Iran became ever more vital to the major multinational companies. In August, cables to its home offices from Aramco indicated that the Saudi Arabian government was seriously considering limiting production for political reasons.[63] That summer, Mobil, Exxon, and Socal publicly urged a change in American policy toward the Arab world; they did so to show the Saudi Arabian government that they were sympathetic to its feelings.[64]

On September 1, 1973, Tripoli radio announced the nationalization of 51 per cent of the assets of the remaining totally foreign-owned oil companies—mainly the majors—operating in Libya.[65] Now the majors could no longer assist Bunker Hunt. They tried to boycott Libyan oil, but oil was in short supply, and Libya had too many willing buyers.

By the fall of 1973, the "moderate" Gulf countries were calling for a drastic revision of the Teheran Agreement, and they arranged for negotiations to begin in Vienna on October 8. The companies reconvened their London Policy Group and sought to prepare common negotiating ground. On October 6, the fourth Middle Eastern war broke out, but, war notwithstanding, on October 8, 1973, OPEC-company discussions began in Vienna as scheduled. The oil industry's negotiating team, however, adjourned the negotiations on October 12, indicating that they had to consult consuming governments regarding the steep OPEC demands.[66] That day the board chairmen of the companies that were parents to Aramco sent a memorandum to President Nixon explaining that the terms demanded by OPEC at Vienna were "of such magnitude that their impact could produce a serious disruption of the balance of payments position of the Western world." The free world had no spare capacity, which meant that the negotiating position of the companies was weak. The company chairmen noted that market forces had pushed crude-oil prices up.

A significant increase in posted prices and in the revenues of the producing countries appears justified under these circumstances; but the magnitude of the increase demanded by OPEC, which is in the order of a 100 percent increase, is unacceptable. Any increase should be one which allows the parties an opportunity to adjust to the situation in an orderly fashion. Accordingly, the companies are resisting the OPEC demands and they are seeking an adjustment of them which can be fair to all the parties concerned.

The four oil executives indicated that the United States policy of open support of Israel was making the companies' position extremely difficult and that, if the support were to be increased, "a major petroleum supply crisis" was probable. Their memorandum concluded that

. . . much more than our commercial interests in the area is now at hazard. The whole position of the United States in the Middle East is on the way to being seriously impaired, with Japanese, European, and perhaps Russian interests largely supplanting United States presence in the area, to the detriment of both our economy and our security.[67]

Apparently, Nixon had not read the memorandum when he ordered further aid to Israel,[68] but it is doubtful whether he would have altered American policy if he had. In any case, following Nixon's order, the Arab cutback and embargo that had been predicted were announced. The owners of Aramco, fearing the loss of the world's most valuable oil concession, dared not defy Saudi Arabia's wishes. They informed Washington of the actions they were taking to carry out the embargo. They were neither ordered nor advised by the American government to do otherwise.[69]

On October 16, before the embargo had been announced, OPEC for the first time had bypassed the negotiating table and unilaterally raised posted prices. The companies learned from the newspapers[70] that the host nations' share per barrel (Saudi Arabian marker crude) had been increased from $1.69 (as of June 1, 1973) to $3.03, and the posted price from $3.01 to $5.12 a barrel. While companies were taken aback, the home governments, absorbed in the issues of the Arab-Israeli war, seemed unaware of the significance of OPEC's actions. When the October 18 embargo followed on the heels of the price announcement, it provided added distraction, and the price increase remained unchallenged.

The companies were alarmed over the rising price levels because they knew that it would take time before the increases could be passed on to the consumer. Moreover, the action of OPEC created new uncertainties. On December 19, 1973, Frank Jungers, chairman of the board of Aramco, had an informal talk with Saudi Arabia's Shaikh Yamani about "the devastating effect of unbridled price increases."[71] "Devastating effect" notwithstanding, the Gulf nations pushed prices up again to $11.65 a barrel—to become effective January 1, 1974; the government share would be $7 a barrel.

Regardless of the reaction portrayed in the press, the corporations—far from glorying in the new high prices and potential windfall profits—reacted with dismay at the destabilization of the oil markets. Historically, multinational corporations had sought stability and order in world oil markets, and conditions were now far from stable. When the embargo ended March 18, 1974, the steep prices remained. Strengthened by their achievement, OPEC countries turned to speeding up the schedule for government ownership of the oil fields. A Shell official later said that, in his view, the high prices opened the way for the participation agreements and "precipitated" the loss of the company's concessions.

On January 29 and February 20, respectively, Kuwait and Qatar signed participation agreements with oil-producing companies operating in their nations; the governments' share was 60 per cent, instead of the earlier 25 per cent. In April, Agip (a subsidiary of the Italian state company ENI) and Occidental Oil made arrangements with Libya on the basis of an 81 per cent share. In the same month, Exxon and Mobil signed agreements with Libya for 51 per cent Libyan participation, in the light of the Kuwait and Qatar decisions, the Agip and Occidental arrangements, and Libya's complete takeover of the concessions of Texaco, Socal, Atlantic Richfield, and Royal Dutch/Shell, after they had refused to compromise on Libya's 51 per cent demands.[72] In June, 1974, Saudi Arabia would obtain a 60 per cent participation agreement.[73] By the fall of 1974, the new government in Venezuela was discussing plans for full nationalization of the oil companies there. The Saudi Arabian government announced in October that it hoped for complete nationalization by early 1975.[74]

These actions seemed to herald basic changes in the industry: The old concessions would be replaced by host government participation; the companies would no longer own their producing properties in the major exporting countries; production levels would be set by governments; security of crude-oil supplies would be endangered; prices, unpredictable. The multinationals were now to be buyers of oil.

But it was in fact unlikely that an arm's length relationship would emerge. The oil corporations still retained their ties and were still negotiating with host states. For one thing, most host nations still needed technical aid in exploration and production. As early as December, 1973, Aramco officials and Saudi Arabia were discussing the formulation of a new, "unique" relationship between that corporation and its host country.[75] In June, 1974, an Exxon vice-president pointed out that no host government oil company had ever made a major oil find. "No one in Government ever drilled at 3,000 feet of water. . . . Maybe some day governments will get all the expertise and we will be pushed out"; but for the time being, the oil companies still had the expertise.[76] Discussions on long-term supply contracts, service contracts, and management contracts went forward.

At the end of 1974, it was clear that the revenues to host producing states would continue to rise sharply. Temporarily, the oil companies' profits had also soared, although it was by no means certain that, as the new arrangements evolved, they would remain so high—the oil companies' "share of the pie" seemed certain to become smaller. The oil-company executives, therefore, continued to worry about security of oil supplies, about their reduced ability to earn profits from their producing operations, about the high cost of oil, and about the nature of the emerging arrangement.

IV

In the meantime, throughout 1974, the United States companies were urging Washington to expedite the approvals needed for the Alaskan pipeline and for off-shore drilling. In Europe, the companies were similarly pressing for extensive exploration in "non-vulnerable" regions, particularly the North Sea. Companies sought sources for oil outside the control of the OPEC nations.[77] They were also displaying more interest in diversified energy supplies: tar sands, shale oil, coal, coal gasification, coal liquefication, natural gas, and uranium. And some companies began to diversify into other industries (thus Mobil's interest in the retail stores of Montgomery Ward).

The oil companies, which had been on the defensive in their relations with the oil-producing nations from 1970 through 1974, now came under attack from the consuming countries, mainly in the form of criticism for their high profits. They felt themselves to be falsely maligned—profits were needed for the major investments required by the industry, inflated profits would not last, and the future was very uncertain. Government—home, consumer, and exporting—intervention seemed bound to increase.

Yet from the corporate standpoint, the multinationals still had a major role to perform. Each had vast resources and expertise in a highly complex industry and extensive organizations. When consumer governments planned to enlarge their role in the oil industry, the perennial industry representative, John J. McCloy, said, in February, 1974, that he felt the "State Department should be very importantly involved." McCloy believed that the best formula was a combination of private oil companies with "knowledgeability [and] good bargaining ability together with the strength of government. . . ." But neither McCloy nor the companies were prepared to let the governments act as chief negotiators. As McCloy put it:

I don't see how the [consumer] Governments can do the negotiating. There are so darn many intricacies with regard to freight rates and all the different kinds of crudes and everything that is involved. You have to have people who are not just passing through a government dealing with that.[78]

Other company representatives also stressed the dangers of the consumer governments' undertaking negotiations. "I think," declared a former Socal vice-president, "to the maximum extent that it is feasible there should be a segregation between the political and the economic problems."[79] Although there never was and never will be a separation between oil and politics, corporate thinking perceives a distinct economic function for the multinational oil company—that of carrying on the tasks of exploring, producing, transporting, refining, and marketing oil and, in the process, of balancing world supply with demand.

REFERENCES

[1]Mira Wilkins, *The Emergence of Multinational Enterprise: American Business Abroad from the Colonial Era to 1914* (Cambridge, Mass., 1970), pp. 62-74, 82-85; *idem, The Maturing of Multinational Enterprise: American Business Abroad from 1914 to 1970* (Cambridge, Mass., 1974). Both have data on the history of U.S. oil companies abroad. On Jersey Standard, see Ralph W. Hidy and Muriel E. Hidy, *Pioneering in Big Business 1882-1911* (New York, 1955); George Sweet Gibb and Evelyn H. Knowlton, *The Resurgent Years 1911-1927* (New York, 1956); and Henrietta Larson et al., *New Horizons 1927-1950* (New York, 1971). For Royal Dutch's history, see F. C. Gerretson, *The History of the Royal Dutch Company*, 4 vols. (Leiden, 1953-1957); on Shell, Robert Henriques, *Marcus Samuel* (London, 1960). For a general survey of multinational enterprises in oil, Raymond Vernon, *Sovereignty at Bay* (New York, 1971), pp. 27-37. Edith Penrose, *The Large International Firm in Developing Countries: The International Petroleum Industry* (Cambridge, Mass., 1968), contains short histories of the major companies.

[2]Henry Longhurst, *Adventure in Oil, The Story of British Petroleum* (London, 1959) and *A Short History of the Anglo-Iranian Oil Company* (London, 1948).

[3]Federal Trade Commission, *International Petroleum Cartel* (Washington, D.C., 1952), for example.

[4]*Ibid.*

[5]My view here is in sharp variance with that of H. G. Morison, Assistant Attorney General in the Truman Administration. See his Memo, June 24, 1952, printed in U.S. Senate, *Committee on Foreign Relations, Subcommittee on Multinational Corporations, The International Petroleum Cartel*, 93rd Congress, second session, 1974, pp. 7 ff. (henceforth cited as *Cartel Docs.*). Morison paid no attention to the vast economic and political changes from 1928 to 1952 that influenced the policies of the oil companies. He stressed the companies' consistent "as is" policy in the 1930s and failed, I believe, to recognize how marginal the major companies' control over the world oil industry actually proved to be.

[6]Wilkins, *Maturing,* Chap. 9.

[7]*Ibid., passim;* corporate *Annual Reports;* Marquis James, *The Texaco Story* (New York, The Texas Company, 1953); Craig Thompson, *Since Spindletop* (Pittsburgh, The Gulf Oil Company, 1951).

[8]Testimony of William P. Tavoulareas, U.S. Senate, Committee on Foreign Relations, Subcommittee on Multinational Corporations [the Church committee], June 6, 1974 (still unpublished).

[9]*Petroleum Press Service,* 36 (December, 1969), 458; Neil H. Jacoby, *Multinational Oil* (New York, 1974); Raymond Vernon, "Competition Policy Toward Multinational Corporations," *American Economic Review,* 64 (March, 1974), 279.

[10]The U.S. market was protected, and avoided the market price decline that occurred in Europe and Japan.

[11]On the 1914 acquisition by the British government, see E. H. Davenport and Sidney Russell Cooke, *The Oil Trusts and Anglo-American Relations* (London, 1923), pp. 11-23, and J. C. Hurewitz, *Diplomacy in the Near East and Middle East,* I (Princeton, 1956), 278-281. With the stock purchase, the government obtained the power to appoint two directors to the board; they were not to interfere in "ordinary commercial management," but could veto the proposals of others. The British government's investment of £2,000,000 enabled the company to add to its refinery at Abadan in Iran and expand production.

[12]Certain nations even passed laws barring companies controlled by foreign governments: Mira Wilkins, "Multinational Oil Companies in South America in the 1920s," *Business History Review,* 48 (Autumn, 1974), 439. See also Michael Posner, *Fuel Policy* (London, 1973), pp. 50-51.

[13]Interviews and corporate correspondence. For a recent statement, see John J. McCloy testimony before U.S. Senate, *Hearing Before the Committee on Foreign Relations, Subcommittee on Multinational Corporations, Multinational Corporations and United States Foreign Policy,* 93rd Congress, second session, 1974, part 5, pp. 274, 281 (henceforth cited as *Church Committee Hearings*).

[14]Compare, for example, Elizabeth Monroe, *Britain's Moment in the Middle East, 1914-1956* (London, 1963), pp. 105-106 with P. H. Frankel, *Mattei* (New York, 1966), pp. 34-35.

[15]This is the impression given by corporate executives.

[16]Jean Rondot, *La Compagnie Française des Pétroles* (Paris, 1962) and Frankel, *Mattei.*

[17]Quotation is from February 28, 1944, recommendations of the National Oil Policy Committee, which included representatives of Jersey Standard, Gulf, Socony-Vacuum, and a number of independents; printed in *A Documentary History of the Petroleum Reserve Corporation*, prepared for use of the U.S. Senate Committee on Foreign Relations, Subcommittee on Multinational Corporations, 93rd Congress, second session, 1974, p. 79.

[18]Testimony of W. W. Rodgers, U.S. Senate, *Special Committee to Investigate the National Defense Program, Petroleum Arrangements with Saudi Arabia*, 80th Congress, first session, 1947-48, part 41, p. 24923.

[19]Wilkins, *Maturing*, p. 320.

[20]*Ibid.*, 290; *Cartel Docs.*, pp. 14-15.

[21]Wilkins, *Maturing*, pp. 321-22; letter, American Independent Oil Company to John Foster Dulles, October 15, 1954, *Church Committee Hearings*, part 7, p. 250.

[22]Wilkins, *Maturing*, p. 322; *Cartel Docs.*, pp. 58, 27, 49.

[23]*Cartel Docs.*, pp. vii, 51 (action memo, August 6, 1953); *ibid.*, pp. 56-58, 82-83; M. Tiger and L. G. Franko, "E.N.I." (mimeographed), Centre d'Études Industrielles (Geneva, 1973).

[24]Eden quoted in Elizabeth Monroe, *Britain's Moment in the Middle East 1914-56* (London, 1963), p. 113; Shoshana Klebanoff, *Middle East Oil and U.S. Foreign Policy* (New York, 1974); data from Angus Beckett.

[25]Figures in *Survey of Current Business* indicate large U.S. capital withdrawals from Venezuela.

[26]U.S. Legislative Reference Service, *Problems Raised by the Soviet Oil Offensive*, 87th Congress, second session, 1962.

[27]Testimony of John J. McCloy, *Church Committee Hearings*, part 5, pp. 258; 256-57; list of clients in 1961 in *ibid.*, part 6, pp. 290; Wilkins, *Maturing*, pp. 299, 388.

[28]Interviews in Iraq and India, 1965; Peter Odell, *Oil and World Power* (Baltimore, 1970), p. 163. McCloy testimony, *Church Committee Hearings*, part 5, pp. 259, 275.

[29]Wilkins, *Maturing*, p. 369; El Paso Natural Gas, *Annual Reports*; *Petroleum Intelligence Weekly (PIW)*, April 22, 1974, pp. 3-4.

[30]Statement and testimony of Piercy, *Church Committee Hearings*, part 5, pp. 214, 196, on reasons companies capitulated.

[31]Testimony of Schuler, Irwin, and Piercy, *Church Committee Hearings*, part 5, pp. 78, 80, 111, 112, 146, 214-15; *New York Times*, December 16, 1970.

[32]*Church Committee Hearings*, part 5, *passim*. Testimony of J. D. Bonney (Socal), in *ibid.*, part 7, p. 383; testimony of William P. Tavoulareas, *Church Committee Hearings*, June 6, 1974 (still unpublished).

[33]*Church Committee Hearings*, part 5, p. 215, and part 6, pp. 234-35. The non-U.S. signatories included CFP (France), BP (Britain), Shell (British/Dutch), Petrofina (Belgium), Gelsenberg (Germany), Hispanoil (Spain), and Arabian Oil (Japan).

[34]*Ibid.*, part 5, pp. 81, 249, 19, 81, 147, 215, 253, and part 6, pp. 224-30.

[35]*Ibid.*, part 5, p. 113, and part 6, p. 62.

[36]*Ibid.*, part 5, pp. 266-67, 85, 117-21, 148-49, 152, 156, 159, 167-68.

[37]*Ibid.*, part 5, pp. 149, 151, 163, 221, 87, 88, 134, 136, 215-16, and part 6, pp. 67 ff.

[38]*Ibid.*, part 5, pp. 82 ff., and 118 ff. (Schuler testimony), 217 (Piercy testimony); see also McCloy testimony, *ibid.*, pp. 264-69.

[39]Piercy of Exxon stated, "We made every effort to minimize posted price increase and to assure some security of supply." *Ibid.*, part 5, p. 217. See also Socal's J. D. Bonney's impassioned testimony on price increases, *ibid.*, part 7, pp. 377-78 ("We were part of a company group dedicated to minimizing the posted price increases that the Libyan government was trying to force on us").

[40]Industry sources have suggested the differences between Exxon and Shell. Exxon's most important Middle Eastern investment was in Saudi Arabia, so it wanted its relationships with Saudi Arabia to remain as satisfactory as they had been over the years. Shell, with no interests in Saudi Arabia, had no special relationship to safeguard there. The difference in interests may well have been responsible for the reported variations in negotiating posture.

[41]*Church Committee Hearings*, part 5, pp. 149, 216.

[42]*Ibid.*, pp 100, 149, 216.

[43]*Ibid.*, p. 150.

[44]Testimony of George Piercy, *ibid.*, p. 217, and McCloy's memos, part 6, p. 297 ff.

[45]*Ibid.*, pp. 21-22, 27.

[46]*Ibid.*, pp. 39-41, and part 6, pp. 320-21.

[47]Corporate cables and memos, in *ibid.*, part 7, pp. 504-09.

[48]For an excellent description of concessions in the Middle East, see George Lenczowski, *Oil and State in the Middle East* (Ithaca, 1960), pp. 63 ff. I have drawn on a wide variety of sources for concessions elsewhere.

[49]Wilkins, *Maturing*, pp. 225-30, 255-56, 270-72.

[50]*Ibid.*, pp. 317, 321.

[51]See charts in Geoffrey Chandler's article in *Petroleum Review*, June, 1974.

[52]Wilkins, *Maturing*, pp. 367-68.

[53]Monroe, *Britain's Moment*, p. 115, notes that the British government was "inextricably confused with the [British Petroleum] company in all local minds."

[54]Wilkins, *Maturing*, p. 351.

[55]See George Lenczowski's article in this issue and "The Middle East: U.S. Policy, Israel, Oil and the Arabs," *Congressional Quarterly*, April, 1974, p. 8.

[56]Geoffrey Chandler, "The New Energy Prospect," Address given January 31, 1974.

[57]*Church Committee Hearings*, part 5, p. 216.

[58]*PIW*, April 22, 1974.

[59]*New York Times*, April 19, 1973.

[60]*Ibid.*, August 14, 1973.

[61]*Church Committee Hearings*, part 5, p. 94.

[62]*Ibid.*, part 5, pp. 231, 178, and part 6, pp. 268, 318.

[63]*Ibid.*, part 7, pp. 541-42.

[64]*Ibid.*, pp. 510-12.

[65]*Ibid.*, part 6, pp. 318-19, and *New York Times*, September 2, 1973.

[66]*Church Committee Hearings*, part 5, pp. 216-17 (statement of George Piercy).

[67]Copy of the memo of October 12, 1973, in *ibid.*, part 7, pp. 546-47.

[68]Jack Anderson's column, *Miami Herald*, July 20, 1974.

[69]Robert Stobaugh's article in this issue covers corporate responses to the embargo. A copy of Aramco's October 21, 1973, internal cable on embargo implementation went to the U.S. ambassador. *Church Committee Hearings*, part 7, p. 515, prints the cable.

[70]Statement of Piercy in *ibid.*, part 5, p. 217.

[71]Statement of Shell executive, Geoffrey Chandler, February 15, 1975. Jungers' cable to home office, December 19, 1973, printed in *Church Committee Hearings*, part 7, p. 532.

[72]BP, *Annual Report 1973; Wall Street Journal*, April 17, 1974; *New York Times*, April 19, February 12, March 19, 1974; *Church Committee Hearings*, part 6, p. 319 (Libyan 1974 expropriation).

[73]*New York Times*, August 5, 1974.

[74]*Miami Herald*, October 29, 1974.

[75]See cable from Jungers, December 19, 1973, in *Church Committee Hearings*, part 7, pp. 532-33.

[76]Emilio Collado testimony before Church Committee (Vol. X, unpublished). Perhaps Pemex's recent oil find in Mexico may make part of this statement obsolete.

[77]After the 1974 Mexican oil discoveries, for example, companies rushed to obtain concessions in nearby Central American nations.

[78]*Church Committee Hearings*, part 5, pp. 273-74, 281 (McCloy testimony).

[79]*Ibid.*, part 7, p. 384 (Parkhurst testimony).

ROBERT B. STOBAUGH

The Oil Companies in the Crisis

ON OCTOBER 18, 1973, Frank Jungers, then president of the Arabian American Oil Company (Aramco), was informed that Radio Riyadh had just announced King Faisal's decision to cut by 10 per cent Aramco's production of crude oil. Without asking for further details, Jungers ordered an immediate cut, not just of the 10 per cent required, but of a little more for good measure.[1] Three days later, Shaikh Ahmed Zaki Yamani, Saudi Arabia's minister of petroleum, summoned Jungers to a meeting, and there the 10 per cent turned out to have been only the beginning.[2] For, at the meeting, Yamani announced a cut of about 25 per cent from the level programmed for October. He also announced an embargo on shipments of oil to certain countries, especially the United States, allocations to selected countries, which he termed "exempt," of a quantity of oil equal to the average daily amount shipped to them between January and September, 1973, and the allocation of the balance of Saudi production on a pro-rata basis to whatever countries remained.

Thus began the Saudi oil cutback and embargo, which lasted from October 18, 1973, to March 18, 1974. Other Arab nations joined in the cutback and embargo. But Saudi Arabia, the world's largest exporter and the only country holding proven reserves sufficient for important increases in oil production, was clearly the leader in the movement and the key to its success.

Jungers' action on October 18 of immediately cutting production even a little more than the radio broadcast had indicated was characteristic of the companies' behavior in following the instructions of the Arab nations. As the vice-chairman of Aramco's board later explained, "the only alternative was not to ship the oil at all. . . , obviously it was in the best interests of the United States to move 5, 6, 7 million barrels a day to our friends around the world rather than to have that cut off."[3]

The Companies and the Producing Nations

Of all the connections to be found in the world oil industry, the ties between the Saudis and Aramco (completely owned by four American-based companies—Exxon, Mobil, Standard Oil of California, and Texaco) were probably the closest. This closeness is exemplified in their mutual efforts to arrange Saudi participation in the ownership of the company's properties. Both sought an arrangement that would insulate

179

Aramco-Saudi relations from events in the other producing countries, for Aramco executives had reported that Yamani believed some of these countries—Libya and Kuwait, for example—would soon come under the control of radical groups. The radicals would then take steps to which the companies might have to comply in order to meet their oil needs.[4] According to an Aramco vice-president, the Saudis were willing to leave Aramco's parents in control of most of the crude oil, "especially if we could fuzz up the deal somehow" to prevent ready comparison with arrangements in other countries.[5] The close ties between Aramco and the Saudis are also reflected in Yamani's obvious assumption that Aramco would be willing to comply with the embargo and his references from time to time to Aramco as "the Saudis' largest asset."[6]

Nevertheless, the companies and the Saudis were not entirely united. At the "instruction" meeting of October 21, Yamani warned Aramco that it would be dealt with harshly if all details of the program were not met, and its subsequent actions were subject to strict surveillance by the Saudi Arabian government.[7] Captains of ships loading Saudi crude were required to sign affidavits stating their destination and to report by cable to Aramco when they arrived. As an additional check, Saudi diplomats monitored public records of oil imports by country of origin. For example, when tankers arrived at Exxon's refinery in New Jersey in February, 1974, with Saudi oil, Aramco's president was called by the Saudi government within hours of their arrival to explain the apparent violation of the embargo. Fortunately for Exxon, it was able to prove that one of its cargo tankers, the *Berge Duke*, had been loaded prior to the embargo and had sailed to the company refinery in Aruba, where it was anchored for several months because the Aruba tankage was being used for Venezuelan crude. The decision then was made to reload the crude into smaller ships for delivery to New Jersey—the *Berge Duke* was much too large for the New Jersey harbor facilities. These were the tankers the Saudis had questioned.

The cutbacks imposed by the Arab countries reduced the availability of Arab oil from 20.8 million barrels a day in October, 1973, to 15.8 million by December. But the Arab cut of 5 million barrels, of which 2.3 million was Saudi, was partially offset by slight increases in production in Canada, Iran, Indonesia, Nigeria, and the Communist countries. The December low for the world was therefore only 4.4 million, or 7 per cent below the October high of 59.2 million (Table 1). However, in terms of international trade in crude oil, the net loss was quite considerable—about 14 per cent.

Crude-oil availability began to climb again after December. The average amount available was 56.1 million barrels daily from December, 1973, through March, 1974—about the same as it had been a year earlier (55.9 million barrels daily for the same period).[8] These numbers understate the real economic impact of the cutbacks, however, because world oil consumption had been growing about 7.5 per cent yearly and international trade in oil about 11 per cent yearly.[9]

The Arabs did not act as a unified group. Although Table 1 shows that Iraqi production was reduced between October, 1973, and December, 1973, Iraq neither endorsed nor enforced a cutback, nor did it specifically embargo the United States or the Netherlands. The Iraqi reduction had actually been the result of damage to its loading facilities; by December, 1973, its output was about 20 per cent higher than it had been

Table 1

Average Daily Availability of World Crude Oil, Selected Periods

(millions of barrels daily)

	December, 1972, through March, 1973[a]	October, 1973[b]	December, 1973[c]	December, 1973, through March, 1974[a]
ARAB:				
Saudi Arabia	7.1	8.6	6.3	7.0
Kuwait	3.3	3.5	2.6	2.7
Libya	2.3	2.3	1.8	1.9
Iraq	1.8	2.2	1.9	1.9
Abu Dhabi	1.3	1.4	1.2	1.2
Bahrain, Dubai, Qatar, Oman	1.2	1.3	1.0	1.0
Algeria	1.1	1.1	.9	1.0
Morocco, Tunisia, UAR, Syria	.4	.4	.1	.2
Arab Total	18.3	20.8	15.8	16.8
NON-ARAB:				
United States	11.1	10.8	10.8	10.8
Communist states	9.1	9.7	9.8	10.1
Iran	5.8	5.8	6.0	6.1
Venezuela	3.3	3.4	3.4	3.3
Nigeria	1.9	2.1	2.2	2.2
Canada	1.8	1.7	1.8	1.8
Other Western Hemisphere	1.8	1.8	1.8	1.9
Indonesia	1.2	1.3	1.4	1.4
Western Europe	.4	.4	.4	.4
Other Non-Arab States	1.4	1.4	1.4	1.5
Non-Arab Total	37.6	38.4	39.0	39.3
WORLD TOTAL	55.9	59.2	54.8	56.1

[a] Equal to quantities *produced* from November through February.
[b] Month immediately preceding time when embargo began to be felt by importing nations.
[c] Month with lowest availability for world total.
Source: Statistics from governments and reported in *Oil and Gas Journal*, various issues, modified by addition of 1.7 million barrels daily of natural-gas liquids in the United States. Oil is assumed to be "available" one month after production.

in October.[10] Kuwait and other Arab countries also differed from Saudi Arabia in that they did not instruct the companies regarding the countries that should receive priority on deliveries.

The lack of a concerted policy among the Arabs and the willingness of non-Arab OPEC countries to increase production during the embargo are not surprising. The disunity among the OPEC countries is reflected in one of the discussions that Yamani had with Aramco representatives in May, 1973, on participating in Aramco ownership.[11] Yamani said that he could not sign an agreement with the Aramco parents at that stage because critics might use it against him should Iran, Libya, and possibly Nigeria act differently. The Saudis were not the only ones to wait for other countries to act: Yamani said that the Kuwait assembly would also not take action until the companies' arrangements with Libya were resolved. Libya's indecision was of special concern to the Arab world; and Yamani was convinced that "Libya does not know what they want." The Algerians were prodding the Libyans to nationalize all oil properties. The Libyans were in turn attempting to influence Nigeria against signing an agreement with the oil companies that would specify the degree of its participation in the ownership of the companies, while the Saudis were attempting just the opposite. Even within each government, there was a lack of unanimity. For example, when the Libyans were considering nationalizing their oil fields, some doubted whether they could run the operations without the companies, and its government leaders were divided on the question of allowing the Russians to help run the operations.[12]

Not all the rivalries between these nations concerned their oil policies. Kamal Adham, a top political adviser to King Faisal, gave Jungers a lengthy explanation for the Saudi acceleration of its military procurement program. The Saudi government, according to a report by Jungers to Aramco's parents, wondered why the Iranians had undertaken this vast military expansion, since the Saudis were all certain "the Shah knew that such expenditures were fruitless as far as the Russians were concerned, thus they must be only for one purpose, namely control of the Gulf."[13]

While it seemed certain that the oil companies would obey the instructions of the producing nations, it seemed equally certain that the companies would be responsible for allocating the available oil not covered explicitly by these instructions. This responsibility, when coupled with their operation of the physical facilities, placed the companies in the key role of managing the cutback and embargo. The most important of the managers were Aramco, its four American parents, and the three other so-called "majors"—Gulf Oil, based in the United States, British Petroleum, based in the United Kingdom, and Royal Dutch/Shell, based partly in the United Kingdom and partly in the Netherlands.

The Companies Are Warned

Although Jungers had not expected quite such a severe cutback, neither he nor the executives of Aramco's parent companies were really surprised when it happened. Just a few days before the "instruction" meeting of October 21, Aramco had provided the Saudis—at their request—with data that showed Aramco exports by grade to various countries for the first nine months of 1973.[14] The Saudi oil technocrats used this information to help them decide on a detailed plan of action. But the request for the list had not been Aramco's first indication that the Saudis might reduce oil production. Two significant meetings had taken place in May of 1973.

At the first meeting, in Riyadh on May 3, King Faisal delivered an unequivocal message to Jungers and two of his lieutenants, declaring that it was absolutely mandatory for the American government to do something "to change the direction that events are taking us in the Middle East today," and for the Arabs' American friends in

Aramco to do something to change the posture of the American government. They were told that a simple disavowal of Israeli policies and actions would go a long way toward quieting anti-American feeling in the Middle East.[15] In an explanation accompanying his message, the King said that Saudi Arabia had a deep friendship with the United States, but the Zionists together with the Communists were on the verge of causing American interests to be expelled from the area. Faisal found it was almost inconceivable that a democratic government would have so little concern for the interests of its people.

After the audience with the King, the three Aramco executives paid a courtesy call on Kamal Adham, who elaborated on the King's message. Adham said that the Saudis, along with other Arab countries, believed that Israel had no reason to want to enter into meaningful negotiations with the Arabs, and that therefore they would not do so unless forced into it by the United States. He was quite sure that Egypt's Sadat would begin hostilities of some sort, even though the Egyptians themselves considered the military situation hopeless. By doing so, Sadat thought he would marshal opinion in the United States for initiating a settlement.[16]

The second significant meeting between King Faisal and Aramco representatives took place on May 23 at the International Hotel in Geneva, where the King was resting after visits to Cairo and to Paris.[17] The King had been given a "bad time" in Cairo because Sadat had put pressure on him to provide greater political support. The King brought along three advisers, one of them Yamani, to the Geneva meeting. Aramco was represented by three executives, including Jungers, and the four Aramco board members that represented its four American parents.[18] The King's message was blunt—Saudi Arabia was in danger of being isolated from its Arab friends because of the failure of the United States government to give the Saudis positive support. The King was not going to let that happen. If the trend continued, Aramco would lose everything. The oil companies must inform their government leaders and the American public promptly regarding their true interests in the area.

The companies took various actions that included a public statement by the board chairman of Standard Oil of California (Socal), expressing the need for the United States to conduct a foreign policy that would assure reasonable access to foreign oil, and an advertisement in the New York Times by Mobil. Later in the year, the Saudis indicated great satisfaction with Aramco for its pro-Arab stand and with the parents for being more pro-Arab than they had been before. An unidentified Saudi representative told J. J. Johnson, Aramco's vice-president for government relations, "We hope to reward you," a statement that Johnson interpreted as being the Saudis' promise to allow future growth in production.[19]

However, in spite of these actions by Aramco and its parents, Yamani warned Aramco in early August that its production might be restricted to 7 million barrels daily, down from 8.4 million in August and the 9 million that had been scheduled for November.[20] By that time, there had also been numerous other signs that a production cutback was possible. The parent companies asked Aramco to prepare an estimate of the effects on the Aramco financial position of three different levels of Saudi production: 7.8 million, 9.4 million, and 20 million barrels daily.[21] Given this gradual buildup of tensions, one can see why Aramco and its parents were not taken by surprise when the cutback and embargo were announced. Executives of other major oil companies, although not warned as explicitly as Aramco had been, had also heard enough rumors in 1973 not to be surprised when it happened.[22]

The Inertia of the Consumers

The governments of the oil-importing countries had also been given plenty of warning. For example, Geoffrey Chandler, an executive of Royal Dutch/Shell, said before a Parliamentary Committee in May, 1973, and again publicly in June, 1973, that an agreement to share supplies in the event of shortage "seems urgently required."[23]

The United States government was particularly well briefed. Long before the Geneva meeting of May, 1973, Aramco and its parents had been serving as a conduit for passing both general and quite specific information from Saudi Arabia to Washington. Socal's chairman Otto Miller testified that, on occasions when he talked with President Nixon, he (Miller) "would generally make a comment or two about the problems of the Middle East and the importance of it."[24] Miller also talked with a number of people in the Defense Department and in the State Department, including Secretary of State Rogers. Aramco provided a briefing service for officials going to the Middle East;[25] in January, 1973, for example, James Akins, then a State Department officer on loan to the White House staff (later ambassador to Saudi Arabia), telephoned Mike Ameen,[26] the Aramco Washington lobbyist,

[to get] Ameen to tell Yamani it was very important that he, Yamani, take Ehrlichman [who was planning to visit Saudi Arabia] under his wing and see to it that Ehrlichman was given the message that Saudis' people love you but your American policy is hurting us.[27]

After the Geneva meeting of May 23, 1973, the messages to the American government became even more specific. On May 30, the four parent-company executives and an Aramco vice-president who had made the Geneva visit to King Faisal went to Washington to report the "time running out" message given to the group by the King.[28] They delivered it to four persons in the White House, five in the State Department (including Assistant Secretary Joseph Sisco), and three in Defense (including Acting Secretary William Clements). Later in the summer, they communicated their increasing apprehension regarding the possibility of a cutback and embargo to the government "at every opportunity that [they] had."[29] They also assigned an Exxon representative to brief the State Department routinely on new developments.

On October 12, after the outbreak of the October war but before the embargo, John J. McCloy, the famous New York lawyer who had served in several high government positions and was at this time representing the four Aramco parents, delivered a memorandum from the chairmen of Aramco's four parents to General Alexander Haig for transmittal to President Nixon and Secretary Kissinger. The memorandum told about the "anger" and "great irritation" in the Arab governments over American support of Israel, and it warned that the Saudis would impose a cutback; this could have a "snowballing effect that would produce a major petroleum supply crisis." They further asserted that Japan and Western Europe "cannot face a serious shut-in" of Persian Gulf oil because the world oil industry was operating "with essentially no spare capacity."[30]

In spite of these many warnings, the American government seemed to continue to discount the seriousness of the threat. For example, the group of oil executives who visited Washington on May 30 reported encountering skepticism that drastic action was imminent. Some government officials, they said, "[believe that] his Majesty is calling wolf when no wolf exists except in his imagination."[31] On October 12, after the war had started, a State Department cable to a number of United States embassies abroad played down the possibility of a cutback. It stated that it was not clear Yamani was speaking for the Saudi Arabian Government when he informed the American companies about possible Saudi production cuts.[32]

The American government's attitude expressed in meetings of international organizations also contributed to the unpreparedness of the other oil-importing countries. In committee meetings in the Organization for Economic Cooperation and Development (OECD), for example, the United States advocated an oil-sharing agreement in the event of an emergency, but it also took the firm position that any sharing should be based on "water-borne imports" rather than "vital needs" (the French approach) or total energy requirements (the Japanese approach).[33] From the viewpoint of the OECD countries with small or nonexistent domestic oil production—hence highly dependent on "water-borne imports"—the American position was unacceptable. Water-borne imports accounted for only 14 per cent of American energy supplies as compared with a much higher percentage for most other OECD countries, for example, 80 per cent for Japan. Hence a cut of 20 per cent in water-borne imports would reduce American energy supplies by only about 3 per cent, but Japanese energy supplies by 16 per cent.

Even if the United States position had been more realistic, however, it is still possible that no OECD agreement would have been reached. Before the embargo, the governments of consuming nations, especially Japan, were fearful of forming a group to share oil in an emergency because of possible retaliation from the Arab oil producers. During the embargo, the OECD oil-sharing committee tried to play down the crisis atmosphere and any evidence of controversy. At a meeting on October 25 and 26, just a week after the cutback was announced, Australia and Canada, supported by the United States and the United Kingdom, managed to block a statement to the press expressing "hope" that an agreement on an OECD-wide oil-sharing scheme would be reached by the end of 1973. The committee agreed only to report monthly on oil stock levels, and then they adjourned for one month.[34]

The lack of preparation on the part of the United States and the OECD left the oil companies as the only entities capable of managing the cutback and embargo. However, the companies were not a cohesive group—quite the contrary, as subsequent events would show.

The Companies as Managers

The rivalry among American companies was demonstrated first of all in the negotiations of 1973 over "participation," that is, over the takeover by the OPEC nations of part or all of the ownership of the producing assets within their boundaries. The management of Socal suspected that Mobil, historically a company short of crude oil, would break ranks with the other three Aramco parents in order to obtain a greater share of the oil that the Saudis were to make available. Mobil owned only 10 per cent of Aramco, whereas the other three parents had 30 per cent each. Otto Miller, the chairman of Socal during the negotiations, later told a Senate Subcommittee: "I think that probably by hook or by crook they [Mobil] might end up with more than what they own."[35] This issue surfaced on September 12, 1973, in a meeting of the four companies' chief executive officers and their representatives on the Aramco board. Exxon, Texaco, and Socal favored a course of action that was unsatisfactory to Mobil, whose representatives informed the group: "we are putting you on notice of our reserving the right to make our case individually with SAG [Saudi Arabian Government] if the group effort fails."[36]

The oil companies, saddled with the responsibility for allocating the available oil, faced a formidable task. I know of no evidence to suggest that the parent companies met during the embargo to decide upon, or even to discuss, the problems of allocation; the principal efforts, it appears, were made by each company individually. Each operated a complex chain of facilities, starting with the production of crude oil and ending with a petroleum product delivered to consumers. Adding to the complexity is the fact that each produced a hundred or more types of crude oil, and each crude is somewhat different from the others, both in terms of impurities (such as sulfur) and the composition of the hydrocarbons. Each refinery is consequently also somewhat different from the others; if a refinery designed to operate on a light Libyan crude oil were used to process a heavy Venezuelan crude, its capacity would be reduced, its equipment subject to corrosion, and its products of inferior quality.[37]

Managements normally solve the resulting problems with the aid of computers, which can take into account the capabilities and constraints inherent in the system, such as availability of tankers, quality of crude oil, capacity and processing capabilities of refineries, and products available for trade from other firms. The computer programs also take into account any constraints specified by management, such as allocations of products to a given country, and they indicate how available crude oil should be allocated to maximize profits. During the oil crisis, however, the computer was often bypassed because instructions from the producing nations had removed much of the companies' freedom of action, and management wanted to exercise maximum control over those few decisions that were left to them. The resulting problems of supply and distribution were enormous. One manager told me that he rebalanced his company's worldwide distribution system once a week during the height of the oil crisis. His company, which operates dozens of refineries and over a hundred ships, was faced with continuous changes in both production levels and demand schedules, and also, from time to time, changes in the quotas set by the Arabs.

The power of the companies to control the oil flow had already caused uneasiness among the consumers in the months prior to the embargo, and this turned into down-right panic during the embargo's early stages. Especially prevalent was the conviction that companies would give priority to their home countries. This belief was reflected in statements by both Europeans and Japanese in the early months of the embargo that American companies were diverting "the bulk of their supplies to meet their U.S. shortfall,"[38] and it continued to reverberate around the world long after the embargo had ended. It was equally widely believed that since the Arabs had established priorities giving some countries proportionately more oil than others, this would actually result in some countries receiving more oil in total than other countries. A report by the European Economic Commission stated:

The Middle East crisis has not affected all member countries in the same way, since at the start of the crisis two countries—and later three—enjoyed a privileged position (France, United Kingdom, and Belgium).[39]

Information obtained from interviews with both companies and governments concerning the rules used by the majors to allocate oil among the consuming nations revealed that by and large the companies sought to distribute the impact of any shortages more or less equally among the users in different markets. A directive of one American major rationalized this goal as serving three purposes: minimizing the risk of litigation and reprisal by the governments of the consuming countries; maintaining political neutrality; and maintaining the company's role in serving as a buffer and providing flexibility of supply from many sources.

Of course, implementing such an approach requires a definition of "demand" in each market, and that definition could vary considerably with each company. Nevertheless, their underlying rationale did seem to be to "equalize suffering," no doubt in part because it could be construed as a policy of maximizing profits over the long run. For supplying oil primarily to the markets that seemed most lucrative at the moment probably would have increased the risk of government reprisals and could eventually have adversely affected profits.

Some of the majors issued written directives. For example, officials in one company issued a two-page memorandum calling for the allocation of the company's supplies of crude oil on the basis of quantities supplied for the first nine months of 1973, and they subsequently worked out an international allocation system for oil products after headquarters had received more information. To enforce its policy, the company set up an Emergency Supply Task Force comprised of high-ranking officers. The initial directive, adopted on November 6, had to be revised on March 26 to take into account the fact that supplies in prospect for the United States were considerably smaller than had first been anticipated. Another major established guidelines of a more general nature: 1) contractual obligations to provide crude and products to non-affiliates were to be respected, although volumes were to be reduced where *force majeure* provisions of a contract applied; 2) the remaining petroleum supplies were to be distributed on "as fair a basis as possible" to affiliates, taking into account for each consuming country such factors as inventory position, indigenous supplies, and the outlook for industry supply and demand. Other companies used less formal systems. For example, two majors reported that they issued no policy directives, but simply made specific allocation decisions on the basis of their evaluations, made in a series of discussions among the top management, of alternative sources of energy and relative dependence on foreign oil supplies.

In addition to the perceived risks from governments in not allocating oil "fairly," oil executives felt a great deal of pressure from the company's managers in marketing areas, and this reinforced their efforts to "equalize suffering." A vice-president in charge of the Far East region, for example, would not, without strenuous protest, allow the European region to be given an abnormally large share of the pie. Similarly, the manager of a French subsidiary would attempt to protect his interests by ensuring that France would receive a "fair" allocation compared with, say, West Germany or the United Kingdom. The importance of internal organizational pressures is illustrated by the actions of the chief executive officer of one American major, who compelled his area managers to hammer out an allocation scheme in nonstop face-to-face negotiations held at headquarters.

The so-called "independent" oil companies faced a simpler task in allocating oil than did the majors. First of all, they were smaller, and therefore they had fewer persons involved in the allocation process. Second, with fewer sources of supply and fewer national markets, their choices were more limited. Instead of operating with general guidelines and allocating internationally, they could make ad hoc decisions on each specific allocation problem. For example, one company cut off its shipments of Libyan crude to the United States and did not replace them from other sources since its American refinery was designed to process low-sulfur Libyan crude and its alternative source oil was too high in sulfur content. As a result, the United States received no oil from this independent, while other nations received even more from it than they had before the embargo.

The Advice and Pressure of the Oil-Importing Nations

In addition to the pressures that emanated from their own distribution network and from their customers, practically all oil companies were exposed to—and sometimes experienced—pressures from governments. So far as the United States government was concerned, there is no evidence that any attempt was made to direct oil to the American market during the embargo period. Perhaps American officials realized, as John Tillotson had put it three centuries earlier, that "they who [are] in highest places, and have the most power, have the least liberty, because they are more observed."[40] But, whatever the reason, at the three meetings Secretary Kissinger held with American oil executives between October and December, 1973, although he asked a number of questions about supply and how the oil companies were handling it, his only request was that the companies keep the existence of the meetings strictly confidential.[41] Another oil executive, whom I interviewed, told a slightly different story; he said that Kissinger had asked the companies "to take care of Holland." Another American company executive reported that he was told on visits to the State Department and the Federal Energy Office in late October to share available supplies among consuming governments. The testimony of John Sawhill, the head of the Federal Energy Administration (the successor of the Federal Energy Office), before a Senate Subcommittee generally confirmed this: the United States government, he said, urged the international oil companies "to bring as much as possible" into the United States, but at the same time, "[to recognize] the interest of all of the countries of the world in having some kind of equitable share of the world's supplies."[42]

British government policy required that crude-oil deliveries from all companies,[43] including those headquartered there (British Petroleum and Royal Dutch/Shell), not be reduced. The Secretary of Trade and Industry told the House of Commons on November 15, 1973, "There is no need to introduce any scheme of rationing on the information available to us," and referred to "assurances from Arab states."[44] Prime Minister Heath called executives of British Petroleum to a meeting to ensure that the United Kingdom would receive preferential treatment because it was on the Arabs' "priority" list. BP executives, undaunted, maintained that they intended to treat all nations—including Britain—equally. Each would receive an immediate 10 per cent cut in supplies, and a subsequent 15 per cent cut. Heath reportedly exploded, an episode sometimes referred to by British civil servants as Heath's "temper tantrum." BP executives remained adamant, saying that laws and the contracts for delivery written under those laws took priority over instructions from their shareholders (a reference to the fact that the government owned 48 per cent of BP's stock, although by tradition did not concern itself with operations); they suggested that Mr. Heath have a law passed if he was not satisfied. Three days later, Heath decided that it would not be wise to pass a law dealing with oil supplies, because of possible adverse effects on Britain's relations with its European allies.

Royal Dutch/Shell was also briefly a target of the prime minister's wrath. Heath was reported to have been very angry when he learned that Royal Dutch/Shell was applying a uniform percentage reduction to all customer nations, but he took no action. In answer to a question about his company's relation with Mr. Heath, G. A. Wagner, chairman of Royal Dutch/Shell, stated that when the pursuits of national interests by different nations produce conflict, the governments should sort it out for themselves.[45]

Although political leaders in the British government apparently expected preferential treatment for the United Kingdom, its civil servants were more sympathetic to the "equal suffering" policy of the majors. This is illustrated by their complex negotiations with Exxon over supplies for a recently expanded refinery in Britain. The new refinery capacity was designed to run on Saudi Arabian crude and to replace products that previously had been shipped to Britain from European refineries. But when the Arab cutback came, Exxon based its allocation to each country on crude-oil imports of a prior period and hence the British share was lower than it would have been if the new refinery capacity had been put into operation earlier. When Exxon reported its crude-oil allocation to the government, the British civil servants pointed out that Exxon was a major international company with many sources of crude oil, that they expected the United Kingdom to receive at least its "fair share" (and perhaps a bit more), and that—not incidentally—it is the government that decides who gets oil rights in the North Sea. To reinforce the point, Lord Carrington, a senior official in charge of international affairs at the Ministry of Trade and Industry, reminded Kenneth Jamieson, Exxon's chairman, in December that Exxon had served the British for eighty years and had 25 per cent of the market. Exxon increased its planned shipments to Britain. The extent of the effect these pressures had is unclear, however, because simultaneously the total output of Arab oil was increased.

France, another large consumer on the Arabs' priority list, was probably the best organized government in the world when it came to dealing with the oil industry. For five decades it had allocated domestic markets among the companies, and it had played an international role through its support and partial ownership of the company often called the "eighth major"—Compagnie Française des Pétroles (CFP). In addition, the government completely owns Elf-Erap, a company that primarily serves France, but also has several foreign sources of crude oil and several national markets.

French government officials, who had available a list of oil deliveries scheduled by each company from each producing nation, instructed the companies to deliver as scheduled all oil from nations whose production was not cut (Nigeria, for example) and to deliver the oil designated for France as a priority nation (from Saudi Arabia, for example). Only oil from the nations that had reduced output but did not have a priority list (Kuwait, for example) could be cut back.[46] The French officials seemed certain that the companies would obey their instructions. The government owned 100 per cent of Elf-Erap and 42 per cent of the voting shares of CFP, where it had two board members and approved the selection of the president. In addition, any foreign company not in compliance would face the possibility of losing its license to operate in France. France also expected special treatment by Algeria to offset whatever cutbacks it might suffer. The Algerian government instructed both the CFP and Elf-Erap to send their entire output to France, thereby cutting off deliveries of Algerian oil by these companies to other nations. Algeria also ordered non-French companies to maintain their shipments of Algerian oil to France, thereby routing still more Algerian crude to that country.[47]

In spite of all these directives, on November 21, 1973, the majors announced December delivery cuts for France of 10 to 15 per cent. Although the government had never veered from its original instructions to the companies, the companies, even the two French companies, had also never promised to follow them. The only difference between the French and foreign companies, so far as the government was concerned, was that it had access to the records of the former but not of the latter. The CFP allocated its available supplies on the same pattern as the majors. When disagreement arose between the government and the CFP about planned deliveries to a neighboring nation, the CFP won the dispute and delivered the oil. Elf-Erap also continued to deliver oil to its non-French customers, although the amounts were small because France had always comprised most of Elf-Erap's market. Nor did the Algerian government's policy help France as much as might have been expected. Sonatrach, the Algerian national oil company, refused to give the French companies all of the oil indicated in the Algerian government's announcement and instead sold it on the open market.[48]

In spite of the gap between official instructions and company performance, an open break did not develop between the French government and any company. By February, 1974, the crisis in supply was clearly past, and negotiations between the companies and the government began to center on price controls. An industry group, composed of CFP, Elf-Erap, and the local subsidiaries of the majors, jointly brought up the subject of the need for higher prices. The common front of the French-owned companies and the foreign-owned subsidiaries was further emphasized when a complaint from Air France, the state-owned airline, charged that the French and foreign-owned oil companies were dividing Air France's oil purchases among themselves along geographical lines, rather than competing for the airline's business. The general manager of CFP simply replied that "CFP had ceased to deliver fuel to Air France within France because the prices the airlines offered were too low."[49]

During its negotiations with the oil companies, the French government shielded the stresses and strains from the French public. Jean Charbonnel, Minister of Science and Industrial Development, sought to allay any fears by announcing in November that France had a 90-day stock of oil and that supplies were coming in normally. He also maintained that the foreign companies were diverting oil from France to other countries where prices were higher,[50] reinforcing the myth that the CFP and Elf-Erap, being under his direct control, were the salvation of France in the crisis.

But Britain and France were not the only nations to put pressure on the companies. For example, Italy also told the companies that it expected full, normal deliveries of crude oil;[51] Italy and Spain restricted exports to the United States, Norway, and Sweden,[52] and Belgium licensed exports.[53] The companies feared that many other restrictions would be placed on the oil trade in Europe.[54]

On the other hand, the reader hardly needs reminding that the oil companies were not powerless in their negotiations with the consuming nations. For example, during the embargo some companies told the Dutch and Italian governments, among others, that if they blocked exports of oil products to the United States, crude oil would be shipped elsewhere.[55] In fact, one company that I interviewed routed oil bound for Italy to its British refinery, because the British government allowed it to export products. Belgium provides another example: in March, 1974, Petrofina, headquartered in Belgium, joined ten other oil companies in boycotting Belgium in order to obtain higher prices for their refined products in that country.[56]

The Netherlands was on the Arabs' embargo list and was therefore in a precarious position. The relationship between the Dutch government and the oil companies has not yet been fully revealed. Prior to the embargo, the Netherlands was receiving some 71 per cent of its oil from Arab sources.[57] The chairman of Royal Dutch/Shell mentioned on television that it would provide all nations (presumably including the Netherlands) with the same percentage of the supplies that they had received in September, 1973.[58] But the government still felt insecure and instituted rationing in January. Rationing was abandoned in February, however, when the government announced that supplies were running only 15 per cent short.[59]

Apparently the West German government placed little pressure on the companies over supplies. Instead, it took measures to restrict consumption, including a ban on Sunday driving and the enforcement of speed limits. The government then relied on "free-market" prices, hoping that this would bring in the oil demanded by consumers.[60]

On the other side of the world, the situation in Japan, more heavily dependent on imported oil than any other major industrial nation, was confused. Japanese sources claimed that they were suffering from severe shortages because oil was being diverted by companies to their home countries, especially the United States. The overall cut in oil deliveries to Japan was reported in the Japanese press to have been 30 per cent,[61] and some Japanese reports gave an even higher figure. In late November, 1973, the petroleum association of Japan, which represents the domestic refining industry, predicted that arrivals of crude oil in January would be 50 to 70 per cent below "normal" levels; and for the period December through March, it expected that arrivals would be down 36 per cent.[62] Meanwhile, however, there was little evidence of any actual diversion, and supplies held up remarkably well. In November, 1973, Japan announced its new pro-Arab policy, and it was placed on the Arab's priority list.

The Japanese government then turned its attentions to the companies. The head of MITI summoned the majors to a meeting to warn them about diverting non-Arab oil from Japan. The companies replied that they were allocating oil as fairly as possible and that they would gladly hand over to the governments the thankless task of deciding who got what.[63] In fact, as the embargo continued, the Japanese government seemed to have been pleased with the results—executives of one major told me that Japanese government officials, upon reviewing the company's allocation system during a visit to New York, said that they were satisfied with it. In addition, Japanese firms lobbied continually at the headquarters of the majors; the president of one American major told me that for a time during the embargo he was visited by a contingent of Japanese almost every week.

How the Nations Fared

Measured by the consumption levels of oil prevailing a year before the embargo, Japan fared the best of the three major industrial areas. The United States was second, and Western Europe, third (Table 2). Among four selected countries in Europe, the United Kingdom fared best, followed by France, the Netherlands, and West Germany. But perhaps a more appropriate guide is found in a comparison of projections for oil consumption during the period, had the embargo not taken place, with actual performance during the embargo. Although the percentages change, the ranking of the various markets remains the same (Table 3).

Table 2

Estimated Change in Consumption of Petroleum, Selected Markets, January Through March 1974, as a Percentage of Year-Earlier Period

Japan	10
United States	− 7
Western Europe:	−12
United Kingdom	− 8
France	−11
Netherlands	−17
West Germany	−20

Source: Table A-1 in Appendix.

Table 3

Estimated Change in Consumption of Petroleum from Forecast, Selected Markets,
January Through March, 1974

	As a percentage of forecast of petroleum consumption	As a percentage of forecast of total energy consumption
Japan	− 3	− 2
United States	−11	− 5
Western Europe:	−19	−12
United Kingdom	−12	− 6
France	−21	−15
Netherlands	−22	−12
West Germany	−26	−15

Source: Table A-1 in Appendix.

Furthermore, when the effect of changes in oil consumption on total energy supplies is taken into consideration, the ranking of the countries still remains about the same, except that France ranks much lower—it changes places with the Netherlands and equals West Germany. However, several words of caution: these data reflect consumption, and hence they vary from data on deliveries by the amounts of certain inventories. In addition, the accuracy of those data involving the Netherlands and its contiguous neighbors is open to some question—a great deal of oil normally passed through the Netherlands en route to other countries that were not on the embargo list. During the embargo, both the Netherlands and the Arabs let Arab oil pass through Dutch ports. The Dutch agreed to this "pass through" after the Arabs agreed not to demand dockside inspections to make certain that oil was not being diverted to Dutch refineries.[64] In addition, there were unconfirmed reports that an inordinately large amount of fuel was being consumed in northern France; presumably, overland shipments, not reported in official statistics, were being made to the Netherlands.

The data available on a company-by-company basis cover only the intra-company shipments of the thirteen firms headquartered in the United States—five majors and eight selected independents. These intra-company shipments represent 80 to 90 per cent of the total international shipments of these firms. The intra-company shipments of crude oil by the five majors averaged about 9.2 million barrels a day during the embargo period, a drop of some 6 per cent from the year-earlier period; the comparable figure for the eight independents was 2.0 million barrels daily, down 14 per cent from the year-earlier period.[65] Because Arab crude was more important for these American companies than for the world as a whole, the percentage reduction in supply for these companies was larger than that for the world (which, as indicated in Table 1, was about equal to the year-earlier output).

Regardless of whether one compares the embargo period with year-earlier figures or with forecasts for the embargo period, the five American majors supplied Japan more generously in terms of net imports than they did either the United States or Western Europe. Japan received one per cent more oil during the embargo period than during the year-earlier period, but 11 per cent less than the forecast (Table 4). The rest of the world received relatively more than any of the three major industrial areas. Within Europe, Britain and France fared better than West Germany or the Netherlands. In fact, the five American majors supplied France with 9 per cent more oil during the embargo period than during the year-earlier period; in comparison, the French companies increased their supply by 2 per cent over the year-earlier period.[66] However, a judgment on France's treatment by the French companies requires data that I do not possess on the world-wide availability of oil for these companies.

Table 4

Net Supply of Crude Oil Plus Products to Selected Markets by the Five U.S. Major Oil Companies,[a] Embargo Period Compared with Year-Earlier Period and Forecast for Embargo Period
(thousands of barrels daily unless otherwise noted)

	Base Period (December, 1972, through March, 1973)	Embargo Period (December, 1973, through March, 1974)	Difference between Embargo Period and Base Period, as a per cent of Base Period		Difference between actual and forecast during Embargo Period as a per cent of forecast
			Actual	Forecast[b]	
Japan	1,682	1,701	1	13	−11
United States	2,603	2,290	−12	20	−27
Western Europe:	3,870	3,606	−14	8	−20
United Kingdom	873	927	6	5	1
France	430	467	9	12	− 3
West Germany	1,021	908	−11	8	−18
Netherlands	367	233	−37	7[c]	−41
Rest of Importing Countries[d]	1,745	1,877	8	8	0

[a]Exxon, Gulf, Mobil, Socal, and Texaco.
[b]Assumed to be the same as compounded growth rate for consumption of all petroleum between 1968 and 1973 as reported in *BP Statistical Yearbook, 1973*, p. 21.
[c]Benelux.
[d]Includes imports of crude oil and products minus exports of products for all areas except Japan, United States, Western Europe, OPEC nations, Canada, and Caribbean nations.
Source: Federal Energy Administration, *U.S. Oil Companies and the Arab Oil Embargo: The International Allocation of Constrained Supplies* (Washington, D.C.: U.S. Government Printing Office, 1975), Appendix II.

As for the eight American independents, they supplied Japan with 37 per cent more oil during the embargo than they had during the year-earlier period and 21 per cent more than the forecast, far more than they gave any other area of the world (Table 5). Britain also received more than it had in the year-earlier period and more than the forecast. But the characteristic of the independents lies in the wide range of the treatment given to various nations. Net deliveries to France, for example, were cut from 47,000 barrels a day to 6,000 barrels a day, while, in the case of the Netherlands, the independents exported more petroleum than they imported.

Table 5

Net Supply of Crude Oil Plus Products to Selected Markets by Eight U.S. Independent Oil Companies,[a] Embargo Period Compared with Year-Earlier Period and Forecast for Embargo Period
(thousands of barrels daily unless otherwise noted)

	Base Period (December, 1972, through March, 1973)	Embargo Period (December, 1974, through March, 1974)	Difference between Embargo Period and Base Period, as a per cent of Base Period Actual	Forecast[b]	Difference between actual and forecast during Embargo Period, as a per cent of forecast
Japan	119	163	37	13	21
United States	1,269	1,097	-14	20	-28
Western Europe:	870	753	-13	8	-19
United Kingdom	257	322	25	5	19
France	47	6	-87	12	-88
West Germany	232	196	-16	8	-22
Netherlands	48	-4	-108	7[c]	-107
Rest of Import- ing Countries[d]	168	102	-39	8	-44

[a]Amerada Hess, Arco, Continental, Getty, Marathon, Occidental, Phillips, and Standard (Indiana).
[b]Assumed to be the same as compounded growth rate for consumption of all petroleum between 1968 and 1973 as reported in *BP Statistical Yearbook, 1973*, p. 21.
[c]Benelux.
[d]Includes imports of crude oil and products minus exports of products for all areas except Japan, United States, Western Europe, OPEC nations, Canada, and the Caribbean nations.
Source: Same as Table 4.

In addition to the quantities supplied to different nations, statistics also reveal the extent to which the Arab embargo was in fact observed. The data only allow us to explore this question for the United States market, and then only for imports of crude oil; refined products remain a mystery because of the difficulties in determining the precise sources of crude for substantial quantities of these products imported by the United States. There was a substantial rise in American imports of crude oil between the period December, 1972-March, 1973, and the period immediately prior to the embargo, August-September, 1973—from 3,190 thousand barrels daily to 4,114 thousand barrels daily (Table 6). From August-September, 1973, to the height of the shortage in the United States (January-February, 1974), American crude-oil imports dropped 30 per cent. Statistics for March are not available by source at the present writing, making it impossible to calculate an average for the embargo period. More than 98 per cent of the Arab oil that the United States had been receiving was lost—imports of Arab oil dropped from 1,200,000 barrels a day in August-September to only 19,000 a day during January-February. (The 19,000-barrel average could have represented the load of one large tanker; as mentioned above, the oil from at least one large tanker, which was loaded in Saudi Arabia before the embargo, was indirectly delivered to the United States during the embargo period.)

Table 6

U.S. Imports of Crude Oil by Country of Origin, Base Period Compared with August-
September 1973 and January-February 1974
(thousands of barrels daily unless otherwise noted)

	(1)	(2)	(3)	Difference Between Columns 2 and 3	
Country of Origin	Base Period (December, 1972, through March, 1973)	August-September, 1973	January-February, 1974	Absolute	Per Cent
TOTAL	3,190	4,114	2,885	−1,229	−30
Total, Arab States	730	1,203	19	−1,184	−98
Total, Non-Arab States:	2,460	2,911	2,886	− 45	− 2
Venezuela	516	633	493	− 140	−22
Canada	1,116	1,082	960	− 122	−11
Nigeria	334	524	462	− 62	−12
Iran	133	229	430	201	88
Indonesia	126	237	275	38	16
All other Non-Arab	206	206	246	40	19

Source: U.S. Department of Commerce, *Imports Commodity by Country*, monthly, various issues, and news release
CB 74-83, April 8, 1974.

Total American supplies of non-Arab oil dropped very slightly, only about 2 per
cent, from August-September, 1973, to January-February, 1974, but increased by 17
per cent, if the December, 1972, to March, 1973, base period is used as the reference.
But if the analysis is carried further, one finds that substantial cuts were made by some
non-Arab suppliers, principally Venezuela, Canada, and Nigeria. Some of the Venezuelan oil was diverted from the United States to provide more oil to other Latin
American countries;[67] and the Canadians reduced exports to the United States in order
to make up for their own reduced supplies of Arab oil.[68] I have no data on the destination of Nigerian oil diverted from the United States, but it may have been sent to the
Netherlands to help make up for that country's heavy dependence on Arab oil.

These cutbacks of non-Arab oil coming into the United States were, however, almost entirely offset by increased imports from other non-Arab sources, primarily Iran,
but also, to a lesser extent, Indonesia and several other countries. (Production in Iran
and Indonesia during this period increased by an amount greater than that represented
by the additional American imports.) Imports of Iranian crude oil almost doubled, increasing by 201,000 barrels a day.

Table 7 shows the pattern for American imports, broken down by the five American majors and eight selected American independents. The Arab embargo, if measured
solely by supplies of Arab oil, was at least 91 per cent effective for both groups of companies. It was even more effective when one excludes the Arab oil delivered to the
United States that was loaded prior to the embargo. The majors increased American

Table 7

U.S. Imports of Crude Oil by the Five U.S. Majors and Eight U.S. Independents by Country
of Origin, Embargo Period Compared with Year-Earlier Period
(thousands of barrels daily unless otherwise noted)

Country of Origin	Base Period (December, 1972, through March, 1973)	Embargo Period (December, 1973, through March, 1974)	Difference between Embargo Period and Base Period	
			Absolute	Per Cent of Base Period
MAJORS				
TOTAL	1,360	1,137	−223	−16
Total, Arab States	395	35	−360	−91
Total, Non-Arab States:	965	1,102	137	14
Venezuela	182	188	6	3
Canada	302	199	−103	−34
Nigeria	214	358	144	67
Iran	41	59	18	44
Indonesia	152	196	44	29
INDEPENDENTS				
TOTAL	657	467	−190	−29
Total, Arab States	211	19	−192	−91
Total, Non-Arab States:	446	448	2	neg.
Venezuela	73	66	− 7	−10
Canada	209	153	− 56	−27
Nigeria	46	96	50	109
Iran	39	47	8	21
Indonesia	14	23	9	64

Source: See Table 4.

imports of non-Arab oil by 14 per cent (from 965,000 barrels to 1,102,000 daily).
Imports from Nigeria increased sharply (by 144,000 barrels daily) along with smaller
increases from Indonesia and Iran. This more than offset the large drop of 103,000
barrels daily in imports from Canada. On the other hand, imports of non-Arab oil by
the independents were about the same during the embargo period as during the base
period. These companies imported more from Nigeria and less from Canada, and their
imports from other countries changed little.

The five American majors appear to have diverted about 31 per cent of their
crude-oil shipments during the embargo period to countries different from what their
non-embargo destination would have been; a comparable figure for the eight independents is 27 per cent. Table 8 shows the diversions of the American majors (in terms of
crude oil only; Table 4 includes crude oil plus products). As would be expected, shipments of non-Arab oil to the United States and the Netherlands were increased to replace Arab oil; that oil, in turn, was shipped in greater quantities to Japan and the
United Kingdom to offset reductions of non-Arab oil. Relatively little diversion oc-

Table 8

Supply of Crude Oil from Arab and Non-Arab Sources to Selected OECD Countries by the
Five U.S. Majors, Embargo Period Compared with Base Period
(thousands of barrels daily unless otherwise noted)

	Base Period (December, 1972, through March, 1973)	Embargo Period (December, 1973, through March, 1974)	Difference between Embargo Period and Base Period, as a Per Cent of Base Period
United States			
Arab	395	35	−91
Non-Arab	966	1,104	14
Total	1,359	1,137	−16
Japan			
Arab	661	817	24
Non-Arab	898	689	−23
Total	1,558	1,506	− 3
United Kingdom			
Arab	516	794	54
Non-Arab	148	62	−58
Total	663	857	29
France			
Arab	410	442	8
Non-Arab	46	31	−33
Total	455	472	4
West Germany			
Arab	721	721	0
Non-Arab	117	109	−7
Total	837	827	−1
Netherlands			
Arab	701	18	−97
Non-Arab	131	430	228
Total	831	447	−46

Source: See Table 4.

curred either to or from France and West Germany; imports into those countries by
the majors consisted mainly of Arab oil both before and during the embargo.

Conclusion

The consumption of petroleum (Tables 2 and 3) and the imports of the five American majors (Table 4) seem generally to support the view that shortages were being
equalized in all markets—most countries experienced shortages, but no one much more
than the others. There is some apparent variance among countries, however; what can
be said about its causes?

One obvious possibility is that the companies deviated from the "equal-suffering" rule in order to serve the markets with the highest prices. But the admittedly fragmentary evidence does not support this hypothesis. For example, prices in West Germany were not controlled and so were higher than in France;[69] yet the oil companies in general, including the American majors, supplied France more generously than West Germany. Moreover, price differences between the various countries persisted, suggesting that some sort of allocation process was applied. In February, 1974, for example, Iranian crude oil was imported into the United States at an average value of $7.55 a barrel, compared with $8.61 in West Germany, and the average value of Nigerian crude delivered to the United States was $11.86, compared with $14.70 for Japan.[70] In neither case was the difference in freight cost likely to have accounted for much of the difference in price.

Another possible explanation for the observed variations in the market is that the oil companies favored large customers over smaller ones, or ones with a favorable long-term outlook over those with an uncertain one. But these possibilities also gain little support from the data.[71]

Still another explanation for the differences in treatment might be that government influence did have its effects. But the data do not conclusively support this hypothesis either. To be sure, the United Kingdom and France, both relatively well-supplied by the American majors, were on the priority lists supplied by some Arab countries, and both countries also applied government pressure on the oil companies. Yet, in terms of total energy supply from all sources, France seems to have been supplied below the average for the rest of Western Europe, particularly the United Kingdom and the Netherlands. Even the relatively large supplies of oil received by the United Kingdom could equally well be explained as an attempt on the part of the oil companies to compensate for a coal strike that had simultaneously occurred in Britain. Furthermore, Japan, apparently the best served of all the markets, does not appear to have applied any more government pressure than did the other countries. Nevertheless, whatever the exercise of government pressure might have achieved in general, it is clear that pressure from home governments on firms headquartered within their own boundaries did not play a major role in determining the allocations of supply.

The most plausible explanation for the differences in treatment during the embargo seems to be a combination of random events and imperfections in data and in the measurement of shortage. Different companies undoubtedly used different bases from which to compute their allocations. Differences in inventory levels among countries and aberrations in computed averages caused by the bunching or scattering of tanker deliveries would also undoubtedly have caused deviations from the anticipated norm in deliveries within a precisely defined time period.

Given the circumstances, the companies on the whole did an efficient job of allocation during the oil crisis. As a Federal Energy Administration report concluded, "It is difficult to imagine that any allocation plan would have achieved a more equitable allocation of reduced supplies."[72] Nevertheless, the experience of the American majors in helping to manage the shortages of world oil during the embargo seems to have changed their view of the function that they might appropriately play in the future. Prior to the embargo, it would have been difficult to find an American company official who thought it was the responsibility of governments to allocate oil during a crisis— these officials would have taken for granted that the companies would have that task. After the embargo was over, however, a number of company spokesmen, presumably worn by pressures and stung by the criticism that had resulted from their handling of the situation, began to claim that it was the responsibility of governments to determine the allocations.[73] The companies seem to have learned an important lesson, succinctly phrased two and a half centuries ago: "Those who in quarrels interpose must often wipe a bloody nose."[74]

REFERENCES

[1]As reported in *Fortune*, February, 1974, p. 58.

[2]This meeting is described in a cable of October 21, 1973, from Jungers to Aramco in New York and transmitted to the four parent companies, used as Exhibit 12 in U.S. Senate, *Hearings before the Subcommittee on Multinational Corporations of the Committee on Foreign Relations, Multinational Petroleum Companies and Foreign Policy*, 93rd Congress, June 20, 1974, p. 515.

[3]*Ibid.*, p. 418.

[4]*Ibid.*, p. 427.

[5]*Ibid.*, p. 531.

[6]*Ibid.*, p. 418.

[7]Federal Energy Administration, *U.S. Oil Companies and the Arab Oil Embargo: The International Allocation of Constrained Supplies* (Washington, D.C., 1975), p. 2; and company interviews.

[8]See Table 1.

[9]*BP Statistical Review of the World Oil Industry, 1973* (London, The British Petroleum Company Limited, 1974), p. 21.

[10]*Oil and Gas Journal*, January 28, 1974, p. 191; and March 25, 1974, p. 158.

[11]*Hearings, op. cit.*, June 20, 1974, p. 504.

[12]*Ibid.*, p. 504.

[13]*Ibid.*, p. 507.

[14]*Ibid.*, p. 515.

[15]*Ibid.*, p. 507.

[16]*Ibid.*, p. 507.

[17]*Ibid.*, p. 504.

[18]*Ibid.*, p. 504.

[19]*Ibid.*, p. 531.

[20]*Ibid.*, p. 432; *Oil and Gas Journal*, November 19, 1973, p. 101; and *Middle East Economic Survey*, November 2, 1973, p. 1.

[21]*Hearings, op. cit.*, June 20, 1974, p. 431.

[22]Company interviews.

[23]Geoffrey Chandler, "*The Energy Prospect*," an address to the Parliamentary and Scientific Committee, House of Commons, May 15, 1973, p. 2; and "Energy: The Changed and Changing Scene," Keynote Address, Institute of Petroleum, Summer Meeting, 1973, p. 11.

[24]*Hearings, op. cit.*, June 20, 1974, p. 446.

[25]*Ibid.*, p. 423.

[26]*Ibid.*, p. 423.

[27]*Ibid.*, p. 423.

[28]*Ibid.*, pp. 412-13.

[29]*Ibid.*, p. 429.

[30]*Ibid.*, pp. 546-47.

[31]*Ibid.*, p. 509.

[32]U.S. State Department Confidential Cable 202315, October 12, 1973, p. 3.

[33]U.S. State Department Confidential Cable 23307, September 4, 1973, p. 2.

[34]U.S. State Department Confidential Cable 214639, October 31, 1973, p. 2.

[35]*Hearings, op. cit.*, June 20, 1974, p. 456.

[36]*Ibid.*, pp. 528-29.

[37]*Hearings, op. cit.*, October 11, 1973, p. 46.

[38]*The Economist*, "Multinational Business Report," December 1973, p. 1.

[39]Directorate-General for Energy, Commission of the European Communities, "Chronological Summary of Developments in the Oil Crisis Which Have Affected Community Supplies," XVII/110/74-E, p. 6.

[40]John Tillotson (1630-1694), in *Reflections*.

[41]*Hearings, op. cit.*, June 20, 1974, p. 450.

[42]*Hearings, op. cit.*, June 5, 1974, p. 1129.

[43]*Petroleum Intelligence Weekly (PIW)*, November 5, 1973, p. 1.

[44]*New York Times*, November 6, 1973, p. 58.

[45]*Transcript*, "Dutch TV Interview with Mr. G. A. Wagner on 4th December, 1973," p. 9.

[46]*Middle East Economic Survey*, November 5, 1973, p. 1; interviews with company officials and government officials.

[47]*Middle East Economic Survey*, November 23, 1973, p. 14.

[48]Company interviews.

[49]*New York Times*, November 9, 1974, pp. 41-43.

[50]*New York Times*, November 22, 1973, p. 61.

[51]*PIW*, November 5, 1973, p. 1.

[52]*PIW*, October 29, 1973, p. 5; October 15, 1973, pp. 3-4; and December 3, 1973, p. 4.

[53]*PIW*, November 19, 1973, p. 10.

[54]*Ibid.*

[55]*PIW*, November 5, 1973, p. 2.

[56]*PIW*, March 18, 1974, p. 4.

[57]*Middle East Economic Survey*, November 2, 1973, p. 1.

[58]See reference 45.

[59]*The Economist*, January 26, 1974, p. 87.

[60]*The Economist*, December 8, 1973, pp. 84-85.

[61]*The Economist*, "Multinational Business Report," December 1973, p. 1; cf. also the article by Yoshi Tsurumi in this issue.

[62]*The Economist*, December 1, 1973, p. 103 and December 8, 1973, p. 84.

[63]*The Economist*, December 1, 1973, p. 103.

[64]*New York Times*, January 4, 1974, pp. 43, 49.

[65]Federal Energy Administration publication referenced on Table 4.

[66]Schvartz Commission, "Sociétés Pétrolières Opérant en France," Report to French Parliament, III, p. 413.

[67]*PIW*, January 14, 1974, p. 12.

[68]*PIW*, November 12, 1973, p. 9.

[69]Company interviews.

[70]National import statistics of each country with local currencies translated into American dollars at then-prevailing exchange rates.

[71]For example, with an independent variable of "imports in the period one year before the embargo period" and a dependent variable of "imports during the embargo period divided by imports during the year-earlier period," an R^{-2} of .02 was obtained for data covering the five American majors and a .11 for data covering the eight American independents.

[72]Page 10 of Federal Energy Administration publication, reference on Table 4.

[73]For example, see statement of Emilio G. Collado, director and executive vice-president of Exxon, to the Subcommittee on Multinational Corporations of the U.S. Senate Committee on Foreign Relations, June 6, 1974, p. 5.

[74]John Gay (1688-1732), *Mastiffs*, p. 401*b*.

APPENDIX
Table A-1

Estimated Inland Consumption of Total Petroleum, Selected Markets, First Quarter, 1974,
Compared with First Quarter, 1973, and Forecast for First Quarter, 1974[a]
(millions of barrels daily unless otherwise noted)

| | | Petroleum | | | Effect of Changes in Petroleum Supply on Supply of Total Energy |
	First Quarter 1973	First Quarter 1974	Difference between First Quarter, 1973, and First Quarter, 1974, as a per cent of First Quarter 1973 — Actual	Difference between First Quarter, 1973, and First Quarter, 1974, as a per cent of First Quarter 1973 — Forecast[b]	Differences between actual and forecast during First Quarter, 1974, as a per cent of forecast	Difference between actual and forecast during First Quarter, 1974, as a per cent of forecast
Japan	4.78	5.26	10	13	− 3	− 2
United States	18.24	16.88	− 7	5	−11	− 5
West Europe[d]	14.09	12.39	−12	8	−19	−12
United Kingdom	2.31	2.13	− 8	5	−12	− 6[e]
France	2.68	2.39	−11	12	−21	−15
Netherlands	.58	.48	−17	7[f]	−22	−12
West Germany	2.90	2.31	−20	8	−26	−15
Rest of World[g]	19.94	22.90	15	8	6	2

[a] Inland consumption excludes fuel provided for ships, i.e., bunker fuel. Such fuel is equivalent to about 6 per cent of inland consumption both for Japan and for Western Europe on the average.

[b] Assumed to be the same as compounded growth rate for consumption of all petroleum between 1968 and 1973 as reported in *BP Statistical Review of the World Oil Industry, 1973* (London, The British Petroleum Company Limited), p. 21.

[c] Assumes importance of oil in total energy supply as calculated from *BP Statistical Review of the World Oil Industry, 1973*, p. 16, as follows: United States, 47; Japan, 80; West Europe, 64; West Germany, 59; France, 72; United Kingdom, 52; Netherlands, 54.

[d] Includes all OECD European countries except Ireland and Turkey, for which data are not available in source.

[e] The coal miners' strike reduced the consumption of coal. Thus, actual consumption of energy in the United Kingdom was 13 per cent below forecast instead of 6 per cent. The change in total energy consumption of the United Kingdom was calculated from page 2 of *Energy Trends*, February, 1975, a statistical bulletin of the Statistics Division of the Department of Energy of the British Government.

[f] Benelux.

[g] Includes quantities of bunker fuel not included in totals for United States, Japan, and Western Europe.

Source: All oil volumes except those for United States and rest of world are from *Provisional Oil Statistics by Quarters* (Paris: Organization for Economic Cooperation and Development, 1975), p. 15. Metric tons were converted to barrels by using a conversion factor of 7.4 barrels. The conversion factor was calculated from *BP Statistical Review of the World Oil Industry 1973*, pp. 20, 21, and 24. U.S. data are from Bureau of Mines as reported in Federal Energy Administration, *Monthly Energy Review*, February, 1975, p. 8. Data for rest of world were obtained by subtracting totals for United States, Japan, and Western Europe from total oil available for world, which in turn was assumed to be the amount of oil produced in the respective December, January, and February periods. Production data were obtained from "Worldwide Crude Production" reports in *Oil and Gas Journal* (e.g., see January 28, 1974, p. 191).

Part 5: THE INTERNATIONAL ORGANIZATIONS

ZUHAYR MIKDASHI

The OPEC Process

In SEPTEMBER, 1960, five leading oil-exporting countries—three Arab (Iraq, Kuwait, and Saudi Arabia) and two non-Arab (Iran and Venezuela)—founded the Organization of Petroleum Exporting Countries (OPEC) as an intergovernmental, interregional organization, the purpose of which was the protection of their common oil interests, especially prices and oil-export revenues. The transnational oil companies, which had the power to fix oil-export prices prior to September, 1960, reluctantly accepted the OPEC governments as partners in the pricing process.

The immediate issue that had brought the OPEC governments together that September had been their refusal to accept the cuts in posted prices that had been initiated for tax-reference purposes by the transnational companies in August. The OPEC governments began by passing resolutions and issuing statements protesting the companies' action and calling for the restoration of prices to their pre-August levels. But neither these strenuous governmental protests nor subsequent protracted negotiations with the transnational enterprises restored prices completely. The small increases that the OPEC governments did obtain, however, were largely offset by increases in the price levels of their own imports. Their failure to negotiate successfully can be attributed in part to their lack of bargaining power—which was in turn due primarily to the surplus conditions in the market—and in part to the structure of the world oil industry dominated by a closely knit, vertically integrated oligopoly, to the competition between this oligopoly and the new independents, and to monopsonist practices throughout the world, but especially in Japan.

It was not until October 16, 1973—thirteen years after this initial meeting—that the host governments of the oil-exporting countries wrested from the transnational companies the right to set and change the prices of their oil exports in the light of their perceived collective interests. In the meantime, both market conditions and political circumstances had become more propitious. By that time, there were thirteen member countries spanning four continents. The majority (seven) were Arab countries (Iraq, Kuwait, Qatar, Saudi Arabia, the United Arab Emirates, Algeria, and Libya); six were in West Asia (Iran, Iraq, Kuwait, Qatar, Saudi Arabia, and the United Arab Emirates); four in Africa (Algeria, Libya, Nigeria, and Gabon as associate; it was to become a full member later); two in Latin America (Venezuela and Ecuador); and one in Oceania (Indonesia). Their exports accounted for over 85 per cent of the world oil trade.

That the transnational companies tolerated in October, 1973, what they had refused to tolerate in September, 1960—namely, the role of deciding crude-oil export prices—was the result of particular market and political circumstances. The years 1972-73 had witnessed oil shortages as world demand expanded beyond planned production, refining, and transport capacities. The shortages in this "sellers' market" had more recently been increased by the October hostilities between Egypt, Israel, and Syria. On October 11, 1973, Israel bombarded the eastern Mediterranean oil terminals that loaded Iraqi and Syrian oil, thereby denying the world market a flow of close to one million barrels of short-haul crude. Several Arab oil-exporting countries—this time acting outside the framework of the OPEC organization—were subsequently to constrain world oil supplies still further by gradually reducing their oil exports until the cutbacks reached, late in 1973, a peak of about four million barrels per day.

The OPEC nations were in no way responsible for the outbreak of war in October. The countries engaged in hostilities were Israel, Egypt, and Syria; although the latter two countries were members of the Organization for Arab Petroleum Exporting Countries (OAPEC), they were not members of OPEC. Cutbacks in oil supplies and the boycott of certain destinations were debated within OAPEC, but not within OPEC. Nor was there a consensus among OAPEC members with respect to the "cutback" policy. Iraq, with its "radical" government, was an important dissenter, and Saudi Arabia, the most important Arab producer, adopted the cutback only as a so-called "defensive" measure. The Saudi Minister of Petroleum and Mineral Resources explains his government's strategy thus:

The October war broke out and it immediately became clear that it was politically necessary to use oil in defending the legitimate rights of the Palestinian nation and three Arab countries—Egypt, Syria and Jordan—whose territories continued to be occupied by Israel in flagrant defiance of law, justice and the United Nations. There were several alternative courses of action in which oil could have been used to serve the Arab cause: Saudi Arabia chose one course and insisted upon it in the face of the other differing views. It insisted that oil was to be used only to attract the attention of the people and governments of the West to two facts: first, the Arabs had been deprived of certain rights—a situation to which the West had shown indifference and sometimes aggravated by siding with the aggressor; second, Western economic interests in the Arab world were far more than in Israel, if indeed there were any in Israel.

Saudi Arabia adamantly refused to use oil as a weapon to disrupt Western economies or as a punitive device. Therefore, the basis upon which the Arab producing countries reached agreement was that oil production should be reduced and not stopped altogether. As for the ban on exports to the United States, the late King Faisal's Government actively tried to avert it. However, the American military assistance to Israel during the war, unprecedented in its volume and nature, changed the course of the war and prevented the Arabs from achieving their legitimate goal—the recovery of their occupied lands. The natural reaction to this American action was the imposition of an embargo on oil exports to the United States. Saudi Arabia also tried to ward off the ban on oil exports to Holland and urgent contacts were made, and the mediation of a Western power was sought, in an attempt to correct the hostile Dutch attitude. When these efforts failed, the Saudi ban on oil to Holland was imposed, rather belatedly.

To those who watched the events during the period of the oil measures, it was clear from the successive amendments introduced to these measures that the Arab producers were not seeking to inflict damage on the Western economies. The objective was simply to remind the West of the Arab problem and of its own interests in the Arab world.[1]

However, although OPEC members were not responsible for the war, they nevertheless made capital use of its impact. When the war broke out, the OPEC delegations had been in Vienna negotiating with the transnational companies for higher export prices, demanding increases of up to $5.60 per barrel for Arabian crude. The companies were reluctant to agree to these demands, but, when they took no action, the OPEC delegations decided that the company representatives were stalling for time, and ended the negotiations. This action gave them the opportunity they were looking for: they could now set prices unilaterally, without prior agreement with the oil companies, since their position was by now much strengthened by both the actual and the impending shortages in world oil supplies. The October increases were followed by another set of increases in December; taken together they amounted to a four-fold increase in the price of oil.

OPEC's collective efforts in 1973-74 appear on balance to have contributed to both more rapid and higher price increases than would otherwise have been the case if the transnational oil companies had still been in a position to influence pricing decisions—if only by delaying their responses to proposed price increases. Under purely competitive conditions—which can scarcely be said to exist in the real world—prices could have risen to even much higher levels during the 1973-1974 reductions of oil flows had the oil-exporting countries offered their oil at auction when buyers were scrambling for urgently needed supplies. On the other hand, prices could have been lower in early 1975 had competitive auctions been used, since this was a period of slackening world demand for energy. OPEC's pricing arrangements in the period from 1973 to 1975, therefore, represented a compromise between those seeking greater price increases and those content with smaller price increases combined with long-term maximization of benefits. Since the beginning of 1975, OPEC members have attempted to stabilize the 1974 oil-export price level, largely by accepting voluntarily (and uncoordinated) cutbacks in their oil exports.

OPEC's creation and survival can best be interpreted as a response to a perceived collective threat: the transnational oil companies were threatening the prosperity of the developing oil-exporting countries by unilateral cuts in oil-export prices and other acts regarded as actual or potential threats to their welfare. To understand its success in achieving this goal and its role in the oil industry today calls for a review of OPEC's structure and method of operation. The following sections attempt to offer such a review and to scrutinize the sources of weakness and strength in OPEC's cooperative efforts.

OPEC's Structure and Function

OPEC was created in 1960 as a political entity by ministers and diplomats, primarily because at the time of its establishment, and throughout most of the nineteen-sixties, few of the countries involved had either national petroleum companies or the managerial expertise from which to draw on to staff the organization. In the nineteen-seventies, however, once full control over the oil industries had been achieved, the OPEC conference, the secretariat, and the economic commission became as concerned about the "business" of oil as they had already been with its "politics." The business of oil called *inter alia* for expert skills in devising optimal pricing, production, and financial policies. For these, the OPEC governments came to rely on the technical staff of their newly formed national petroleum companies and of their central banks. OPEC delegations began to be composed of national corporation executives and bankers rather than the politicians and bureaucrats of earlier years. This new personnel soon proved its competence in both "battling" and "cooperating" with the transnationals.[2]

The OPEC organization also became increasingly concerned about the dynamics of the world energy market.[3] Whereas previously it had negotiated only with those few transnational enterprises that stood between its members and the world market, beginning in the nineteen-seventies, it began to deal with a multitude of actors and a rapidly changing situation. This new environment elicited increasing expertise and business acumen from the OPEC nations.

In addition to the meetings of OPEC itself, the national companies of the member countries held other meetings among themselves. Although these companies were state owned, they were in closer contact with markets than were the governments. It was the companies that initiated price cuts in response to sagging demand—a move that was either disregarded or supported by individual governments without recourse to OPEC. They also individually or collectively (e.g., through OAPEC in the Middle East, and ARPEL [Asistencia Reciproca Petrolera Estatal Latinoamerica] in Latin America) established their oil-related enterprises dealing with exploration, tankers, petrochemical refining, financing, and so on. The growth in these enterprises and in the expertise available to their management had a beneficial influence on shaping governmental policy in the direction of pragmatism and profitability.

Bearing in mind this shift in character over the years, the main features of the organization were—and continue to be—the following: 1) OPEC is an intergovernmental organization restricted exclusively to countries heavily dependent on oil exports for foreign exchange and economic development; 2) membership is in practice limited to developing countries; 3) decisions are reached by a ministerial conference, require a unanimous vote, and are subject to the approval of all member governments; 4) political matters are at least formally outside the scope and purview of the organization.

By the mid-nineteen-seventies, the organization was comprised of four groups: a secretariat, a conference of ministers, a board of governors, and an economic commission. The secretariat acts as the administrative arm of the organization, has a permanent headquarters, and operates on a regular basis. The three other groups respectively determine policy, oversee the secretariat, and act as an economic advisory group (notably on prices). The secretariat is run by professionals with a loyalty to the common objectives of the organization; its chief officer, the secretary general, is appointed for a two-year term by the conference on a rotating basis from each of the member countries. The other three branches of the organization are composed of delegates from member governments, who represent national interests and convene periodically to make decisions and recommendations.

The staffing of the secretariat is restricted to OPEC nationals, and these are almost always on loan from government civil service or, more recently, from the management of the national oil companies. It is difficult to see how, in the absence of a tenure system, the OPEC staff can have a large measure of freedom of action to work for the "common interest"; it is not even clear that a tenure system could offset national loyalties. Moreover, a permanent secretariat staff can easily become bureaucratic, expansionist, and inefficient. Recognizing these problems, the ministers' conference agreed in 1974 to reform the secretariat's structure and staffing. One study commissioned by OPEC called for the creation of a "think tank" attached to the secretariat, and staffed with qualified experts recruited from both rich and poor non-member states.[4]

Descriptions of OPEC range from assertions that it is no more than a shallow organization, a feeble overlay that can only slightly improve communication among member countries, to the assertion that it is a powerful autonomous entity with a momentum of its own, whose leadership can generate common policies and strategies that otherwise could not be achieved by the countries acting individually. Those of the former persuasion regard it as little more than an information bureau or a research center, while those tending to the latter view find it acting as an effective cartel.

A number of OPEC policy-makers had in fact looked to the cartel model in shaping OPEC's functions. One of the acknowledged "fathers" of OPEC, the Venezuelan minister of mines and hydrocarbons, Perez Alfonzo,[5] sought in 1960 to design the organization's structure and statutes to permit the governments to operate as a cartel. The instrument for this cartel function was to be the joint prorationing of oil production in the member countries. Libya also viewed OPEC in 1972 as a countervailing force along the lines of a cartel to counter both the transnational companies themselves and other cooperating groups in the industrial countries,[6] since in that year the remarkable shift in bargaining power in favor of the OPEC countries allowed them to acquire effective control over their oil industry either through participation or through a full takeover of the transnational operations.

The Saudi minister of oil and minerals was credited with introducing the participation system to the Middle East; he perceived its success in late 1972 as being "a historic event and a fundamental turning point in the international oil industry, that will make [OPEC] prominent in the world and give its member countries political strength to be taken into account in the world balance of power."[7] In 1974, however, the Saudi minister questioned the allegation that OPEC was a "selfish cartel" on the grounds that the two countries with the largest oil reserves and oil exports (Saudi Arabia and Iran) were producing beyond their domestic needs in order to meet the requirements for energy of the world economy. In October, 1974, the Minister said:

In Saudi Arabia we can easily cut down by about 5.5 million barrels a day, and live in comfort and meet our requirements inside and outside Saudi Arabia. In Iran, even with all the huge development programs, they can easily cut down by 2 million barrels a day, and they will meet their requirements both internally and internationally. And this is also the case with so many other [OPEC] countries. This is the difference between OPEC as a cartel and any other cartel. . . . What we are doing right now is something foolish if I take if from the narrow angle of our selfish interest in Saudi Arabia. We are producing 5.5 million barrels a day just to help the economy of the world, to help not to have another price increase. . . .[8]

If one applies the definition of export cartels used by Western countries (for example, the Webb-Pomerene Act of the United States),[9] the OPEC organization does not pass the test. An export cartel not only must include rigid agreements on prices, but also related agreements in such key areas as production control and market sharing. The export cartel must also be responsible for monitoring the activities of its constituent members with a view to policing violations and penalizing violators. The OPEC member governments do not perform any of these cartel functions. Their agreement on oil-export prices is strictly voluntary, and does not carry with it sanctions or rewards. Moreover, the agreements leave to the discretion of each member government the setting and changing of prices within a range considered reasonable by OPEC members. A close scrutiny of OPEC's statutes and resolutions shows that the Organization does not have supranational powers. Member countries do not delegate to any central body their policy- or decision-making powers. Indeed, they jealously guard their sovereignty, and consider their freedom of action to be paramount.

This relatively low level of authority that member countries provide the OPEC organization has so far not proven a source of weakness. On the contrary, it has enabled the organization to weather dramatic changes in world oil conditions and in its members' policies. It has enabled OPEC to survive periods of political division, of economic rivalry, and of mutations in international relations. Flexibility and responsiveness to changing circumstances have throughout characterized the OPEC organization. OPEC's development as an institution has been slow; its capacity for maintaining itself without the constant support of individual governments is limited, as is its ability to take initiatives.[10] Nevertheless, it has acted over its fifteen-year life in different capacities that have been largely determined by changes in oil-market conditions and in the perceptions of the leading member governments regarding the viability of collective action. In periods of a buyers' market (as in the sixties), OPEC was primarily a clearing house and a relatively weak negotiator vis-à-vis transnational companies. In periods of a sellers' market (as in the seventies), OPEC has been able to exercise greater authority.

The Effectiveness of the Cooperative Bond

One possible description of OPEC is that it is a "structured oligopoly," i.e., an organization where leadership in setting oil-export terms belongs to the few. This leadership might be exercised by a country controlling large productive resources (for example, Iran and Saudi Arabia) or by one that has the "right" perception of the balance of market forces at work, or the ability to act as a market "barometer," or as a catalyst (for example, Algeria or Libya). In any case, the effectiveness of the cooperative bond is a function of the conditions and objectives of the member governments, of the policies and actions of the other protagonists, and of the circumstances.

Diversity—whether economic, political, or cultural—has generally been a source of rivalry and divisiveness, while conditions and objectives have been the basis for solidarity. Among the economic differences that have loomed large in intra-OPEC rivalry have been the wide variations in resource endowment, in requirements, and in expectations. A country with very large oil reserves—such as Saudi Arabia—is more concerned with the economic obsolescence of that resource, should its exploitation be

constrained by price-raising policies as compared with a country of moderate oil reserves, whose concern would be to husband that resource and to produce it at a price
level higher than could be considered optimal by the country with vast oil reserves.

The requirements of the member governments for oil revenues and the urgency of
their acquiring these revenues vary considerably. The populous countries, with relatively low per capita incomes and vast needs for socio-economic reforms, with ambitious developmental programs, and with pressing military expenditures (e.g.,
Algeria, Indonesia, Iran, Iraq, and Nigeria) actively push for larger incomes. This
pressure is concentrated on price increases (rather than volume increases) whenever the
countries concerned perceive that they can exploit their oil reserves in a fixed period of
inelastic demand. Countries whose economies have a limited absorptive capacity (e.g.,
the United Arab Emirates, Kuwait, Libya, and to some extent Saudi Arabia) are reluctant to increase their oil exports in order to hold depreciable currencies until they
can be profitably invested. Individual plans for production control exist in a few OPEC
countries, but production programs covering them all have not yet been devised because of the divergences among OPEC members and the practical problems of implementation.

The price issue has often been a source of controversy among OPEC members, especially since 1973; differences usually center on the "optimal" size and timing of price
changes. Saudi Arabia is known to be relatively moderate in its demands for price increases. It in fact threatened, at an emotional and acrimonious OPEC conference in
June, 1974, to reduce prices on its own oil by two dollars a barrel and to back up this
reduction with an increase in production of three million barrels a day should the other
OPEC members institute a price increase. That posted prices were kept unchanged at
that time in the face of demands for higher prices can be attributed to the great influence of Saudi Arabia based on its vast oil reserves and production.[11]

The role of the national oil enterprises in the OPEC countries represents another
source of disunity. Some of them aspire to become transnational, and are willing to
play according to the rules laid down by such enterprises (this is most notable in the
case of the National Iranian Oil Company). Others want to remain agents of national
governments and are primarily concerned about their domestic role in industrial development (as in the case of the Iraqi National Oil Company).

Economic differences are not the only source of division and rivalry within OPEC;
there are important socio-political differences among the member nations as well. The
OPEC countries belong to a wide array of cultural, religious, linguistic, and racial
groups. Their systems of government vary all the way from parliamentary democracies
(e.g., Venezuela and Kuwait) to conservative, paternalistic, autocratic monarchies
(e.g., Iran and Saudi Arabia), to revolutionary military governments (e.g., Algeria
and Iraq). Differences in regional and East-West alliance structures are also pronounced. OPEC members have at various times broken diplomatic relations with one
another owing to territorial and political disputes (Iraq vs. Kuwait; Iran vs. Libya
and/or Iraq). Their actions, as a consequence, have more often than not been far from
unified.[12] The exception has been the regional Organization of Arab Petroleum Exporting Countries (OAPEC); representative of its unity is the oil cutback its members

imposed on the Western countries both in late 1973 and in early 1974 as a "defense" protest over the Israeli occupation of Arab territories in 1967.

But, despite their important differences, OPEC member governments have come to realize the possibilities of cooperative action. They are aware that unless they remain together on the oil-pricing issue, they may well fail separately, and that their protagonists, the industrial powers and the major companies, will regain the power they held over the oil industry until recently.

While agreements on coordinating oil-export prices have been relatively more effective than collective production programming, at no time has there been uniformity in pricing methods and in allowances for location and quality differentials. Differences in government levies, in fiscal systems, in oil-production costs, in export prices, and in freight rates continued in 1974-75, at times exceeding one dollar per barrel for a given delivery point on similar quality crudes. Moreover, OPEC governments have offered extensive credit terms for up to five years to some developing countries, such as India, a practice which amounts to indirect discounting of prices. In the future, such price discounts might be provoked by a drastic cut in demand and could affect some members more harshly than others. This could place additional strain on OPEC solidarity, especially if the demand cuts hit countries with the most pressing needs, and with sizable reserves capable of sustaining larger production.

The transnational companies already made one attempt, early in 1975, in this direction when they cut production in Abu Dhabi (the leading constituent member of the United Arab Emirates) from 1.65 million barrels a day in July, 1974, to 550,000 barrels a day early in February, 1975. The cutback was a serious one for Abu Dhabi, owing largely to its relatively heavy commitments in domestic projects for socio-economic development and foreign aid. Abu Dhabi reacted by reducing its oil-export prices in the hope of encouraging exports and raising its export revenues. Officials of the United Arab Emirates were convinced that the companies' tactics vis-à-vis Abu Dhabi represented a deliberate attempt to destroy OPEC unity.[13] *The Economist* apparently agreed when it remarked that such tactics could backfire by inducing OPEC countries to "take their revenge for attempts to break their unity by picking on Abu Dhabi."[14]

In order to assist Abu Dhabi, OPEC member countries had two options: one was to institute OPEC-wide production programming; the other was to tolerate a pre-agreed price reduction or "realignment" of Abu Dhabi crude to render it comparatively attractive for additional lifting by oil companies. OPEC members opted, on February 28, 1975, for the latter approach as being the more effective one, since it would not require complex administrative machinery or self-discipline by member countries for its efficient operation. The major reserve countries and large producers—notably Saudi Arabia, Iran, Kuwait, Libya, and the United Arab Emirates—agreed to absorb large reductions in 1974-75, rather than permit a disastrous collapse of the OPEC price structure. By mid-1975, Saudi Arabia was content to allow reductions of oil-export prices to take effect gradually through a decline in the purchasing power of oil paid for with depreciating currencies.

OPEC and the Importing Countries

The oil-exporting countries, thanks to concerted action and to propitious economic and environmental factors, dramatically increased their profits from oil production and trade in 1973-74. Their success effected a world-wide redistribution of wealth in their favor, largely at the expense of the industrialized oil-importing countries, though the developing oil-importing countries were also affected. Estimates place the additional oil bill for 1974 over 1973 for the industrialized Western developed countries at about $60 billion and for developing countries at about a tenth of that figure.

While the whole world was affected by commodity booms—some favorably, others adversely—the least developed countries with limited export potential suffered most, particularly several countries of the African Sahel (Senegal, Mali, Niger, and Chad), Ethiopia, and the Indian subcontinent. Nevertheless, OPEC continued to symbolize for them the emancipation of the developing world from its dependency upon the industrial nations. To these developing countries, OPEC was often seen as a trade union facing the industrial powers as the employers, and it thus offered a model for other resource-rich developing countries to emulate. It inspired governments of copper-exporting countries, which set up the Intergovernmental Council of Copper Exporting Countries (CIPEC), of bauxite-exporting countries, which formed the International Bauxite Association (IBA), of phosphate-exporting countries, and of other primary exporting countries as well.[15]

To keep old friends and win new ones—especially after the surge in oil prices—the OPEC countries made use of their surplus income from oil to offer grants, interest-free or concessional loans, and other forms of aid both to developing countries and to a few Western nations. In addition, development aid was disbursed or committed by several national and multinational funds serving developing countries. Examples are the Kuwait Fund for Arab Economic Development (cap. $3.38 billion), the Saudi Development Fund (cap. $2.8 billion), the Abu Dhabi Fund for Arab Economic Development (cap. $500 million), and the Iraq Fund for External Development (cap. $170 million). Among multinational funds are the Islamic Development Bank (cap. $2.4 billion), the Arab Fund for Economic and Social Development (cap. $340 million), the Arab Bank for Industrial and Agricultural Development in Africa (cap. $275 million), the Arab Fund for African Development (cap. $200 million), and the Arab International Bank for Foreign Trade and Development (cap. $70 million).[16] In addition, the OPEC countries have offered direct assistance to individual oil-importing countries, and indirect assistance through contributions to the World Bank, to the IMF oil facility, and to the United Nations Emergency Fund. According to estimates disclosed by the Organization for Economic Cooperation and Development (OECD) and *The Economist*, the oil-exporting countries were, in 1974, the world's largest donors in terms of aid given as a proportion of gross national product, averaging proportionately five times more than the major industrialized members of OECD. *The Economist* commented that "oil exporters are mainly poor but generous."[17]

The burdens of oil-price increases were unevenly distributed among the oil-importing countries. Among the developed countries that together imported some 90 per cent of OPEC's oil in 1973-74, Japan and Western Europe were more affected than the United States. Japan's oil imports represented about 80 per cent of its primary energy consumption, as compared with about 60 per cent for Western Europe, and about 18 per cent for the United States.

Though less dependent on imported oil than its OECD partners, American leaders insinuated, early in 1974, that as a last recourse they would resort to military action to reduce oil-import prices. President Ford, speaking in September, 1974, to the World Energy Conference in Detroit, declared: "Throughout history, nations have gone to war over natural advantages such as water or food. . . ."[18] This was later repeated by Secretary of State Henry Kissinger in his *Business Week* interview of January 13, 1975. Echoing this American saber rattling, West German Finance Minister Hans Apel remarked that, in the absence of "normal weapons" and "when nations are hopeless, when they don't see any further way out, when they have to fear the destruction of their social wealth or their democratic structures, then everything might happen," including what he termed "military reprisals" against the oil producers.[19]

While OPEC's successes were largely attributed to its price-raising activities of 1973-74, its increases were sufficiently sudden to provoke the creation of a "counter cartel" of buyers, called the International Energy Agency (IEA), late in 1974, set up under the effective leadership of the United States government. The IEA was formed as an autonomous body by fifteen governments of the rich Western countries and was restricted to members of the Organization for Economic Cooperation and Development (OECD), which included all the developed non-Socialist countries. IEA's members accounted in 1974 for 80 per cent of the world's oil imports. The functions of IEA were to attenuate the effects of collective actions by the OPEC countries and to reduce Western dependence upon them. These functions included the building of huge oil stockpiles, sharing in total production and imports in periods of oil supply cutbacks, the conservation of energy, and the joint development of alternative energy sources.

The human, industrial, and financial resources at the command of the IEA countries dwarf those of OPEC. Where there is bargaining or confrontation between the IEA and OPEC, it is only realistic to expect the balance to be in favor of the industrialized countries. The IEA is therefore a potentially powerful cartel, not only because it includes several of the rich industrial countries, but because it represents a bold plan for supranational cooperation. Member governments have agreed to subordinate national sovereignty and oil-company independence to the supranational agency's authority in the event of oil shortages. The IEA treaty provides for an automatic demand restraint and allocation of existing oil resources, should any country or group of countries sustain a reduction in their oil supplies of 7 per cent or more in relation to a designated base period. In other areas of cooperation, IEA's decisions are based on proportional representation, in which the EEC and the United States, for example, have equal votes.

OPEC's Performance

An assessment of the performance of an intergovernmental organization is a difficult exercise. One possible assessment might be on the basis of perceived costs and benefits. Using these as the standard of performance, the member governments would probably consider OPEC a beneficial creation. Expenditures incurred by the governments in connection with financing OPEC have averaged about $1.5 million per annum, a "drop in the bucket" in terms of the additional oil revenues that OPEC countries have received. Revenues for the five founding members (Iran, Iraq, Kuwait, Saudi Arabia, and Venezuela) have jumped from $2.25 billion in 1960 to about $80 billion in 1974. Not that one should attribute this spectacular rise in revenues entirely to the OPEC organization: individual governmental initiatives, the growth in demand, and the shortages in world supplies also had important roles in the surge of oil revenues. Nor would OPEC have been the relatively successful organization it was had it not been for certain characteristics of the world oil industry—namely its oligopolistic structure, the vertical integration of the major companies, the foreign ownership of major oil concessions in developing countries, and the absence of acceptable commercial substitutes for oil in the short and intermediate terms. Nevertheless, the creation of OPEC has provided a starting point for its members to gain the authority to decide their oil-export prices by winning effective control over oil production. Armed with the principle of permanent sovereignty over natural resources enunciated by the United Nations General Assembly and a remarkable show of solidarity, the OPEC countries were able to acquire majority or total control of their oil operations during the early seventies. This assumption of national control could not have been the relatively smooth operation it was had the host governments tried to operate independently.

The absence of readily available commercial substitutes to natural crudes combined with the growth in demand for oil has alerted the member governments to the scope of the opportunities available for exploiting market forces, at least in the short and intermediate terms; they have realized that in theory they can price their oil close to the value of substitutes. Greatly increasing prices is not without risk, however; there are other "costs," which in the long term might include such problems as the effective allocation of production resulting from demand shortfalls, possible injury to key trading partners, stimulating the search for alternative energy sources that would render OPEC oil superfluous, and possible political or military intervention by the United States. On balance, however, at least for the short term, OPEC leaders perceive the advantages to outweigh the disadvantages.

OPEC's achievements should not prevent us from seeing its problems and failures. It has encountered many difficulties in developing additional institutional bonds among members. Specifically, attempts to establish common institutions and projects approved in conference resolutions have often subsequently bogged down in the unwillingness of members to delegate national authority to a central body. Among these projects have been: 1) a high court of justice to arbitrate in oil and related affairs among members; 2) an OPEC-wide production prorationing system; 3) a mutual-aid fund to support member countries should their observance of OPEC resolutions lead them to suffer punitive action by major companies; 4) an OPEC developmental bank to help oil-importing developing countries; and 5) a code of uniform petroleum laws.

Summary

Beginning in the nineteen-sixties, the spectacular growth of oil as a major source of energy in industrial societies after World War II, the emergence of developing countries as major oil exporters, and the rise of national consciousness in these countries had alerted them to the opportunities for collective action. To several of them, the conditions obtaining in the nineteen-seventies represented an opportunity to take advantage of these factors and to achieve higher levels of socio-economic development. They perceived their best strategy to be in the maximizing of oil receipts over a period of time that would be sufficient to permit the speedy implementation of their development programs. To these practical considerations, one should add the psychological one of "emancipation." As eloquently phrased by a former American lawmaker, "United in OPEC, they set out to redress the imbalance between cheap oil and costly imports, and also, in the psychological sense, to redress centuries of colonialism and exploitation."[20]

The leaders of the rich industrial countries were—and remain—generally reluctant to adjust economically and psychologically to this changed balance in the division of wealth accruing from the oil trade. Some even toyed with the idea of military occupation of oil fields in order to restore the status quo. On the basis of the dramatic changes of 1973-74, one should view with some skepticism the prospect that the richer nations will willingly share their wealth and power with the less fortunate members of the international community.

OPEC might well be seen in retrospect to have been a key contributor to the changing of the character of the international system, and as a catalyst in the formation of new political relations both among nations and between nations and major "transnational" enterprises. It has already had its influence both as a protagonist in the international system and as a model to be emulated by other resource-rich underdeveloped countries. The remarkable fact about OPEC is that it proved to be a reasonably effective force for arriving at consensus, narrowing the differences in policies among member governments, and contributing to collective action.

The emergence of the importance of the OPEC national oil companies appears to be a development for the future. While the OPEC organization was dominated until the early seventies by civil servants, whose concern was to negotiate for larger governmental income from foreign concessionaire companies, by the mid-seventies, these companies had moved in to take a greater share in OPEC's decision-making processes with the objectives of seeking profitability and growth, acquiring modern technology and managerial capacity, increasing international markets, and stabilizing business relations with the outside, including the major foreign oil enterprises. By mid-1975, some OPEC national oil companies had already entered into joint ventures with the major oil companies; for example, the National Iranian Oil Company was exploring for oil in the British sector of the North Sea in cooperation with British Petroleum and was linking up with American and European enterprises in their domestic markets. Clearly, their attitude is no longer one of helpless domination by the economic enterprises of the Western world.

REFERENCES

[1]Ahmed Zaki Yamani, Saudi Arabian Minister of Petroleum and Mineral Resources, "Oil and International Politics," address at a meeting of the American Society of Newspaper Editors in Washington, D.C., April 18, 1975, reproduced by *Middle East Economic Survey [MEES] Supplement*, April 25, 1975, p. 3.

[2]For personalities and statistics on the OPEC national oil companies, see the series of articles on Petromin of Saudi Arabia, the National Iranian Oil Company, the Iraq National Oil Company, and Indonesia's Pertamina in *The Petroleum Economist* of London, with the Pertamina article appearing in March, 1975, pp. 99–101.

[3]See "OPEC National Oil Companies to Hold Regular Consultations for Coordination of Pricing and Marketing Policies," *MEES*, August 24, 1975, pp. 1–4.

[4]See "OPEC May Woo Outside Experts to Bolster Vienna Staff," *Petroleum Intelligence Weekly* [PIW], January 6, 1975, p. 5.

[5]Interviewed by author in Caracas on June 6, 1970.

[6]See Libyan Arab Republic Ministry of Petroleum, *Libyan Oil 1951-1971* (Tripoli, 1972), p. 163.

[7]Speech by Ahmed Zaki Yamani, President of the OPEC XXX Conference (Extraordinary) held in Riyadh on October 26, 1972; text published in *OPEC Bulletin*, 2 (November 15, 1972).

[8]Ahmed Zaki Yamani, quoted in *MEES Supplement* October 25, 1974, p. 9.

[9]See "Should Export Cartels Be More Closely Controlled," *OECD Observer*, October-November, 1974, pp. 30-32.

[10]Zuhayr Mikdashi, "Cooperation Among Oil Exporting Countries with Special Reference to Arab Countries: A Political Economy Analysis," *International Organization*, 28:1 (Winter, 1974), p. 26.

[11]See, for example, "Saudi Arabia Versus the Rest at Quito OPEC Conference," *MEES*, June 20, 1974, pp. 1–5.

[12]See Zuhayr Mikdashi, *The Community of Oil Exporting Countries: A Study in Governmental Cooperation* (Ithaca, New York, 1972).

[13]See *MEES*, March 1, 1975, p. 1; and *Newsweek*, March 10, 1975, p. 21.

[14]"A squeeze on Abu Dhabi today could be a squeeze on Britain tomorrow," *The Economist*, February 8, 1975, p. 69.

[15]The author analyzes these organizations in a forthcoming book, *Business and Government in Resource Industries: A Multinational Perspective*, to be published under the auspices of Harvard University Center for International Affairs.

[16]See "Large Increase During 1974 in Arab Development Funds," *Bulletin of the American-Arab Association*, published by the American-Arab Association for Commerce and Industry (New York), November, 1974, pp. 1–2. See also "Venezuela Offers Cheap Oil As Aid to Central Americans—Coffee Price Stability Also Aim," *International Herald Tribune*, December 17, 1974, p. 4.

[17]"Petro-Aid Takes Off," *The Economist*, February 15, 1975, pp. 72–73.

[18]*Business Week*, September 28, 1975, p. 116.

[19]"Oil War Not Barred by Bonn Aide," *International Herald Tribune*, October 24, 1975, p. 1.

[20]Senator J. W. Fulbright, "The Clear and Present Danger," speech given on November 2, 1974, at Westminister College, Fulton, Missouri on the occasion of the centennial anniversary of the birth of Winston Churchill; printed in *MEES*, November 15, 1974 (emphasis added).

ULF LANTZKE

The OECD and Its International Energy Agency

As THE PROBLEMS associated with energy continued to build up in the early nineteen-seventies, the industrialized countries turned repeatedly to the question of developing a set of common policies. There were in the industrialized world two organizations where efforts were made to coordinate energy policies: one was the European Economic Community (EEC), the other the Organization for Economic Cooperation and Development (OECD). Though the problems seemed pressing enough, neither of these organizations had managed to make significant progress by 1973. The necessity for greater coordination was recognized; the actual achievements were few.

So far as the EEC was concerned, the lack of success may be attributed to the fact that the individual member countries relied on quite different structures and very different strategies to secure their energy supplies—differences that expressed the unique historical developments of the various countries involved. Thus, for example, in Germany and the United Kingdom, tradition favored a more domestically oriented energy policy, while remaining countries in the Community tended to be more import-oriented in their policies. In addition, the economic policies applied in the various countries of Europe were very different, ranging from the fairly tight controls characteristic of France (based on a 1928 law) to the more obviously market-oriented energy policies of the Federal Republic of Germany. In the efforts made to harmonize these quite distinctive national policies, the basic differences could never be overcome.

In the broader framework of the OECD, national differences produced even more difficult problems. Although Japan and Europe were shifting from indigenous sources of energy to international supplies in the early fifties and sixties—depending increasingly on Middle Eastern and African oil—they continued to regard energy as being principally a question of domestic policy. At first their industries had adapted to the change by building up international systems of supply, initially using the multinational oil companies as their principal vehicles, and later including the independent companies to achieve a secure supply. Those countries that were not host to one of the multinational oil companies tried to build up their own national companies; ENI in Italy, Elf-Erap in France, and Deminex in Germany are examples of this tendency. But the various national authorities were never convinced that these major structural changes required that an equivalent international effort be made to harmonize energy policies within the Western world itself.

217

This situation remained unchanged even after the clear indications of the late sixties and early seventies that the world oil industry was shifting from a buyers' to a sellers' market, with the consequent fragility and uncertainty this shift implied for the security of supply and for the maintenance of favorable price conditions. By 1968, the United States could no longer be regarded as a secure supplier for Europe and Japan, since it was consuming domestically all that it produced. Indeed, the United States was becoming a competitor for the oil supplies that were coming from the Middle East and Africa.

In 1971, new price agreements were negotiated in Teheran and Tripoli. Their terms clearly indicated that the position of the producing countries vis-à-vis the oil companies had altered; the sellers' position had been strengthened to the point that the individual producing countries no longer had to compete among themselves for a larger share of the market. The negotiations on questions of ownership and control continued in the years that followed, and these also indicated clearly that the bargaining position of the producing countries was being strengthened as the market structure rapidly changed in their favor.

The industrialized countries were relatively slow to react to these changes. Given the very complicated structure of the market, analysis took a great deal of time, and transformation of analysis into policies was even more time-consuming. Thus, changes in policies came about only gradually and, again, always on a national basis. The efforts of Japan and Germany to strengthen the security of their oil supply in the Middle East and Africa by involving their own national companies directly are examples of what went on during this period. Attempts at the harmonization of energy policies on an international level had not gone very far by October, 1973.

The Eve of the Oil Crisis

The machinery available in the OECD for cooperation on energy policies was relatively weak when the crisis broke in October, 1973. In organizational terms, there were two committees available to deal with the problem—the oil committee and the energy committee. The work of the former centered mainly on the exchange of views among governments on major developments in the oil market; the latter was principally a forum for exchanging information on energy questions more generally, including the periodic review of national energy policies. The object of such exchanges, however, was not the achieving of harmony; indeed, the oil committee never really got down to influencing oil developments and policies, though the need for common action was discussed in the committee prior to the crisis.

There were, however, numerous activities in the OECD that, while not having great influence at the time, had some potential importance for the future. The European member countries of the OECD, for example, had agreed to a system of sharing in the event of an emergency. However, this scheme for the equitable sharing of oil within Europe could only be put into operation if there was unanimous agreement among the participating members. Should this agreement be obtained, formal oil sharing could be organized, and an international industry advisory body could be activated to coordinate government and industry action for the European market.

In addition, the OECD had recommended that the European member countries maintain a ninety-day emergency supply. Supplies had accordingly been built up over the years, and these stockpiles did help somewhat in the management of the crisis during the winter of 1973-74. Although the ninety-day target had not been met by most of the member countries, stocks had, on the average, reached a level of well over seventy days by the time of the crisis, and this emergency buffer did allow Europe to survive for a considerable period without the risk of economic collapse. The oil committee had initiated a similar emergency system with recommendations for the building up of stocks by the non-European members of the OECD well before the crisis broke; negotiations on these recommendations had not been completed by October, 1973.

In 1972, the secretary-general of the OECD had also undertaken an assessment of the long-term development of the energy market. This might have served as the basis for energy-policy changes for OECD countries had it been completed when the crisis came. As it turned out, it was not available until October, 1974, but consequently it does provide the first basic analysis of probable developments in the energy and oil markets since the quadrupling of oil prices in 1973.

During the Crisis

After the onset of the crisis, activities within the OECD increased markedly. The Middle Eastern war broke out on October 6; within a week, Iraq had nationalized American interests in Basra Petroleum's southern Iraqi production, and three of the eastern Mediterranean pipeline terminals had been virtually shut down. These events were followed on October 17 with the announcement by ten Arab states of a progressive step-by-step production cutback and of embargoes to be applied immediately against the United States, the Netherlands, and Denmark.

The oil committee and its high-level group were responsible for cooperative action in the OECD. Both these groups met for the first time on October 25-26, 1973, in November and December, 1973, and again in March, 1974. The committee concentrated principally on trying to assess the rapidly changing situation (this was accomplished mainly by informal consultations between the chairman and the oil companies); there was also an exchange of information on possible supply shortfalls involving individual countries, and on the demand restraint measures that were being implemented by the member countries. There was a continuous review of the stock situation within the OECD area. In addition, certain of the international aspects of the crisis, which could not be dealt with on a national level, were discussed: the problems of marine and aviation bunkers, for example, were examined.

It soon became clear that a reliable assessment of the world-market situation was not at all easy to obtain; information available to the governments of the industrialized countries, as well as to the oil companies, concerning events in the oil-producing countries was incomplete, contradictory, and confusing. Nevertheless, the exchange of views on actual and expected supplies available to individual member countries provided reasonably accurate information at any given time, though it could not precisely predict future developments. The assumptions regarding the supply situation were thus constantly being revised.

No attempt was made by the member countries to coordinate restraints on demand; national policies followed distinctly different lines, and measures varied from systems of direct allocation to arrangements for indirect cooperation between governments and the oil industry. However, the exchange of information at least permitted the member countries to draw on the experience of others in the development of their national policies.

Had the European sharing scheme proposed by the OECD been put into effect, it could have guaranteed that all countries would have shared the burden equally, thereby eliminating the consequences of the embargoes imposed by the Arab oil exporters on certain European countries. The scheme could not be activated, however, in part because trust in the flexibility of the oil market was too strong to achieve the unanimous agreement that was necessary. Other considerations, such as a concern lest the oil exporters be offended, may have added to the reluctance, but lack of information must be regarded as the really decisive element.

In sum, the industrialized countries did not recognize that world-wide economic interdependence would guarantee that isolated embargoes directed against certain countries could have damaging consequences for all of them. There was a tendency on the part of those countries given preferential status by the Arabs to capitalize on the advantages that came with that treatment. As a result, all OECD countries were obliged to fall back on uncoordinated national emergency measures that followed different patterns, involving the use of quite different measures and resulting in additional distortions.

Still, it is possible to detect certain positive elements in this otherwise dismal picture. First, the exchange of information among the industrialized countries was relatively effective, and to some extent it compensated for the lack of multinational cooperation. Most countries tried to take into account the importance of international economic interdependence in their individual policies and to maintain their exports of oil products, using restraints only in proportion to the restrictions on their own energy consumption. The availability of information provided the national governments with a uniform picture of the rapidly changing situation in the oil market on which they could rely. Second, the international oil companies proved to be reasonably flexible in their policies. They managed to adopt supply patterns of a kind that allowed for more or less equal distribution, thus avoiding the deterioration of individual economies that would have followed a breakdown of oil supply.

All told, however, cooperation among the industrialized countries was in no way comparable to the unity of the oil-producing countries. Agreement among the latter was exceptionally well maintained with respect to prices, and relatively well maintained with respect to production. It was only the general capacity of the economic system of the Western world to adjust to these imbalances that prevented catastrophe for the system itself, as well as for individual economies.

The Basic Interest in Energy Cooperation

The energy crisis underlined the interdependence of the Western economies, and it demonstrated—and is today still demonstrating—the extent to which the smooth functioning of the world economy depends upon regular energy supplies. It also made clear that the energy problem could not be solved by national policies alone, and that energy problems were indeed closely linked to policies of a more broadly political nature.

Europe's basic position is underlined by the fact that its energy consumption in 1974 fell just short of the equivalent of 1,200 million tons[1] of oil, or about two-thirds of the United States figure. Out of this 1,200 million tons, oil itself accounted for 720 million tons, some 700 million tons of which had to be imported. In absolute terms, these imports are higher than the total imports of the United States and Japan combined. While Europe is less dependent on imports than Japan, it is more dependent than the United States. The prospects for altering this situation are, on the whole, good, if a determined policy is followed; the present 65 per cent import dependence could certainly be reduced to 40 per cent. In order to do this, however, quite a number of constraints would have to be overcome, beginning with obstacles involving environmental controls and ending with the need to achieve greater harmony among the various national energy policies of the European countries. Because the European energy-supply situation is closely bound to the international energy-supply-and-demand situation, decisions taken in this field in the United States and Japan will have a direct bearing on supply conditions in Europe. These are obviously compelling reasons for European countries to seek international cooperation with at least these two other major consumers.

Moreover, because European economies are so interrelated, not only among themselves, but also with the rest of the world, there is a strong need for Europe to maintain an international division of labor based on the avoidance of trade restrictions between individual countries. It is hardly conceivable, however, that trade relations can be maintained undisturbed if, at the same time, conditions of energy supply do not follow analogous patterns. European interests are clearly served by strengthening international cooperation.

Even more generally, the situation in the Middle East, because of both geographical proximity and economic and historical links, affects Europe as a whole. This situation has been reflected in a series of European initiatives, including the European/Arab dialogue, the Common Market Mediterranean policy, and the EEC policy of trade preferences extended to Mediterranean and Middle Eastern countries.

Though cooperation in the energy field is no doubt desirable, there is still some question whether Europe is prepared to coordinate its various national policies and to integrate them into a more internationally cooperative system. The difficulties are many. Europe is itself divided into energy-rich and energy-poor countries. The fact that the energy-rich countries—once the United Kingdom, Germany, Belgium, and France—now include Norway and the Netherlands but no longer include Germany, Belgium, and France will not make things easier. Difficulties arise also from the different approaches to energy policy, ranging from those that depend on publicly administered markets to others that are comparatively unregulated. Nor is the general political situation in Europe particularly encouraging to those who look for greater integration. In Germany's revision of its national energy program, accomplished in October, 1974, that country strongly endorsed international cooperation. On the other hand, energy production policies in countries such as Britain and Norway, which now have new production possibilities, tend toward a more national character. While this is entirely comprehensible, it certainly does not improve conditions for a common European approach.

Japan, one of the major consuming areas, is the most dependent on the smooth functioning of the world oil markets. In Japan, energy consumption of all kinds in 1974 totaled the equivalent of about 350 million tons of oil, oil itself accounting for 260 million tons of this amount. All of it had to be imported. The prospects for changing this pattern of supply are very slight, the only real possibility, for the moment, being nuclear power—an energy source that generates deep emotional reactions in Japan, as it does elsewhere. Japan, given its dependence on the smooth functioning of the world oil market, can choose between one of two strategies. Either it can seek special arrangements with the oil-producing countries, a policy that it pursued to some extent up to the beginning of the crisis, where, through the use of substantial resources, it sought to secure a foothold in the Middle East, or it can acknowledge that for the longer term its security of supply can be guaranteed only if new forms of energy sources are developed that provide a broader energy-supply pattern than it presently has. This longer-term aspect of supply security calls for close cooperation with the rest of the industrialized world. Japan's energy policies will have to provide for the stabilization of oil supplies in the short and medium term, while accepting the need to develop greater diversification for the longer term through new technologies.

This duality of needs becomes more apparent if we look at Japan's general economic interests. Japan owes its economic recovery largely to its greatly expanded exports. It is, therefore, especially dependent on a stable division of labor and on the maintenance of free conditions of trade. A sustained period of tension in the energy market would almost certainly lead to restrictive trade practices, which would have negative effects on the Japanese economy. International energy cooperation is virtually imperative for the solution of Japan's difficulties.

As for the United States, it consumed energy equivalent to about 1,750 million tons of oil in 1974, five times Japan's consumption, and about one and a half times the consumption of all of Western Europe. About half of this (800 million tons) came from oil, and 300 million tons of this oil was imported. If no resolute government measures are taken, the growth in energy consumption will have to be met largely by oil imports. Conversely, a positive and determined policy on the opening up of higher-cost indigenous sources could provide a comparatively high degree of independence from the world market. Even if the original objectives of "Project Independence" were too ambitious, there is no doubt that the American potential in unexploited indigenous resources is considerable. In view of the sheer size of the American market, a policy of unlimited imports would have a decisive impact on European and Japanese supply problems, since it would destroy any hope of restoring some sort of equilibrium to the world market in crude oil.

The implementation of a strong plan to develop domestic energy sources is, however, meeting with considerable difficulty. There are environmental problems in virtually every conceivable area: nuclear power, oil exploration on the American continental shelf, the development of shale oil, the substantial increase in coal production. Still, the United States remains the country best equipped to solve its energy problems with its own resources. Because the American economy is generally somewhat less dependent on world trade, given the size of its domestic market, it has somewhat greater flexibility. However, even the United States would be subject to severe economic depression if the world economy were to take a marked turn for the worse.

The United States, as one of the world's leading powers, must—and does—acknowledge that energy stands at the very center of major new developments in relations between the industrial countries and the developing world. American policies are bound to contribute more than those of others to stable and equitable conditions in the energy market. This responsibility may at times be uncomfortable for the United States and even not welcomed by its partners in the world community, but it cannot be ignored.

The United States must—and does—recognize the links between international economic problems and the larger issues of foreign policy, which involve the economic and political stability of so many states. Although the direct need for the United States to participate in such large-scale cooperation may not be imperative, the indirect consequences of its failure to deal with the energy problem will imperil its alliances all over the world.

Apart from the specific interests of the different regions, one general element favoring extended energy cooperation should have become evident through the repercussions of the crisis common to all the industrialized countries—namely, the indirect consequences of unstable energy supplies on their economies. How far these consequences reach has been clearly demonstrated. The impact of the crisis has been seen in the slowdown of economic growth, the acceleration of inflation, new disequilibria in balance-of-payments relations, difficulties in certain industrial sectors, and substantial shifts in real resources from the industrialized to the oil-producing countries. A number of these problems existed prior to the crisis, but the quadrupling of oil prices aggravated them substantially. Efforts undertaken before the crisis that might have been successful were quickly overtaken by events. The direct relation between stability of energy-supply conditions and the more general problems of economic policy has been amply demonstrated.

The Scope and Objectives of the International Energy Program

Although the industrialized countries were inadequately prepared to cope with the crisis in a coordinated manner, the basic similarity of their interests was always apparent. The initiatives taken to reconcile the differing policies of the Western world seem, in retrospect, entirely natural. The initial step, taken by the United States government in a conference convened on February 13, 1974, saw the beginning of a period of intensive negotiation among the industrialized countries; the objective was to establish a new foundation for energy cooperation within the Western world. These negotiations resulted in an explicit and detailed agreement on an International Energy Program, which was signed by sixteen member countries of the OECD on November 18, 1974. The scope of the agreement is ambitious; it covers every major energy-policy issue. It is all the more ambitious because the cooperation intended under the agreement has had to be built from fairly insubstantial bases, and because it must confront the fact that energy policies in member countries are now divergent.

The objectives of the program, centered on more stable and equitable supply conditions in the world, seek to provide assurances against the disruptive effects of oil emergencies; these involve a greater independence from imported oil, the broadening of cooperative relations among the participating countries and between other producing and consuming countries, and the establishment of an international system to make available more information and greater predictability regarding oil-market developments.

The most developed part of the program relates to the precautions to be taken against the possibility of new emergencies. The basic principles of the system may be summarized as follows: Each country will have to build up its own emergency self-sufficiency in oil supplies to cover an initial period of sixty days, thereby securing, in case of crisis, an inadequate buffer to allow for cooperative political action. In the event of crisis—a term specifically defined as a reduction in the daily rate of oil supplies of the IEA countries as a whole, of a single member, or of a few—a sharing system is activated. The system aims at equitable burden-sharing among the participating countries, including a common reduction of demand on the basis of explicitly identified measures. Finally, the emergency system provides for political consultations that will enable participating countries to undertake joint action to meet the requirements of the crisis.

The longer-term objective of reducing dependence on imported oil has to be tackled by influencing the overall demand and supply situation through deliberate conservation programs, through enlarging available energy resources, and through developing new forms of energy. This section of the program is probably, for the long-term future, the most important; it is also the most difficult. Consuming countries are not accustomed to deliberate policies of energy conservation, and consumers, so long as energy costs were comparatively low, were not in the habit of giving much attention to this subject. The development of new energy resources is expensive, technically complicated, and potentially disturbing to the competitive position of individual economies. As for research and development on new types of energy, these have concentrated in the past on nuclear energy to the almost total neglect of other resources.

It is understandable that, given the complexity of the negotiations, the brief period between February and November, 1974, did not suffice to develop very detailed targets and explicit modes of cooperation. It became necessary to rely on agreements in principle and on procedural arrangements for work programs that would subsequently be developed. In any case, the justification for this part of the program is independent of the current situation in the oil market and of relations between consumer and producer countries. Given the enormous periods of time that must elapse between results in research and development and their large-scale economic applicability, the benefits of energy sources now being developed can only be expected to accrue to future generations.

Greater understanding of what is going on in the oil market is another issue of major political importance. Until the crisis, there were substantial differences in the amounts of information available to individual countries. The host countries of international companies tended to have better access to information, though public antagonism to the activities of the companies had already begun to build up in these countries even then. Because everyone recognizes that the companies are highly efficient managerial instruments for dealing with the problems of vastness and complexity involved in supplying the world with oil, there is considerable respect for their ability. At the same time, a certain amount of ambivalence exists in both governmental and other circles about the role of the multinational oil companies; further, public criticism of these institutions, insofar as it is based on emotion rather than fact, can lead to a loss in their efficiency. It is in the interest of both the governments and the oil companies to resolve these conflicts before they endanger the smooth functioning of the world oil market. The improvement in general information envisaged in the International Energy Program should help to direct future discussions of these problems into clearer and more objective channels.

The improvement of relations between producing and consuming countries is of vital importance to both sides. For a considerable time to come, the consuming countries will be dependent on the producers for securing their oil on reasonable terms; they will have a natural interest in a better integration of the world economy. Producing countries are similarly dependent for their well-being on good relations with the industrialized world. While, for the immediate future, they remain the major outlets for oil, the industrial countries still provide both the means and the opportunities to help them broaden and diversify their existing economic structures. Both sides stand to gain through an amicable development of their relations; both stand to lose by a continuation of adversary or restrictive policies.

Possibilities of Success

The International Energy Agency of the OECD, created by a decision of the council of the OECD on November 15, 1974, is the instrument for the realization of this ambitious energy program. As these words are being written, the Agency is just six months old—far too short a time to judge its effectiveness. The work already accomplished by the member countries is, however, encouraging, and the institutional and operational organization of the Agency favors substantial accomplishment. New elements in the structure of the Agency distinguish it markedly from other international organizations of a similar kind. In contrast, for example, to the EEC, the Agency is not territorially limited, and it can therefore provide for international energy cooperation on a world-wide basis. Only France, among the major industrial countries, has chosen not to join the IEA. In contrast to its parent organization, the OECD, the Agency is not bound to the principle of unanimity. In contrast to the economic bodies of the United Nations, as well as the United Nations itself, the interests are not so varied as to impede concrete progress on the specific issue of energy.

Accordingly, the present combination of institutional features and circumstances encourages achievement: broad geographical scope, narrow functional objectives, abandonment of the rule of unanimity, and avoidance of certain of the old adversary relationships that have restricted the ability of industrialized countries in the past to work together effectively. Moreover, in the emergency scheme, the oil industry continues to be responsible for the system's practical implementation and thus has parallel interests. These factors converge to create confidence that a new oil crisis would be far better managed by the governments than was the case last time.

In the field of long-term cooperation—conservation, development of alternative energy resources, and research—the initial work of the Agency has been equally encouraging. Policies have already been developed that reach beyond the consensus contained in the agreement itself. Concrete conservation targets have been set for 1975, and basic commitments have been undertaken by member governments on targets for the future. Consensus has been reached on a system favoring the development of alternative energy sources, and energy research projects have been developed to a point where their practical implementation is only a question of time. Moreover, a general information system on the oil market is by now well established; the first round of data has been supplied, and its analysis and use is now possible.

The most difficult area so far has been the improvement of relations between the oil-exporting and the oil-importing countries. A preparatory meeting in Paris, from April 7 to 16, 1975, between representatives of the two groups was not able to reconcile their different views concerning the scope of a conference planned for later in the year. The industrialized countries felt that the agenda ought to be restricted, with energy and energy-related questions as the focal point; the producing and developing countries wanted a more general agenda that would include most of the economic issues now being disputed by the developing world and the OECD countries. By the summer of 1975, some bridging of this gap seemed possible, as the United States government showed a greater willingness to broaden the scope of the proposed talks.

But even if this meeting does not produce immediate results, it seems obvious that direct contacts and better relations remain the principal objectives of the Agency. Indeed, the importance of oil to both sides emphasized the primacy of those objectives. In 1974, approximately 1,100 million tons of oil, valued at $90,000 million, were exported from the OPEC countries into the OECD area. This amount represented nearly two-thirds of the oil supplies of the consuming countries and four-fifths of the production of the OPEC countries. Given these conditions, it is difficult to imagine how the parties can fail sooner or later to settle the problems that separate them.

Finally, a more general point has to be mentioned, if one is to take accurate stock of OECD's and IEA's present energy activities. The secretary-general of the OECD has provided in his report, "Energy Prospects to 1985," a valid assessment of future energy developments and the needs of energy policy. This study must be regarded as only a starting point, since the abrupt and far-reaching changes in the supply conditions that developed in the early nineteen-seventies are without precedent. If it is to serve as an instrument for the future development of energy policy, its assessment will have to be regularly updated.

Comparing the situation between the OECD countries—and even within the EEC—prior to October, 1973, with the situation that now exists, one can only conclude that the radical changes in the oil market have led to equally radical changes in the willingness of industrial countries to cooperate on energy. The emergency system has undergone substantial improvement; the basis for long-term cooperation has been established; the idea that oil consumers have common interests has made very considerable progress. One must realize, however, that the real tasks still lie ahead, and that the IEA will be successful only if member governments remain determined to achieve larger results.

At present, there is some danger that public awareness of the problem may diminish. Such a development would be disastrous, because in five or six years the results of a "business as usual" attitude would be keenly felt. The present excess of oil production should not mislead anyone into assuming that the situation has not changed in basic ways. Most of the potential elements of crisis remain; in the longer term, the additional elements of a more general energy-supply crisis are becoming apparent. It is, therefore, of vital importance that the underlying roots of the difficulties in energy supply—short term and long term—be kept constantly in mind. If the energy problem is not to constitute a source of permanent unrest in international relations, the different regions of the world will have to undertake a continuous effort at cooperation in solving the world's energy problems.

REFERENCE

[1] Figures in this paper are expressed in millions of tons in order to give the reader an easier grasp of the order of magnitude involved. To convert tons per year into a rough estimate of barrels per day, divide the tons-per-year figure by 50.

Part 6: SYNTHESIS

KLAUS KNORR

The Limits of Economic and Military Power

THE PRESTIGIOUS *Strategic Survey*, published annually by the International Institute for Strategic Studies in London, begins its analysis for 1973 with the statement: "The successful use of the oil weapon by the Arab states in connection with the Middle East war of October produced the greatest shock, the most potent sense of a new era, of any event of recent years." This event initiated and portended a vast redistribution of wealth and income internationally—a subject treated in other parts of this volume. It also expressed shifts in the international distribution of power and influence, and related modifications in the perceptions and policies of key international states. Some of these changes had been in the making for some time, but they were fully recognized only in the glare of crisis confrontation. To probe these changes is the purpose of this essay.

I

It is not clear, in retrospect, why the use of the oil weapon—that is, the use of monopoly economic power for coercive political purposes in matters of high policy—should have come as so much of a surprise. Rapidly rising oil prices had·already indicated the presence of a growing monopoly power, and the employment of the oil weapon in support of the Arab cause against Israel had been referred to repeatedly by Arab spokesmen for many months before October, 1973. But because some Arab countries had been unsuccessful in enforcing an embargo on oil shipments during the Arab-Israeli war of 1967 and because the extent of Arab unity was believed to be insufficient to impose a successful embargo in the future, the Arab warnings were disregarded. The failure to foresee the Yom Kippur War also greatly contributed to the surprise. Its precipitation by Egypt and Syria disturbed a Middle Eastern status quo that had been frozen into immobility since the 1967 war; it fused and strengthened Arab unity and prepared the stage for invoking the oil sanction. Since the sanction was not invoked until October 17, when Israeli armies had gained a bridgehead on the west bank of the Suez Canal, we can infer that the chief Arab leaders did not think the oil weapon would be sufficiently effective without the solvent of war, and that war was perhaps even a precondition for the degree of Arab cohesion that would permit the OAPEC action.

229

Whether the use of the oil weapon in 1973 was effective in achieving all the Arab objectives is not an easy question to answer. But it certainly achieved one of them—namely, to call the attention of the industrial countries to the shift in international economic power that had taken place and to consequent Arab expectations that this shift should lead to a more cooperative attitude toward Arab goals in its conflict with Israel. Employment of the oil weapon combined with the Yom Kippur War made clear to all the new urgency of the Arab cause and the increase in Arab resources and determination.

The surprise of the embargo and the sudden recognition by the industrial nations of their economic vulnerability induced a state of panic in these countries, at least at first. Of course, they had already supported the United Nations Resolution 242, which called on Israel to give up the territories conquered in 1967, but such declarations were mostly *pro forma*. Furthermore, most of these states had virtually no influence on the settlement of the Arab-Israel conflict, and the effectiveness of the oil embargo in these cases could produce for the Arabs little more than a further step in the diplomatic isolation of Israel. Still, panic was a major factor in the refusal of the West European countries (with the exception of Portugal) to cooperate with the United States by facilitating the flow of American arms to embattled Israel. And the cutting edge of the oil weapon was evidently sharp, as the Western European nations and Japan quickly responded to Arab demands for declarations in support of the Arab cause.

Whether the United States was coerced by the oil embargo is more doubtful. Not yet vitally dependent on Arab oil, the United States could be discommoded but not coerced. To be sure, it did its utmost to bring about a negotiated ceasefire, but neither President Nixon nor Secretary Kissinger ever attributed this eagerness to Arab pressure. It is likely that they acted as they did for other sufficient reasons. The material costs of supplying Israel with armies and the risk, however slight, of a confrontation with the Soviet Union were powerful enough incentives by themselves for working toward a ceasefire and an Arab-Israeli settlement. It is possible that the oil embargo added to these incentives, but, if it did, the effect was scarcely more than marginal. When the oil weapon was sheathed, the objectives that the Arabs had announced at the beginning of the embargo were still far from being achieved.

If the world was shocked into a sense that the autumn of 1973 was the beginning of a new era, it was primarily because of the vulnerability revealed by the powerful and rich industrial countries when challenged by a group of supposedly weak, less-developed states, and the meekness with which they had bowed to the challenge. That the powerful victims showed so little indignation was especially startling when one considers that oil had been used as a weapon against them even though most of them had no control whatever over Israel's actions.

The successful coercion of Japan and the large industrial states in Europe also provoked shock because of a faulty conception of economic power that prevailed in the West, both in government and academia. It had become fashionable to argue that a recent decline in the utility of military power, the diminishing importance of high policy, and the rising salience of economic issues were making national economic power increasingly important to the settlement of international disputes. In this view, the bipolar structuring of inter-state politics around the two superpowers had given way to a "pentagon of power" in which, in addition to the great military powers and China, both Japan and Western Europe would be members on the basis not of their military strength but of their economic resources.[1] The jarring events of 1973 suggested that, though economic power properly defined might still have new importance, the vaunted economic power of Japan and Western Europe stood on feet of clay.

It is undeniable, of course, that these nations account for a very large proportion of world trade and a considerable proportion of the world's wealth. But to equate national wealth with economic power is just as fallacious as to equate it with military power. Wealth can be an ingredient of either, but only an ingredient. Power also rests on the structure or composition of resources and on the crucial ability and will to direct their flow, and it was precisely this kind of power that the Arab oil states possessed. They were able to withhold from Japan and Western Europe something of absolutely indispensable value to their economies.

What, then, is the quintessential basis of power? If two actors, A and B, are dependent upon one another for the enjoyment of peace, or trade, or whatever they value, but the degree of dependence is not equal, the power of the less dependent actor A is measured by the degree to which he can tolerate a rupture of the relationship more easily than can B. A's power is absolute if the continuation of the relationship is indispensable to B, but not to A. Such a situation exemplifies what is known in economics as the law of reciprocal demand, which fixes the terms of trade in exchange relationships. It explains the basis of economic power that OAPEC wielded over Japan and Western Europe. Progressively decreasing oil imports would have paralyzed their economic life. Below a small margin, oil imports were felt to be indispensable. These countries could not threaten the oil-exporting states with any economic reprisals of even nearly equivalent power. Short of a willingness to suffer enormous hardship, they would have been helpless in the face of a protracted and severe embargo on oil shipments. In contrast, the United States would not.

However, although the economic power of the OAPEC countries was effective in 1973, this does not mean, as has frequently been argued, that economic power is becoming increasingly important in shaping international relations, or that it is becoming a substitute for military power. Even if the utility of military force has declined somewhat in the resolution of international conflicts, there is no logical reason why a substitute form of power must necessarily come forward. It is possible, and indeed arguable, that nations are less easily coerced than in the past, whatever the form of threatening power. Moreover, it is doubtful that a notable increase in the international use of economic power has in fact taken place, although one can understand how the impression might have arisen. Western Europe and Japan are now much less given to the use of military power than formerly; and the vast expansion of aid to an increasing number of poor countries in recent decades has no doubt enlarged the economic factor in political relations.

To identify clear global trends is not, however, so simple; such trends do not lend themselves to statistical measures, and one's view of them also depends upon how one defines economic power to begin with. If we mean something more than ordinary bargaining strength in trade or other economic negotiations, that is, if we mean the exercise of economic coercion on other than ordinary economic issues (what European diplomats in the nineteenth century called *haute finance*), then it is difficult to believe that economic power is now finding appreciably more use than it did, say, in the thirties or around 1950. The vast expansion of economic aid has only occasionally led to power plays, although it is a typical instance of asymmetrical interdependence and is potentially coercive. Donor influence has no doubt been frequent and often successful, but only so long as it takes place in private (as between officials) and does not affect the basic interests of the recipient country. The cases in which the threat of suspending aid was used dramatically in order to coerce a recalcitrant recipient have not been very numerous, and usually have in the end failed to deter or compel.

The question remains whether economic power of the magnitude represented by the 1973 oil crisis can readily recur, and, more generally, whether or not the elements of economic power can be mustered with sufficient ease to make the economic weapon an attractive form of coercion. My research suggests that the answer is negative.[2] Cases of success are very rare. For example, in 1948 and again in 1957, the Soviet Union cut off both trade with and aid to Yugoslavia, because Belgrade's policies were insufficiently responsive to Moscow's demands. Although Yugoslavia's economic dependence on the Soviet Union had been very great, especially in 1948, and although the economic rupture certainly hurt, the Yugoslav government was nevertheless able to defy its powerful opponent, because other, particularly Western, countries were willing to provide alternative trade and economic aid. Again, in 1960, the United States placed an embargo on trade with Cuba (and induced some members of the OAS and NATO to do likewise for a time); this time, it was the Soviet Union and other Communist states who came to the rescue in trade and economic aid. These two cases are typical of the outcome of attempts at coercing a state through the use of economic reprisals.

The monopoly power of a state that attempts to exploit a relationship of asymmetrical economic interdependence is usually insufficient to achieve coercion. Exports and imports are seldom specific to a particular market or source of supply. Competing countries are ready to function as substitutes. There may be short-term problems, as commodity and service flows are diverted from the most economical channels, but the price of defiance is readily borne. In addition, because yielding to international economic coercion in dramatic conflicts of will usually seems to be regarded by the victim as less acceptable than yielding to military might, the resisting governments can usually count on strong domestic political support. From the point of view of other countries, it is also riskier to aid a country subject to military threat or attack than it is to increase trade with a state subject to economic pressure. The rivalry of great powers and considerations of the international balance of power reduce the threat to the victim country even more. For example, the keen Soviet-American rivalry during the period of the Cold War greatly diminished the political leverage either state could derive from extending economic aid. The pressured country could always threaten to change its patron, and not infrequently did play off one donor against the other.

It is clear, then, that only unusual combinations of circumstances can make economic power succeed when applied to international politics. For instance, the American trade embargo against Great Britain in 1811-12 succeeded only because Napoleon's Continental blockade had already severed British trade with most of Europe and because by chance inclement weather had caused one of the worst crop failures in English annals. The United States' halting of the Anglo-French invasion of Suez in 1956 is also sometimes attributed to economic pressure, in this case, American refusal to bolster Britain's weak currency. But—as is usually the case—several other factors were at work as well. The British public was divided, the execution of the military move was slow, the Soviet government uttered vague nuclear threats, and the very fact that Britain's foremost ally, the United States, disapproved exerted strong psychological pressure. Yet, to the extent that financial leverage did stop this ill-fated enterprise, it was because, in 1956, there was no other country that could have lent effective support to a slumping pound sterling.

To what degree one state can coerce another will, then, always depend upon a rare combination of international and domestic circumstances. The limited effectiveness of the Arab oil weapon in 1973 was no exception. There is at present no single commodity as indispensable as petroleum, especially for the industrialized countries. A high inelasticity of supply and demand, an increasing sellers' market since 1970, and a limited number of key exporting countries that managed to agree on the use of the oil weapon combined to make coercion possible. Its employment is probably not the harbinger of big economic power plays based on other commodities in short supply. If countries exporting bauxite, copper, and other materials can join effectively to establish a high degree of monopoly power, particularly while importing industrial countries are prospering, they may manage to extort higher prices for a time, but probably at the cost of reducing demand and expanding supplies in the longer run. Because the commodities involved are not nearly so indispensable as oil, such temporary market power can hardly lend itself to effective coercive use. Theoretically, food could be used coercively against a society exposed to mass starvation, but who would exploit that kind of leverage internationally for political purpose in the contemporary world? The only other commodity that could—although probably only fleetingly—generate a great deal of leverage is military weaponry. There are so few exporters of sophisticated weaponry that the possibility of switching quickly from one to another source of supply is severely limited, while expectations of war or ongoing hostilities may render the need for imports extremely urgent. Once the October war had broken out and had begun to absorb enormous quantities of expensive armaments, the Soviet Union and the United States did have ample leverage temporarily, but even this leverage was limited because neither superpower could afford to let its clients be defeated.

Two other circumstances contributed to the effectiveness of the oil embargo in 1973. One was the effect of the Yom Kippur War on Arab unity. The other was the lack of preparedness in the industrial importing countries. The oil weapon continues to constitute a potential threat. Yet any reimposition of the oil embargo would take place in a different context. Next time, no surprise or panic would be found in the industrial importing countries, partly because they are now better prepared to cope with the contingency, so that a second OAPEC embargo would have to be far more extensive and prolonged than the first, and partly because the United States and the other industrial importing states can and probably will do more to keep their dependence on Arab oil down to a level that is manageable in an emergency. If they act with determination, the economic power inherent in the use and threat of the oil weapon may turn out to have been ephemeral.

There are also circumstances on the Arab side that operate against a renewed use of the oil weapon. Any such attempt in support of political objectives demands close co-operation among the major Arab oil exporters. This cohesion cannot be taken for granted. In view of the political differences among Arab regimes, it seems unlikely to be forthcoming for any cause other than the Arab struggle against Israel. It cannot even be assumed that the Arab governments will agree on the precise settlements to be achieved with Israel. Even if they did, they would lack the power to coerce Israel directly; Israel is not readily coerced regarding any matter that it perceives to be a threat to its survival. Militarily, Israel is apt to remain formidable for a long time to come, and the only Arab economic pressure that could conceivably help to bring an Israeli government to heel is the protracted economic attrition resulting from the need continuously to match a high level of military mobilization by neighboring Arab states.

The Arab governments could again try to coerce major oil-importing countries to bring pressure to bear by using an oil embargo. But, if the Arabs are to achieve a satisfactory settlement with Israel, the key country is obviously the United States, and the oil weapon is the least effective in that country. It was already not very vulnerable to the last boycott, and it would be even less vulnerable to the next one unless it is careless enough to let its dependence on OAPEC oil increase appreciably. In addition to stockpiling and other emergency measures that have already been taken, it would probably be able to augment its imports from non-OAPEC countries. Only a very severe and prolonged curtailment of production could threaten economic havoc in the United States, and that would come long after other oil-importing countries had suffered much more grievous shortage.

The United States has two important incentives to work toward a settlement of the Arab-Israeli situation, as we have already seen. If it proves impossible to find a solution that satisfies both sides, however, the United States is still unlikely to allow itself to be coerced into compelling Israel to accept a settlement that clearly jeopardizes Israeli security. To do so would compromise America's image of itself as a great power, and it would also run up against the strong identification of many Americans with Israel. The oil weapon would do little to offset these considerations.

Still other considerations restrict the utility of the Arab oil weapon. We mentioned earlier that, though the Soviet Union and the United States could derive substantial power from denying arms to a client state, the coercive use of this dependence is limited by the need to see the client state survive. Interdependence, though frequently asymmetrical, is seldom one-sided; there is usually some mutuality of dependence. When exploiting monopolistic market power for economic gain, the oil-exporting states do not want to raise prices and revenues to the point where they will ruin the economic life of the dependent importing states.

Nor will most Arab governments be eager to drive any major importing country—

including the United States—to economic collapse for political reasons. Where the United States is concerned, at least some Arab states are likely to weigh the possibility that its continued vigor may prove indispensable to Arab independence. Arab societies do not want to fall under Soviet domination, whether imposed by Soviet military strength or their own dependence on Soviet arms. Only the United States can balance Soviet military power in the Middle East, and, while Arab countries currently depend on Soviet arms, economic aid, and technical assistance, their governments can minimize Soviet power so long as they have the possibility of switching to the United States as an alternative source for these things. This need to maintain the United States as a friendly—or at least potentially friendly—power severely circumscribes the use of the oil weapon against it. Finally, should the Arab oil-exporting countries apply their weapon with utter ruthlessness, they cannot be sure that the United States would not resort to some sort of military response, unlikely as this eventuality now seems to be.

Speculation that the major Arab oil-exporting countries will soon acquire additional international economic power through the amassing of huge financial claims on the industrial importing states is also heard. Assuming that oil prices hold up, the consequent shift in the international distribution of income generally will increase the influence of the Arab oil-exporting countries. Oil-importing states with persistent external deficits, especially among the less-developed countries, may then set out to please Arab governments in the hope of attracting aid or credits from them. But it is unlikely that, by using the threat to withdraw investments or the offer to grant credit, the OAPEC countries could exert coercive power on any one of the major industrial countries, let alone on several of them at the same time. Projected export surpluses for those Arab states whose imports will lag behind the value of exports are large (especially for Saudi Arabia), but they are not huge in relation to the combined currency reserves of the OECD states. "Safety net" arrangements are being developed among the industrial countries to aid deficit countries in distress. Moreover, as Arab surplus funds accumulate, the vast bulk can find profitable investment only in the industrial countries. The United States and other strong-currency countries can absorb even large Arab funds with relative ease. As this process goes on, these countries are placed in a correspondingly stronger position to assist or rescue deficit states by recycling Arab capital. As Arab "petro-dollars" are channeled into investments, the investors themselves become hostages to the fact that the liquidity of their assets is limited. Finally, as Arab investments in the industrial oil-importing nations increase, Arab interests in the economic prosperity of these countries will also grow. It seems likely, therefore, that, as in the case of the oil-embargo weapon, increased Arab financial assets will add little leverage in matters of high policy.

Although the economic power that certain Arab countries wield may not be as robust as appears at first blush, it is also far from trivial. The countries concerned are experiencing a major accretion of power as well as of wealth. Following a long period in which they were subjected to the humiliating exercises of power from outside the region, the Arab elites have recovered a sense of confidence, self-reliance, and strength. So long as they remain united, they will be able to claim a larger role in the management of international as well as regional affairs. As the examples of Iran, Saudi

Arabia, Egypt, and Syria demonstrate, they can also finance large imports of modern weapons and become countries of greater military consequence. Indeed, it is by no means unimaginable that the staggering financial resources flowing to some Middle Eastern countries will enable them before long to import the ingredients necessary for the development of nuclear arms.

II

Perhaps the most remarkable thing about the Arab employment of the oil weapon in 1973 was the absence of any military response. Using economic power for political coercion is not a friendly act, and of all the forms of strength, military power has been regarded traditionally as the ultimate one, capable of trumping all others. To be sure, some hints at the use of force were made during the period of crisis. On November 21, 1973, Secretary of State Kissinger warned that the United States would consider counter-measures if the oil embargo continued "unreasonably and indefinitely." On January 6, 1974, Secretary of Defense Schlesinger announced that the Arab states would run a risk of violence if they used their control over oil supplies to "cripple the larger mass of the industrial world," and he added on January 7 that there was a risk that the Arab oil embargo would create domestic pressure favoring an American show of force. On January 9, Kuwait's foreign minister announced that his country was prepared to blow up oil installations should any foreign power invade it in order to secure oil. It is probable that a "crippling" oil embargo would have stimulated a serious consideration of using military force, but it had hardly been considered by the time the embargo was lifted. The statements quoted seemed more in the nature of advice and warning than of genuine threat.

Why this reluctance to resort to force? Only a few decades ago the great powers would have met such a challenge unhesitatingly with force. Even after World War II, European powers readily applied force in the Middle East. When Iran nationalized properties of the Anglo-Iranian Oil Company in 1951, Britain seriously threatened military reprisal. And in 1956, when Egypt nationalized the Suez Canal, Britain and France actually attempted a military invasion, although it was called off chiefly in response to external, especially American, pressure. So why no recourse to force in 1973? We will find that the answer to our question will be elusive. Still, a brief examination of the many possible explanations will at least shed some light on the changes that have occurred in the world power structure.

To begin with, the days of old-style gunboat diplomacy are gone. A threat backed by a modest show of force offshore will no longer intimidate any Arab government and make it change its policy, and no small expeditionary force can readily invade an Arab country, quickly install a new, compliant government, and withdraw. To conquer and occupy any of the populous oil-producing states and defend it against guerrillas would necessitate large forces for an indefinite future. Even the small states on the Persian Gulf would probably be difficult targets: they could demolish most of the equipment for producing and shipping oil, and would probably receive effective support from other Arab countries in the form of regular military forces, guerrilla fighters, and, most seriously, a drastic embargo on oil. Against anything resembling Western imperialism, even small, underdeveloped countries can now generate determined resistance.

On the other side, with the exception of the United States and the Soviet Union, none of the big industrial states any longer possesses military force sufficient for large-scale military action overseas. Japan has only modest capabilities for defense, and the larger NATO countries have a force structure designed only for nuclear deterrence and/or the defense of Western Europe. Thus, none of the industrial societies most vulnerable to the oil weapon possesses the capacity to contemplate military reprisal. It is even doubtful that the United States possessed it in 1973. While American forces in the Mediterranean were probably sufficient to launch such an operation, their ability to sustain it was more questionable in view of the Vietnam debacle, the transition to an all-volunteer force, and the limited training of American soldiers for desert warfare.

Numerous other restraints in the use of force would have come into play. Following the traumatic events associated with the war in Vietnam, military intervention would not have elicited adequate public support in the United States. American allies would have been likely to censure such an action, and the indignation of the Third World would have been profound. Domestic support for—even the very political survival of—those rulers (e.g., King Faisal and the Shah of Iran) favorably disposed toward the United States would have been threatened. Military intervention in any part of the Middle East would have forced other states in the region to proceed to a complete halt of oil exports. Finally, the Soviet Union would have supported the Arab side not only by diplomacy and propaganda, but also by supplying arms and probably by military counterthreats. The risk of Soviet interference would surely have had a discouraging influence. Clearly, the reasons for not launching an American military response were many and overwhelming.

For some time to come, only the United States and the Soviet Union will have the capacity to engage in sustained conventional battles and military occupation far distant from the homeland. Soviet capabilities—while on paper perhaps still below the American level—have been expanding rapidly for more than a decade. This was dramatically demonstrated in October, 1973, when the Soviet Union expanded its Mediterranean fleet to more than ninety ships and supplied the Arab belligerents with a large number of sophisticated weapons and ammunition by using a rapid, continuous airlift as well as sea transport. It was also reported to have alerted several divisions of airborne troops for possible deployment in the Middle East.[3]

The overall capabilities of the United States, on the other hand, have been on the decline. Nevertheless, Secretaries Kissinger and Schlesinger, supported by President Ford, declared in January, 1975, that American military intervention would not be out of the question if the OAPEC states imposed an oil embargo with economically "strangulating" effect on the industrial societies. This, as we have shown, is a most remote contingency. But if it does materialize, the threat should not be regarded as entirely empty. If the Arab governments are convinced of this, it might—in addition to the other conditions we have discussed—further deter the Arabs from the application of extreme economic pressure.

III

What accounts for the unwillingness or inability of the major industrial nations that used to maintain large military forces and to employ them readily overseas to do so now? It is evident that the democratic capitalist societies in Western Europe have recently limited their military effort to deterrence of attack and to preparations for defense, should deterrence fail, and for even these limited functions they rely to a great extent upon the United States. During the last decade, moreover, their defense budgets have been subjected to continual political pressure for cutting, and with substantial success (as a percentage of GNP) in West Germany, France, Italy, and Britain.[4] Recruiting military personnel has also become increasingly difficult. While these same trends have been apparent in the United States as well in recent years, they have not been typical of the remaining parts of the world.

These are recent phenomena, and they have not yet received adequate study. Some scholars speculate that these trends in the Western industrialized countries must be connected with profound changes in the political and economic system. It has been suggested that these societies have undergone progressive democratization with the consequent diffusion of domestic political power. Government authority has weakened and become more dependent upon satisfying rising demands for various public benefits. The politicization of previously less sharply drawn cleavages along lines of religion, ethnicity, class, age, and sex has impaired the sense of national solidarity.[5] Changes in political culture have made these formerly bellicose societies more sensitive to the costs of international war and more skeptical of the benefits to be obtained by it.

The question now arises whether these changes in the character of the Western nations will transform them from militarist societies to strong but peaceful ones, content to maintain only those security forces that are rationally required for deterrence and defense. It may be that some of these countries will move in this direction and become peaceful without weakening themselves to a point threatening to their autonomy. But it is also possible that some will move beyond this point and become more vulnerable to military threats. In the past, it was thought that organized societies could subsist in a fundamentally dangerous world so long as individual states determinedly exercised their military sovereignty for deterrence and defense. Today, awareness of the potential threat, which is rooted in the nature of the international system, seems to have dimmed in the capitalist societies. The end of the Cold War and the prospects of détente have encouraged the public not to look for distant threats. Government leaders find it difficult to provide prudently for the deterrence of any foreign aggression that is not immediate and obvious. That precisely the richest societies claim to be unable to afford defense outlays at past levels, when all other societies seem to have little trouble in doing so, is of course ironic.

One cannot say to what extent these developments will apply to the United States, or even whether they will persist in Western Europe and Japan. If they endure, the nations of Western Europe are likely to experience further shocks of the kind that occurred in the fall of 1973. So far, that shock has generated no visible attempts at mustering some elements of counterpower: for the reasons presented, it seems improbable that the reaction, when it comes, will involve using military strength, nor—despite the signs of local wealth—effective economic power on the part of Western Europe. These countries have no essential economic assets that they can threaten to withhold and on which some powerful nation is dependent. Surely, there is nothing the oil-exporting states can possibly want that they will not be able to buy from countries eager to sell. Moreover, one prerequisite—necessary though not sufficient—for building strength in Western Europe is the progressive economic, political, and military integration within the European Community. The prospects for that, however, are not promising. The energy crisis confirmed the tendency of the Community to fall apart under adversity. So long as the crisis looked ominous, member states acted on the principle of *sauve qui peut*. This propensity to revert to national action under pressure also expressed itself in NATO, where practically all the United States allies refused the American request to use their airfields in supplying Israel.

However, while the democratic industrial countries may find it difficult or unprofitable to use military power, or impossible to develop the kind of economic power enjoyed at present by the members of OAPEC, they can reduce their vulnerability to the coercive employment of economic power by others. Indeed, the recognition of their vulnerability and of the consequent strains that manifested themselves between Western Europe and the United States—easily the most severe strain since the inception of the transatlantic partnership—has evoked a clearer perception of common problems and a renewed disposition to work them out together. In the light of the events of 1973, the recent capacity for cooperation among the countries loosely organized in OECD in matters related to high oil prices, international monetary disequilibrium, and the latent threat of another oil embargo has been remarkable, and is no doubt related to the sense of impotence experienced in the fall of 1973.

IV

According to some recently fashionable hypotheses, the disposition of West European societies to slight the military bases of security is not so important, since the utility of military power in international affairs has been decreasing.[6] Some of the specific arguments in support of this view are quite plausible. First, it is doubtful that strategic nuclear weapons are useful for anything but nuclear deterrence. Second, the spread of nationalism to the less-developed and ex-colonial world forecloses the easy use of modern conventional military power against such countries, no longer so helpless as in years gone by. Third, there has been a crystallization and strengthening of international norms—codified in the Charter and resolutions of the United Nations—which formally prohibit recourse to violence among states for purposes other than self-defense. Other generalizations, however, do not stand up well in the light of the events that took place in the autumn of 1973.

It had been assumed by some, for instance, that the increased economic interdependence would reduce the possibility of a rupture in relationships simply because such ruptures were becoming more costly. According to this argument, most of the growth in interdependence is brought about not by national governments but by actors, including private international organizations and multinational enterprises, whose loyalties and interests were not exclusively national. As a result, some foresaw that the role of national government would very likely contract, and sovereignty would be transferred to newly evolving supranational and transnational authorities. This line of speculation all sounds a bit like the Manchester School in mid-nineteenth-century England, where it was predicted that universal free trade would entail not only vast increases in national production but also the growth of cosmopolitan brotherhood among men, making war impracticable. Although Western and Central Europe came fairly close to conditions of a free-trade regime and even passports were unknown during the second part of the nineteenth century, it did not prevent the catastrophe of World War I.

The current hypothesis of the effects of interdependence is evidently based on the premise that the condition is symmetrical, that is, it is not only beneficial in its global effects, but also to all of the parties. What did the fall of 1973 reveal about these matters?

As between the OAPEC states, on the one hand, and Western Europe and Japan, on the other, interdependence has proved to be predominantly economic. In terms of mutual dependence, this relationship had become extremely asymmetrical. The industrial, oil-importing countries were heavily dependent on a secure (and expanding) inflow of oil, but they were irrelevant to Arab security and almost irrelevant to Arab aspirations regarding Israel; without serious hardship to the Arabs, imports from them could be postponed, or cut back, or replaced by imports from the American and the Communist countries. Such pronounced asymmetry almost invites exploitation. The Arabs reacted to it first by dictating stupendous changes in the terms of trade, and finally by making irresistible political demands.

Turning to interdependence among the Arab countries, an oil weapon, to be effective, requires the ability of the key producers to agree on its use. Although there are very strong transnational ties of religion, culture, and identity—talk of an "Arab nation" is not uncommon—one is also impressed by the many political frictions among these countries and by the resulting fragility of their capacity for joint action. It was far from easy for them to agree both on initiating and on terminating the oil embargo, as it is not easy for them to agree on price policy now. Despite the multitude of transnational bonds, the interests of at least the current governments and elites are in conflict on many issues. The one great unifying impetus is intense hostility to the state of Israel.

International interdependence has grown very unevenly the world over. In terms of military security and economic transactions, it has increased far more among the rich, industrial, and democratic countries than between them and the rest of the world, or than among the Communist states and among the less-developed countries. Within the capitalist group, enmeshment has naturally thickened most among those West European societies that established the Common Market and affirmed their intention eventually to create a political union as well. It is in the experience of those countries that our hypothesis should find its sharpest illustration and strongest support.

Military conflict among these nations seems unimaginable at the present time. This is unquestionably a monumental change from their past behavior. It is doubtful, however, that this salubrious development results from a growing volume of economic or political interaction, or that it will necessarily spread to other parts of the world. It is possible that the abandonment of military threats within this set of societies is rooted in the various changes—more or less common to all—to which we attributed their diminishing willingness to develop and sustain military strength, and that the increase of interdependence among them followed these changes.

What is striking in the economic area—both within Western Europe and in the larger OECD grouping—is the present difficulty of progressing beyond past achievements. The recent disposition of these countries has been to find further advances in economic interdependence burdensome and to revert to nationalist postures when confronted by economic crises. This tendency has been visible for some time, notably in connection with balance-of-payments problems, and it was by no means lessened by the demise of General de Gaulle and by the accession of Britain to the European Economic Community. The energy crisis of 1973 only highlighted the degree to which the centrifugal forces had gained. There was no willingness to share and no willingness to stand together within the European Community. Europeans were less hesitant to cooperate with the United States than with each other, once the immediate Arab threat was over, because they saw their security as depending on that link. Of course, the disposition to cooperate with the United States also had its limits, especially in the period when the Europeans faced the immediate threat of insufficient energy supplies. What the episode demonstrates is that little erosion of national sovereignty has occurred within the European Community. The question now is whether the steps toward economic integration that have been taken so far are a fair-weather phenomenon unable to survive economic adversity.

It is too early to tell whether the capacity of the industrial countries to work together on common problems within the OECD, manifest since 1974, marks the beginning of a new trend or represents simply another isolated episode. The crucial fact seems to be that international interdependence entails costs as well as gains. It may turn out that the benefits are largest in the initial stages of progressive interdependence and that they are subject to diminishing returns in more advanced stages, while the burdens possibly tend to increase with each stage. Increasing interdependence may have the effect of heightening sensitivity in one country to changes in the other over which it has no control. If increasing interdependence is asymmetrical, a society may see itself as more vulnerable to exploitation by the other as interdependence grows. Both among and within societies, interdependence may be compatible with harmony of interest, but it neither presupposes nor engenders it.

Lastly, the energy crisis also shed light on the relationship of the two superpowers, both with one another and with countries of the Middle East. Until the recent thawing of the Cold War, the Soviet-American relationship had been overwhelmingly hostile. Détente changed this, by adding a cooperative dimension to the relationship. While this change might be welcome for many good reasons, the expansion of trade and various cooperative ventures were also advocated on the ground that growing interdependence would promote cooperation and reduce the adversary element in the relationship: the process would foster contacts between more and more Soviet and American individuals, and the fruits of cooperation would become increasingly valuable to both sides.

As I suggested earlier, this is a very dubious proposition. The beginnings of détente touched off excessive expectations in the United States, even though Soviet leaders made it quite clear that a basic adversary relationship was bound to continue. The adversary position of the two powers in the Middle East during the events of 1973 helped to sober American ideas about the realities of détente. The mutual desire for détente survived, but American perceptions of its potentialities became more realistic.

The 1973 crisis also demonstrated once more the limits to the ability of the two superpowers to control events in the Middle East. These limits are set mainly by three factors: One is the disposition already noted of the countries of the Third World to be less susceptible to coercion. The second is the fact that rivalry between great powers reduces the influence of either over a third state. The third is the growing affluence of the oil-exporting countries themselves, which reduces their dependence on economic aid and greatly strengthens their self-confidence as international actors. This last condition may tend to diminish Soviet influence in particular, partly because many Arab countries can now pay cash for arms and partly because the Arabs have managed to muster widespread recognition and support against Israel.

Soviet influence is also limited by the existence of a fundamental asymmetry in the Middle East. While the Soviet Union can only cooperate with the chief Arab governments, and Israel only with the United States, the Arabs and the United States are relatively free to interact. This imbalance made it possible for Secretary Kissinger to act as an effective intermediary in negotiations leading to a cease fire. The imbalance also makes it possible for Arab governments to play off one power against the other in a way that Israel cannot emulate.

V

Our principal conclusions are five: First, increasing international interdependence engenders disadvantages as well as benefits. The problem for any nation is to estimate the yields and the tradeoffs of any chosen policy. Thus, it might well be in the American interest to place limits on its increasing interdependence with the Arab oil-exporting states, as it might be in the interests of some West European governments to reduce and restructure the existing degree of interdependence with these countries.

Second, while the major oil-exporting countries now possess the elements of strong coercive power, the use of that power is highly circumscribed; it depends crucially on their ability to act in concert and on the inability of the industrial oil-importing countries to curtail their vulnerability. Third, it seems unlikely that the application of the oil weapon in 1973 inaugurated a general increase in the coercive use of economic power in matters of high policy. Fourth, although the utility of military power in the economic system seems to have declined somewhat, this is probably not the result of a increase in international interdependence. Fifth, the international distribution of military power has shifted. Compared with the situation just before World War II, military power in Japan and Western Europe has greatly contracted, while in the United States and especially the Soviet Union it has greatly increased. Even more recently, the military power of some less-developed countries, such as China, India, Egypt, and Iran, has also grown. These shifts may prove to be more important for the future course of international relations than the speculations over the increased role of economic power that appeared to accompany the 1973 oil crisis.

REFERENCES

[1]For an enthusiastic exposition of these arguments, see Robert E. Hunter, "Power and Peace," *Foreign Policy*, 9 (Winter, 1972-73), 37-45.

[2]Klaus Knorr, *Power and Wealth* (New York, 1973), Chaps. 6-7.

[3]Elmo R. Zumwalt, "The Lessons for NATO of Recent Military Experience," *The Atlantic Community Quarterly*, 12 (Winter, 1974-75), 456-59.

[4]Stockholm International Peace Research Institute, *World Armaments and Disarmament. SIPRI Yearbook 1973* (Stockholm, 1973), p. 236.

[5]Morris Janowitz, "Toward a Redefinition of Military Strategy in International Relations," *World Politics*, July, 1974, pp. 473-508.

[6]See, for example, Seyom Brown, *New Forces in World Politics* (Washington, D.C., 1974); Stanley Hoffmann, "Choices," *Foreign Policy*, Fall, 1973; Robert O. Keohane and Joseph S. Nye, Jr., "Power and Interdependence," *Survival*, July/August, 1973; Klaus Knorr, *On the Uses of Military Power in the Nuclear Age* (Princeton, 1966); Janowitz, *op. cit.*

RAYMOND VERNON

The Distribution of Power

IN OCTOBER, 1973, as the authors of this volume have repeatedly observed, a group of non-industrialized countries in the Middle East arrived at an agreement to reduce the flow of crude oil to the United States, Europe, and Japan. Soon after, these countries joined with a number of other oil-exporting countries in announcing a huge increase in the price of oil, an increase so great that inflation was given a vigorous push and the international monetary system itself was placed under enormous stress. Yet up to the very moment of this writing, the reply of the rich industrialized countries has been muted and equivocal. So far, there has been neither a credible threat of gunboat diplomacy, nor a significant effort at economic sanctions, nor any other of the traditional responses of seemingly powerful nations.

It is not easy to define and delimit a concept so elusive and ephemeral as the distribution of power. By almost any definition, however, it is evident that power has shifted in important ways since, say, the early nineteen-fifties, when Iran could be soundly chastised for its unwelcome attempts to gain control of its own oil and when Saudi Arabia could be regarded as a reliable ward of American oil interests. To describe the nature of that shift more precisely, one is obliged first to look back into recent history and identify some of the factors that made the 1973 crisis possible. But that is not enough; speculation about future events is also required, particularly about whether the strength displayed by the oil exporters in the October, 1973, crisis represented a solid new fact in the international distribution of power or an aberrant condition that would soon disappear.

The Problem of the Short Term

Projection in this case runs into some special difficulties. From the perspective of mid-1975, the world confronted a problem so large in size and so unprecedented in character that guessing about its solution seemed almost foolhardy. The oil-exporting countries, it appeared, would be accumulating two or three hundred billion dollars of financial assets over the succeeding four or five years, representing an unspent surplus from the payments of the oil-importing nations. According to some versions of the future, to be sure, these unspent funds may prove to be much smaller. No such surplus will develop if the future price of a barrel of oil falls sharply, say to five dollars or so;

245

nor will it occur if the oil-rich countries accelerate their purchases of arms even further. Moreover, if these countries involve one another in a destructive war or are plunged into war by an invasion on the part of the oil-buying countries, the transfer issue is overwhelmed by even more formidable questions.

Force and the threat of force cannot be divorced from any serious speculation about the Middle East, because the present situation is rife with explosive potentials. One is the unfinished war between Israel and the Arab states. Another is the relations among the principal Persian Gulf states, which include several unsettled disputes with a long history. A third is the possibility, slight though it may be, of the use of military power by oil-importing countries.

Barring developments in the Middle East along these lines, however, the problem of transferring unprecedented quantities of resources from the oil-importing countries to the oil exporters remains. There are, in fact, two transfers involved. One is the net transfer of financial resources from oil buyers to oil sellers, which must occur so long as the receipts of the sellers exceed their expenditures; another is the eventual transfer of real resources from the oil-buying countries, as the oil sellers use their piled-up financial assets for the purchase of more goods and services. Of the two transfers, the one entailing real resources is unlikely to pose a serious problem. The size of that transfer is not so very staggering when measured against the economies of the industrialized nations.[1] But the short-term problem of transferring the two or three hundred billion dollars of financial resources—bank accounts, currency, securities, and so on—remains a difficult one.

From a technical viewpoint, any group of well-informed experts can design methods to handle the operation. But the sheer process of elucidating the technical solutions underlines the potential political obstacles that lie in their way.[2] In any such transfer operation, some large industrial countries are obliged to surrender significant portions of their industry to foreigners they regard as potentially hostile, and practically all must subject their foreign-exchange regimes to major influences from the oil-rich nations. Therefore, before one reaches the point where familiar modes of analysis can be utilized, a challenge of disconcerting difficulty will intervene and will have to be overcome.

Once that challenge has been met, the oil-exporting countries can be expected to emerge with a considerably greater measure of power in world affairs than they had before the oil crisis began. Some sources of their added strength will probably be enduring, some more perishable. But taken together, they cannot fail to increase the influence of these countries in world affairs.

The Elements of Power

Of the various sources of strength that the major oil-exporting countries have acquired, one of the most important is the ability to vary the terms at which oil will be available to buyers; so long as prices remain high, offers of selective price reductions and long-term credits can be potent sources of influence. A second source of power for the oil exporters is the ability to buy goods—mainly industrial plants, military supplies, and consumer products—in quantities that may amount to some tens of billions of dollars annually over many years. The third is the ability to use money in all the other ways that money can be used, such as to aid developing countries, to purchase political support in international agencies, and to invest in the industrialized countries. In the period immediately following the crisis, the oil-rich countries rapidly learned to appreciate some of the ways in which their power could be used, and there is every expectation that this process will continue.

Some of these uses of power can be exercised by any of the oil-exporting countries acting individually. If Saudi Arabia were prepared to act without limit as the balance wheel in the world's oil markets, for instance, the price could be held up for some years to come. But the future price of oil is more likely to depend upon the willingness of the oil exporters to act jointly. Even with such cooperation, most prognosticators foresee some decline over the next five years in the power of these countries to determine the price of oil; without cooperation, it could decline more quickly.

On the other hand, the ability of the oil exporters to use their enormous purchasing power to influence the industrialized countries guarantees a more enduring strength. On anyone's guess, the unspent financial assets piled up by the oil exporters will be sufficient to last well beyond the nineteen-seventies, that is, well beyond the period in which the exporters may be able to determine the price of oil itself. As a result, for a long time to come, the industrial countries will be found dispatching one government mission after another to the oil-rich nations, eagerly offering agreements of mutual assistance, promises of technological exchange, and the sale of sophisticated equipment.

It remains to be seen whether bilateral offers of this kind will eventually take forms that entail strong discrimination against other countries. Long-term bilateral barter deals—agreements, for example, to swap so many tons of oil for so many industrial plants or military aircraft—invariably carry discriminatory overtones. If such agreements come to dominate the trading patterns of an industrialized country, they raise serious questions regarding the ability of that country to participate effectively in any system that requires open boundaries and nondiscriminatory treatment for others.

The trading systems that have prevailed among the industrialized countries over the past few decades, based mainly on the principles of the General Agreement on Tariffs and Trade and the European Economic Community, have managed essentially to retain the qualities of openness and nondiscrimination, even though they have been battered and distorted from time to time by acts of restriction and discrimination. Bilateral trading agreements therefore constitute some degree of threat to the system, though the threat is not yet one of very great potency. To be sure, a few countries—notably France and Japan—occasionally seem drawn to quasi-barter arrangements. But, despite occasional dramatic announcements of vast bilateral deals, most arrangements so far have been statements of intent rather than actual swaps of goods and services. There have been general undertakings to provide technology and credit and to apply the good offices of governments in the acquisition of scarce items. Where expensive items, such as nuclear power plants, could be identified for sale, there have been even more concrete affirmations of intent. But barter has not yet become the dominant mode of exchange. A prolonged depression could change the tactics of the industrialized oil-importing countries, pushing them to act on more discriminatory lines in order to promote their exports; but that stage has not yet been reached.

One reason why the reversion to bilateral barter has not occurred more rapidly is probably that many prospective buyers remain uncertain how long the sellers' market in oil might last. If oil promises to be in short supply for only a few years, the damage from undermining the existing system of trade relations may not be worth the assurances of future supplies of oil. Besides, from the viewpoint of the oil-rich countries themselves, the advantages of bilateral barter arrangements are not all that obvious. Such arrangements in the past have usually been thrust upon countries because they lack foreign exchange. Conversely, countries that have ample foreign exchange ordinarily relish the freedom of choice which goes with their plenitude. A long-term commitment to a selected seller is not an attractive prospect for a buyer with the financial means to shop the market.

Some OPEC countries may prove willing to consider exclusive bilateral arrangements, however, because they are apprehensive about their access to the markets of the industrial countries for the oil, petrochemicals, and other products that they expect to be selling in the decade or two ahead. Anticipating that the sellers' market may not last, some may seek to convert their present position of strength into a long-term commitment. But that strategy constitutes precisely the weakness of the discriminatory arrangement from the oil-buying partner's point of view. Also anticipating an end to the sellers' market, the buying partner will be unwilling to relinquish his future maneuverability. In other times and other markets, these and similar considerations have tended to hold bilateral trade agreements in check and have eventually led most governments to recapture the right of flexibility and choice.

Not all the advantages and drawbacks of bilateral trade agreements for the oil-exporting countries can be measured in economic terms, of course. The quest for new political alliances can provide the incentive that makes the restrictions of bilateralism palatable. No doubt some of the oil exporters would welcome new sources of political support, and one or two of the oil-importing countries would enjoy new political alliances as well.

The more lasting source of increased bargaining strength for the oil exporters, therefore, would appear to be in their rapidly accumulating financial assets. These may be used to influence the countries seeking to export industrial and military goods or to influence the countries in which the funds remain as investments. For instance, so long as Israel remains the enemy of the Arab states, such assets will continue to be used to enforce the blacklisting of firms that do business with Israel, as well as for other forms of economic pressure. Such pressures generally have to be applied discreetly, even covertly, because of the obvious risks of backlash. Nevertheless, they can be counted on to have some influence.

Apart from using the funds as a form of economic warfare against Israel, the oil exporters can also threaten to withdraw their assets from any country whose policies seem openly hostile. Withdrawal—even the threat of withdrawal—could produce a drop in asset values and a strain on the balance of payments for any country that was its target. Because the oil exporters already have used the "oil weapon" with some effect in order to achieve their political ends, they are widely assumed to be capable of using the "money weapon" for political purposes as well.

Alliances in the Less-Developed World

Although marginal shifts in the bargaining power of the oil exporters may have their impact, the more important changes in power arising out of the oil crisis may prove to be of quite another kind; they may take the form of new alliances and new perceptions of common interest among groups of nations. One such possibility is that the less-developed areas of the world may find a new basis for uniting in an effective political alignment to be used as necessary in dealing with the rich industrialized areas —Europe, North America, and Japan. Since the beginning of the oil crisis in 1973, the collective strength and cohesion of the less-developed countries, rich and poor alike, have seemed remarkably strong and sustained. In the United Nations General Assembly, in UNESCO, in the two Law of the Sea conferences, and in other international gatherings, the solidarity among the less-developed countries has been impressive. Moreover, although India, Bangladesh, Sri Lanka, and the impoverished African countries face a shattering problem of financing their increased energy requirements, their support of the oil-exporting countries on any international issue of importance to those countries has seem ungrudging.

Some of the reasons for that solidarity are obvious. Two or three billion dollars yearly is about the subsidy needed to finance the increases in energy costs for the poorest countries; but this will apparently not be forthcoming from the traditional aid-givers, the industrialized countries, either on a bilateral or a multilateral basis. Aid of this— and even greater—magnitude is well within the means of the oil-rich countries, and in fact their aid-giving is already on such a scale that the expectations of the impoverished oil importers do not seem wholly unfounded. Commitments of nonmilitary aid by the oil exporters have already reached five or ten billion dollars, depending on how one chooses to count them, and are rising rapidly. Though some of the African countries are beginning to show restiveness over the rate of disbursement, the restiveness appears to be a negotiating tactic rather than a threat to the continuation of the coalition.

Apart from the desire for aid, however, the poor countries also feel a strong pull to the oil-rich countries because of their success in forcing the industrialized world to treat them with deference—an achievement of enormous psychological importance to all developing countries, even to those who suffer from its consequences. As a dramatic illustration of how far and how fast the world has moved from a neo-colonial era, the demonstration of power is something to be savored. The depth of that reaction can be felt in Norman Girvan's article in this collection, an authentic interpretation of the oil crisis as seen through the eyes of many Third World intellectuals.

The oil-exporting countries are likely to try to reinforce their role as champions of the developing world whenever an opportunity arises. As Girvan points out, their strategy of trying to tie the price of oil over the long term to some index of the prices of industrial exports will bring them the added support of the less-developed countries that export other raw materials. Venezuela's willingness to help finance the coffee-exporting countries in their efforts to raise the price of coffee and to help finance development in other parts of Latin America will also be remembered. Measures such as these have fortified the sense of solidarity among Latin American countries in confronting the United States. They help to explain why the members of the Organization of American States exhibit more hostility than approval to any United States effort to bring down the price of oil, even though some of them have been much more affected by the price increase than has the United States.

The stability of a coalition among the developing countries will depend on various factors. The most important of these may prove to be historical and psychological: a basic sense of the rightness and inevitability of a trend that at long last has elevated such countries to a position that has long been merited by their vast numbers and their unrequited needs. But stability may also depend on more mundane factors, such as whether the two or three billion dollars a year needed by the poorer countries as supplementary financing for their oil imports will be regularly forthcoming. Even if they are, there will be other problems. The experience of the United States, the Soviet Union, and France suggests that it is difficult for a donor country, whatever its motivation and experience, to master the art of international giving; if large-scale aid must be given on a continuous basis, resentment and defiance are almost invariably the reactions of the recipient. The probability of that reaction seems particularly high in a situation where the disparity in living standards between donors and recipients promises to grow so rapidly. While the recipients will be building up their obligations, the donors will be building up their assets.

The vulnerability of the coalition in the long run stems also from the fact that the political systems of these countries run a very wide gamut, from absolute monarchy to putative Maoism. The common element in their ideology is not very great; it consists mainly of a certain reserve, sometimes an outright hostility, to the industrialized world. Accordingly, one can only speculate about how long the political and intellectual leaders of South Asia and Africa will find it possible to endure the psychic stress of their position as chronic aid recipients.

Indeed, though the oil-rich countries may see some political advantages in keeping the less-developed world on their side, some are likely simultaneously to try maintaining close ties with the industrialized countries. Whether they place a high value on such ties will depend on many factors, including the political composition of the regimes in power. If, for example, conservative regimes such as those of Iran and Saudi Arabia continue to rule, the importance of these links may be fairly substantial. Even if more radical regimes replace them, the development of similar ties is not to be excluded. Countries such as Algeria and Iraq—even isolated and radical Guinea—have demonstrated that questions of ideology do not prevent them from developing close working relationships with capitalist institutions. The form that these relationships take will also depend in part on how the structure of world industry evolves, and this in turn brings up the emotion-laden question of the future of the multinational enterprise.

The Role of the Multinational Enterprise

The intensity of the debate over the growth and role of multinational enterprises in recent years has distinguished the controversy from the ordinary squabbles over trade, payments, and investment which are the usual fare of international economic relations. Even the term "multinational enterprise" in itself, though apparently devoid of value judgment, has been attacked on the ground that it subtly conditions the direction of the ideological struggle. The intensity of the debate suggests that the disputants regard the multinational enterprise as a symbol for some larger issues—socialism against capitalism, labor against management, Europe against the United States, or the Third World against the industrialized countries.

In this atmosphere, extravagant projections have become the stock-in-trade of discourse. Three or four years ago, it was fashionable to assert that many nations were rapidly becoming so weakened by multinational enterprises as to threaten their national existence. Today, with the oil crisis fresh in mind, it has become equally fashionable to speculate that multinational enterprises are on the way out.

My own expectations do not fit either mold. Under modern conditions of transportation and communication, multinational enterprises have managed to develop some major sources of strength. These strengths do not apply uniformly to all industries—textile firms, for example, derive fewer advantages from the multinational form than, say, chemical firms. Even where the multinational structure provides advantages to an enterprise in a given industry, changes in the nature of the industry can alter these advantages—in some cases increasing, in some cases diminishing them.[3] But for some industries at some stages of their development, the multinational form offers decided advantages.

The dominance of multinational enterprises in the oil industry over the past forty or fifty years was based on a variety of factors. The major companies drew their strength partly from their monopoly power, expressed in the fact that the cost of Middle East oil to them was much lower than the cost of that oil to their competitors. But they also built their strength on the diversity of the markets to which they had gained access, the diversity of their sources of crude oil, the economies of scale in the operation of tanker, storage, and refining facilities, and the economies in the use and reuse of technical, political, and market information. All these advantages played some part in inducing newcomers to adopt the multinational form. The proliferation of these new companies, as I have suggested elsewhere, also contributed to the eventual weakening in position of the original majors.[4]

One of the key decisions that the state-owned oil companies in the exporting nations will have to make in the next decade has to do with the distribution of their oil. They must decide how to discharge their responsibilities for marketing very large quantities of crude oil and oil products. If long-term bilateral barter arrangements prove to be of limited utility, as I think likely, other means will have to be considered. One is to pursue the traditional strategy of the oil industry—acquiring and controlling facilities in key foreign markets for the further fabrication and distribution of products. Another is to develop partnerships and alliances with potential competitors, in order to increase geographical spread and avoid competitive conflict.

Policies of this sort on the part of the oil-rich countries may not be confined to the oil industry. Over the course of the next five years, these countries plan to commit one hundred billion dollars or so to the creation of industrial plants. These facilities will for the most part be large-scale, capital-intensive, energy-consuming establishments, manufacturing such products as plastics, iron and steel, aluminum, and fertilizers, and requiring consistently high rates of production for economic stability. Captive customers in foreign markets can generally provide a critical part of that stability.

So long as the exporting countries consider that they have a sellers' market abroad, the urge to find captive customers can be held in check. But even if these countries feel secure for the present, my assumption is that the feeling will not last for very long. Eventually, perhaps even initially, these new plants will be confronted with the formidable task that faces any seller in a capital-intensive industry—that of making inroads into an established market. Not many industrialized countries are likely to tolerate widespread price-cutting as a means of entry; so the new sellers will face the task of making their arrangements within the established channels of distribution. One obvious method for doing this is for the state-owned companies to create a series of partnerships with existing enterprises in the importing nations. Indeed, I find it difficult to picture how the major part of the industrial output of the oil countries will be marketed without such partnerships.

In projecting the development of partnerships between these new companies and the already existing corporations in the industrialized nations, one is actually describing tendencies that already are visible. The fact that these companies are state-owned has not inhibited them from creating partnerships with privately owned foreign enterprises. So far, the partnerships have been largely for the servicing of home markets. But, if past experience is any guide and if a few early illustrations are indicative of future trends, as I think they are, the ramifications of these partnerships will eventually extend far beyond the home country. Perceptions of common interest and the potential threat of competition are likely to condition the behavior of the managers of the new state-owned oil companies, just as in prior decades they conditioned the behavior of the officials that directed the European state-owned firms, BP, CFP, and ENI. There may be periods of independent market behavior on the part of these state-owned companies as they try to break their way into the existing markets. But, like ENI, their desire for stability seems likely to grow in measure as their market share increases.

If I am right that the oil-rich countries will find themselves under considerable pressure to cultivate new ties with multinational processors and distributors in the importing countries, one is back to considering whether multinational enterprises, acting individually or collectively, will after all figure more prominently in future international relations. My guess is that they will not, that individual enterprises as a rule will carry less weight, and that groups of enterprises will encounter more difficulties, not fewer, in any efforts to act collectively. One needs much more evidence before drawing any firm conclusions on this complicated question, but extrapolations from the oil crisis itself offer some provocative indications.

The oil industry, along with many others, registered a sharp increase in the number of its multinational enterprises after World War II. Studies conducted in about twenty-five other industries so far clearly demonstrate the same strong tendency.[5] Almost without exception, these industries show an increase in the number of enterprises operating on world markets. As a corollary, these industries also display a decline in the relative importance of the established industrial leaders.

An increase in the number of firms in a given industry does not, of course, inevitably reduce the strength of individual firms or reduce their ability to pool their strength in an industry cartel. In fact, there are signs that in some industries, even as the number has proliferated, the leaders have tried to guard the industry against instability. In a few cases, formal cartels have been created. In others, partnerships among producers have become more common. But my guess is that, just as in the case of oil, the increase in the number of multinational enterprises in other products has weakened the old leaders individually and the multinational enterprises collectively.

Nevertheless, the tendency for each individual multinational enterprise to command a smaller share of the world market and the reduced ability of multinational enterprises collectively to control the market are unlikely to solve the problems that many governments see in the operations of multinational enterprises. In a world of open economic boundaries, governments feel more secure when their national economies operate through institutions that are national in structure, on the assumption that they are easier to direct and control. In many commodities, multinational enterprises may be necessary to carry on effective international business simply because of advantages that go with the multinational form. Where that occurs, the existence of such enterprises will remain disconcerting and even threatening to some governments. That reaction is present in some measure in all countries, as recent American responses to the growth of foreign-based investment in the United States demonstrate. So the issues presented by the growth of multinational enterprises are likely to continue with unabated vigor.

Moreover, the oil crisis has exposed one particular situation in which multinational enterprises as a class may keep on exerting considerable influence. The capacity of national governments to take over functions of distribution and rationing in an emergency is limited by their lack of expertise and by their need for time to react to the emergency. The capacity of any group of national governments to take on such emergency functions collectively is even more limited. In such situations, the least cooperative participant usually sets the pace. In emergencies that require the managing of complex international transactions, therefore, it can be expected that governments will respond either slowly or not at all. Unable effectively to monitor, let alone to supervise, the activities of the multinational enterprises, governments may be tempted to leave the problem of management to the enterprises themselves. Whether this management by default will continue for long periods will depend on the technical complexity of the industry and on the national and international politics of control. But multinational enterprises will certainly not suspend their operations while the governmental mills slowly prepare to grind.

The continuing strength of the multinational enterprise will rest largely on its ability to perform its business tasks—developing products and processes, financing operations, and delivering goods and services. Such enterprises may also occasionally draw strength in international markets from the support of governments, as they have from time to time in the past. If that should happen, the political sensitivity of other countries to the manifestation of strength would be especially great.

The oil crisis also provides evidence, however, that governmental use of multi-national enterprises as arms of public power has its limits. In the oil crisis, these limits were swiftly reached. None of the industrialized countries succeeded in obtaining greatly preferred treatment from the oil companies it thought of as its own, though one or two governments tried. On the other hand, oil-exporting governments did not necessarily have their way either; when Arab oil was embargoed to some destinations, the oil companies substituted oil from other sources.

The realization that multinational enterprises cannot be used simply as an extended arm of government is presumably a lesson already half-learned by the governments involved. In the decade ahead, the governments of many industrialized countries seem likely to hold "their" multinational enterprises rather more at arms' length than they have in the past, questioning even more strongly whether their overseas activities are contributing to the national welfare of the home country. Accordingly, the disposition of governments in the industrial countries to support their multinational enterprises in international affairs promises to be fairly subdued.

These developments should make no particular difference to those in the developing countries who base their opposition to multinational enterprises upon ideological grounds, such as the issue of socialism versus capitalism. The issue will remain alive on that level, affecting the national policies of this country or that. My guess is, however, that the ability of the multinational enterprise to survive is likely to rest mainly on grounds that are more empirical in character, namely, on the perception of many countries that the suppression of these enterprises would be costly to the national interest.

Relations Among the Industrialized Countries

Projections of this sort, however, involve some assumptions about how the industrialized world itself is likely to be organized in the next years. They imply the continued existence of an international regime that permits trade to move fairly readily among industrialized areas and permits enterprises to maintain multinational processing and distributing networks very much as they do today. Yet it is possible to picture a set of developments in the relations among the industrialized countries that could alter these assumptions. The markets of the industrial areas may be divided up into tight compartments through restraints on payments or on trade. In that case, enterprises may find it difficult to function easily across national boundaries, and multinational enterprises may lose some of their present prominence.

Most industrialized countries appear to have concluded, however, that the large-scale reestablishment of high trade barriers would be destructive to their national interests. The oil crisis, it is true, revealed some of the costs of interdependence. The exposure of each of the European states to one another and to the outside world was reemphasized. Japan's vulnerability, though always well known, had come to be overlooked by other countries, and that oversight was made apparent in the crisis. Only the American economy's relative self-sufficiency appeared to provide much room for maneuver.

Still, most industrialized countries appear to have concluded for the present that their interdependencies carry more advantages than drawbacks. As was observed earlier, none is rushing ahead with policies such as bilateral barter deals that would separate its national oil market from the markets of the other industrialized countries. Besides, practically all—with the exception of the predictable holdout, France—have joined the International Energy Agency, strong evidence of their desire to be able to cling to one another in times of stress.

Nevertheless, relations among the industrialized countries can easily undergo substantial modifications over the next decade. One possibility, for instance, is that the United States might lose interest in a joint approach with Europe in defense, finance, and trade, partly out of petulance and resentment over its being cast as the archvillain in these relationships. Another possibility is that the United States may prove intractable in the handling of its relations with Europe. Even though the United States in principle may favor the continuation of transatlantic links, it may handle its day-to-day relations in ways that imperil what is left of the Atlantic alliance. Two of those links are the presence of American troops in Germany and the tradition of friendly ties with Britain. Neither of these links seems so robust as to be invulnerable in the mid-nineteen-seventies.

Even if the United States does not take steps that would further weaken its ties with Europe, the European countries themselves might conceivably do so. Europe's choices in the handling of the oil crisis have been reminiscent of its choices in the handling of its military defense. In oil, as in military defense, Europe's basic choice is either to pool its strength with the United States or to face its problems alone. Pooling strength, as some Europeans see it, entails high risks and doubtful benefits. The propensity of the United States to take a hard line with the oil-exporting countries, according to this view, does not serve European interests. Neither does the United States' policy of close identification with Israel. In oil, as in defense, therefore, Europe may see some considerable virtues in pursuing an independent course.

Sharp shifts in policy, such as are suggested here, are not to be excluded from the range of future possibilities. Whether they come about depends in part on whether the separate European states are found capable of developing common policies and common courses of action in foreign affairs. Many Europeans think that the various European states are already developing a capacity for joint political action that is new and important. They see the continual interplay of such men as Giscard and Schmidt to be something more than the ordinary consultations of heads of state, perhaps even something approaching de Gaulle's *famille de patries*. The ease with which European governments consult and debate, though not in the mold of Monnet's European conception and not likely to lead to political union, is regarded as sufficient eventually to produce a joint European will and joint European political action. Other Europeans, however, are far less sanguine and see the present state of European unity as parlous and untenable in the long run.

The studies in this volume were not designed, of course, to lay a basis for assessing the likely pattern of future relations among the rich industrial countries. Moreover, even if the groundwork for such an assessment had been carefully and systematically laid, it could not escape the criticism of appearing too facile. Yet it matters a great deal in the future course of world oil trade whether or not the oil exporters are selling to markets in Europe, Japan, and the United States that are relatively open and whether those countries are prepared to accept a certain amount of continual interdependence among them. So far, the open pattern of past decades seems to be holding. The widespread assumption implicit in some of the papers in this collection is that it will continue to hold for a while longer.

The Ideal and the Likely

The shifts in power that I have sketched in this brief essay appear in retrospect to be limited both in nature and in magnitude. A new group of rich countries is seen as emerging, with both added strengths and added interdependencies. The idea of a powerful Third World bloc, based on the wealth of the newly rich, is thought unlikely. The old group of international business leaders is seen as declining in power, though the place of multinational enterprises in the aggregate continues to be substantial. There are shifts in the position of the United States relative to Europe and Japan, but not so much as to obliterate some of the advantages of America's sheer size.

Though the outline that I have sketched strikes me as plausible and realistic, it would be overstating the case to suggest that it is highly probable; no single pattern can be regarded as highly probable in a situation so loaded with potential crisis. All one can say is that the outline is as plausible as any other, and, in my view, more likely than most. It would be altogether wrong, however, to construe this speculation as an outline that is preferred either by myself or by my colleagues participating in this exercise. It is easy to imagine other distributions of power that would offer greater prospects of equity and tranquility for a troubled world. In fact, the patterns that have been suggested here promise to sharpen some existing problems that already threaten the maintenance of a tolerable and equitable international order. The fundamental dilemma of the modern state, for instance, will still remain: how to retain the national control needed to respond to the political demands of its citizens while guaranteeing the openness of national boundaries that seems indispensable for a nation to achieve acceptable levels of choice and efficiency.

The basic structure of political power and legitimacy in the world will remain a jumble of fearful and quarrelsome nation states, while a considerable part of the economic structure will continue to depend on a cluster of multinational firms with goals fixed in their own survival and growth. So far as the international economy is concerned, nowhere can one see on the horizon a set of institutions with the legitimacy and power capable of speaking and acting for the collective interests of mankind.

REFERENCES

[1] See Hollis B. Chenery, "Restructuring the World Economy," *Foreign Affairs*, January, 1975, pp. 242-63.

[2] For a heroic effort of this sort which succeeds mainly in underlining the difficulties, see K. Farmanfarmaian and others. "How Can the World Afford OPEC Oil?", *Foreign Affairs*, January, 1975, pp. 201-22.

[3]For a full development of this line of analysis, see Raymond Vernon, "The Location of Economic Activity," in John H. Dunning, ed., *Economic Activity and the Multinational Enterprise* (London, 1974).

[4]An elaborate account of the process appears in Neil H. Jacoby, *Multinational Oil* (New York, 1974).

[5]So far, the results of the studies are unpublished. For preliminary results covering four industries, however, see "Competition Policy Toward Multinational Corporations," *American Economic Review*, Papers and Proceedings, LXIV:2 (May, 1974).

IAN SMART

Uniqueness and Generality

I

WHETHER WE TAKE A PARTICULAR "CRISIS" to be unique or representative of a class depends not only on our judgment and the interpretation of evidence but also on our definition of the "crisis" and our delimitation of the class to which it might belong. Defining the "crisis" with which the papers in this collection are concerned is a problem in itself.

An actual or impending "energy crisis" had been advertised long before October, 1973, predicated in part on a belief that oil supplies would be inadequate to meet longer-term demand, either because they would eventually be exhausted or because those in control of production would not want (or be able) to produce oil at a sufficient rate. After October, 1973, a number of associated but distinct collections of events and perceptions existed simultaneously, any one of which could have been described as a crisis in its own right, and some of which were. In the first place, Arab oil producers, with the exception of Iraq, took steps to restrict the production of oil and its export to Western markets, with the explicit objective of enforcing changes in the external political attitudes of the Western countries concerned. In the second place, the members of the Organization of Petroleum Exporting Countries (OPEC) successfully imposed a series of extraordinary increases in the prices of their product, with the stated aim of raising their income and of altering the terms of trade in their own favor. In the third place, the fear that oil supplies might, in the longer term, be insufficient to meet demand was reinforced by the fact that price increases diminished the amount of oil producers would have to sell in order to secure a given revenue.

By the middle of 1974, the politically motivated restriction of exports by Arab producers—the "oil weapon"—was no longer in effect. The perception of crisis remained, however, reflecting primarily the shorter-term effects of increased oil prices, but also the longer-term possibility of insufficient supplies. Where, in this confused territory, are the frontiers of our "crisis" to be drawn? For convenience, the authors of these essays have decided to accept a fifteen-month period, beginning in October, 1973, as the temporal extent of their collective investigation. But whether the collection of events and decisions within those limits is to be regarded as a crisis is as much for the readers as for the authors to decide.

259

Delimiting a class of crises and deciding what we want to know about it is no easier a problem than defining a particular crisis. The efforts of political scientists to classify and analyze crises have generated a substantial literature,[1] but, as is so often the case, the area of consensus has seemed to shrink, rather than expand, in proportion to the volume of writing. Most political scientists would nevertheless agree on three very general characteristics of a "crisis": it causes, or threatens to cause, a considerable degree of change from the preceding or existing situation; the change in question affects something of great value to the people concerned; and the pace of events or essential decisions is greatly accelerated. Others—including, unfortunately, many politicians and several writers on economic affairs—seem to be less concerned that the term should mean anything in particular, provided it is available as a literary or rhetorical convenience.[2]

Taking it as a member of a class, there are four things we might want to learn more about by investigating a particular crisis. We might want to learn something about *when and why crises occur*: about the circumstances in which some category of human affairs characteristically moves from a state that is "normal" to one that is "critical." We might want to learn about *what happens or should happen within crises*: about the characteristic pattern and sequence of events in critical situations or about the "best" way of coping with such situations. Parallel but subtly distinct is the desire to learn about *how people behave in crises*: We might want to learn about the typical reactions of those who find themselves directly involved in a critical situation. Finally, we might want to use crises as a research tool in order to discover *what crises reveal about the nature of politics*, whether within national societies or beyond their borders.

One thing that it is pointless to ask about a crisis is whether it is "unique." On one level, all events and combinations of events are unique. On another level, they are invariably analogous, in however small a measure, to other events or combinations. None of us expects to encounter an exact replica of a past crisis. What we may expect is that the experience or study of a crisis will yield a certain amount of information that can be applied by analogy. Once we have chosen the collection of events or occasions with regard to oil to which we shall attach the term "crisis," we may therefore look for some light on when, why, or if analogous crises may occur in the future. Beyond that, we may ask what, if anything, we can learn from this particular crisis about politics, at least at the international level. Some have seen the events of 1973-74 as the precursor of recurrent crises over oil. Others have held that they point to a wider change in the distribution of bargaining power between the producers and consumers of all natural resources, modifying the patterns of international politics and thereby exemplifying a new type of crisis. Shall we, in fact, ever see the like of this crisis again? And, whether we do or not, should we regard it as an indication of some general evolutionary or revolutionary change in international politics, or merely as an indication of the extent to which the structure of international politics can accommodate ephemeral distortion?

II

The unusual complexity of the events and changes in the international oil market during the fifteen months after October, 1973, has already been made apparent in this volume. Numerous actors were involved, including international and national companies and governments throughout the world. Against the background of what had, in any case, been a growing concern about the long-term sufficiency of oil supplies, these actors found themselves facing a coincidence of two suddenly appearing circumstances: the Arab restriction of oil exports and the OPEC multiplication of oil prices. Even when the restriction on exports had ceased, its memory and the possibility of its recurrence commanded attention. Moreover, for at least three months after the middle of October, 1973, events had followed one another at an extraordinary pace, and some of the changes that occurred, especially in oil prices, were of an extraordinary dimension.

It was this concatenation that, more than anything else, distinguished the circumstances in and after October, 1973, from previous episodes in the international oil market. Some Arab oil producers had, after all, been known to impose relatively brief embargoes on selected Western consumers before, notably in 1956. Some West European countries had faced an oil-supply shortage in 1967, when the Suez Canal was closed again. Individual producers, such as Libya, had previously withheld oil supplies on political grounds. Numerous producers—Kuwait, Libya, Algeria, Venezuela, and, of course, the United States—had previously acted through national legislation to limit or even reduce oil production on economic grounds. OPEC as a whole had already exploited its collective bargaining power to support or raise oil prices, especially in and after 1970. None of the circumstances of 1973-74 was, therefore, intrinsically novel. What was novel was that, in October, 1973, market conditions and international politics combined to bring together such a wide range of circumstances, affecting so many people within such a short space of time.

Among the important reasons for the explosive character of that convergence was that a few countries in the Middle East were in a position in 1973 to act with unprecedented effect simply because they were the dominant suppliers operating in a relatively tight market, while, in the Iranian nationalization attempt in 1951 and the temporary shortages associated with Arab-Israel wars in 1956 and 1967, consumers had had no great difficulty in obtaining alternative supplies. Even then, the effectiveness of the producers' actions in 1973-74—whether OPEC's action in raising prices or the Arab states' action in restricting exports—also depended upon their political cohesion as a group and upon the existence for a sufficient number of producers of a common political motive. Economic circumstances alone might have provoked some increase in the price of oil, but certainly not a quadrupling of prices within twelve months. Nor were economic considerations alone in prompting fears about the longer-term supply of oil. Least of all was it economics that gave the Arab states a motive, as distinct from an opportunity, to employ the oil weapon against Israel and the West. By the same token, it was not general economic forces but particular and very specific political circumstances that brought these three factors together to form the total shape of the 1973-74 oil crisis. The complexity of this crisis combined with the degree to which its complex structure was the product of specific political circumstances is what makes it extraordinary. To that extent, it is to be treated with some caution as a basis for extrapolation.

A further reason for caution is that, even in a relatively few months, the context within which the oil crisis began has been considerably modified. The Arab-Israel conflict, inter-Arab politics, and relations between Arab oil producers and industrialized importers have all evolved. Steps have been taken toward closer cooperation between the major oil consumers of the industrial world. The delicate balance between supply and demand in the oil market, which served as a major premise of the crisis, has significantly shifted. Meanwhile, trepidation over the impending effect on national economies and international monetary arrangements of massive financial transfers to oil exporters has persisted, and even intensified.

Some of the changes that have taken place in the Middle Eastern situation since October, 1973, are themselves traceable to the oil crisis. The disengagement agreements, which removed Israeli forces from some parts of the territory taken from Egypt and Syria in 1967, were negotiated by an American secretary of state moved to a new urgency by the use of the Arab oil weapon. The efforts of Japan and the European Community countries to conciliate the Arab states have had similar roots. On the one hand, all this has served to reinforce the oil weapon; it has been shown to have some effect, and its further employment has thus been encouraged. On the other hand, so long as the threat of the oil weapon spurs the United States to seek further Israeli withdrawals, it may seem wiser to hold the weapon in reserve, rather than to risk the complete alienation of the United States by its repeated use. The eagerness of the main Arab producers in the Gulf area to use the oil weapon may, in any case, have been modified by a shift in the main focus of the Arab struggle against Israel. Saudi Arabia may have been more inclined to deploy the oil weapon in support of Egypt or King Hussein of Jordan—as it was deployed in 1973—than to use it in the interests of a Ba'thist Syria, strongly supported by the Soviet Union, or of a Palestinian movement in which important elements are ideologically committed to the overthrow of monarchic government.

Over all these thoughts, there falls a shadow of uncertainty. Above all, there is the uncertainty of another Arab-Israel war. Such a war would not necessarily bring with it a reimposition of restrictions on the supply of Arab oil to Western markets. Much would presumably depend on the willingness of Western governments openly to support the Arab side, or at least to stand aloof from Israel, which would in turn depend on the circumstances. It would nevertheless be prudent to assume that war will increase the probability of the oil weapon being used, if only in an effort to secure the termination of hostilities in the Arab favor. In war, it will be politically difficult for even conservative Gulf producers to forego the use of the oil weapon, especially when its previous use had such an impact.

The oil weapon's impact obviously depends in part upon the extent to which the West relies upon Arab oil. Here the situation has also changed since October, 1973. First, the international companies and many of the industrialized countries have now built up larger stockpiles of oil. A commitment to maintain stocks (including "shut-in" spare production capacity) equal to sixty days of consumption has been accepted by the governments that established the new International Energy Agency in 1974. Although all have not achieved the agreed levels, probably enough oil and products have been stockpiled within the OECD countries to last between two and three months.

Another change is the new effort that has been made through the International Energy Agency to ensure the efficient sharing of available oil in the case of sudden scarcity. The IEA program, of which that arrangement is a part, has obtained less than unanimous support among the OECD countries and has yet to be reflected in national programs or proved in practice. But its existence and possible effectiveness must already enter into any calculation by Arab exporters of their ability to impose discriminatory pressure upon Western governments.

A third factor is that, since the 1973-74 oil crisis, higher oil prices and the general deceleration of economic activity in the industrial world have combined to slow the growth of world demand for oil; despite cutbacks, world rates of oil production have continued to run ahead of consumption.

Finally, the 1973-74 crisis has had the effect, despite a general economic sluggishness, of accelerating the development of alternatives to Middle Eastern oil. Any substantial production from these alternatives is bound to be many years away, but their prospect may nevertheless carry some psychological weight.

For the moment, however, the industrialized countries—to say nothing of the less-developed nations without fuels of their own—remain critically dependent upon the flow of oil from OPEC members, and especially from the Middle East. The economies of Western Europe and Japan could not withstand a total cessation of Middle Eastern oil supplies for more than a relatively few weeks. Even the United States might eventually be hard pressed without them.

One development that may actually make a relatively prolonged suspension of Arab oil exports to the West more plausible is that the unspent revenues of OPEC governments, still largely in the form of short-term deposits, are continuing to grow.[3] Apart from the fears that the international financial system will be disrupted or that certain producers will soon decide to limit their production on economic grounds, the danger is thus created that Arab producers may attempt for political reasons to use part of their accumulated surpluses to finance a more prolonged suspension of oil supplies to at least some of the OECD countries. Arab producers may well have accumulated surplus funds of some 75 billion dollars by the end of 1975. If redistributed, that amount would in theory be sufficient to sustain the national economies of the whole group at 1974 rates for a period of at least eighteen months, without any further income from oil whatever. If exports to non-OECD states were not cut off, the term might easily be extended to two years. The OECD countries are in no position to withstand a total embargo for anything like such a time.

In practice, it would be difficult for the Arab producers to finance a total embargo for so long. Some of their financial reserves, although technically in liquid form, could not easily be withdrawn quickly. In any case, some OECD countries might respond to an embargo by freezing Arab balances within their own jurisdictions. Against this, Arab governments might first take the precaution of converting some of their reserves into stockpiles of commodities or withdrawing them in the form of gold (as Saudi Arabia withdrew some $100 million of gold from New York in 1974). Even the threat of such withdrawals would greatly alarm many OECD governments, dependent as they are upon the storage of Arab reserves in Western banks. Moreover, if an embargo were selective among OECD countries and effective in the countries where it applied, those that were exempt might be loath to expose their own situations by freezing Arab funds or to give up a lucrative export trade to Arab markets.

It is not easy to summarize these reflections upon a possible recurrence of an oil crisis along 1973 lines. In any case, "recurrence" is not a satisfactory term. One of the dimensions of the 1973-74 crisis—the dramatically higher price of oil secured by OPEC's classic exercise in cartelization—still persists. Another—the longer-term threat that strong economic reasons may prompt certain suppliers drastically to cut production—remains a future danger. As to the third of the main dimensions—the political use of the oil weapon itself—the arguments for fearing its recurrence are delicately balanced. The cautious conclusion should probably be that a major resurgence of this third element in the 1973-74 crisis is, on the whole, unlikely, unless Arab-Israel war is renewed, but that, if war does break out, the employment of the oil weapon is more probable than not, while the possibility that its application might in the future be more protracted is alarming.

III

Not surprisingly, the 1973-74 oil crisis aroused both alarm and ambition with regard to other natural resources. OPEC had apparently carried out an overwhelmingly successful economic coup by operating as an efficient price-fixing cartel. At the same time, the Arab members of OPEC had successfully pursued political objectives by the use of economic sanctions. The importers and exporters of other commodities immediately began to look at their own situations. Whether or not the crisis proved to be unique so far as oil was concerned, might it, they asked themselves, offer a precedent for the exploitation of quasi-monopolistic power over other non-renewable mineral resources, such as metals, or even over ostensibly renewable agricultural resources, such as grain?

There are, however, certain limits on whether any supplier or group of suppliers can act, like OPEC, as an effective price fixer. One typical study asserts four main conditions for the successful manipulation by a group of producers of prices in a world commodity market:[4]

a) the group must control a sufficiently large share of world exports, world production, and, for mineral resources, world reserves;

b) the price elasticity of demand for the commodity in question, including the cross-elasticity with possible substitutes, must be sufficiently low;

c) the price elasticity of alternative supplies of the commodity from other producers must also be sufficiently low;

d) the group itself must be sufficiently cohesive to prevent individual members from pursuing their own advantage through unilateral action in the market.

These criteria are not immune to criticism. In particular, it is not clear that a large share of world production, as opposed to world export, is always a necessary condition of success. Nor, except in the very long term, is the proportion of ultimate world reserves always relevant. What is more troublesome about the criteria, however, is that they are unavoidably imprecise. "Sufficiently" is a flexible standard. The most that can

be said in general terms is that the probability of success increases as the conditions described become more pronounced. Even then, the important element of time is missing.
Price elasticities of both demand and supply may vary significantly between the short
and the long term. Demand may be relatively inelastic in the short term, perhaps because the substitution of an alternative material requires investment in new plants that
will take some time to deliver and install, but much more elastic in the longer term,
when this new equipment and the substitute material become available. In another
case, the short-term elasticity of demand may be relatively high, but demand may
strengthen again as consumers adapt to a higher price (a pattern that has, for example,
been characteristic of the response to increases in governmental taxation of tobacco and
alcohol). Similarly, the short-term price elasticity of alternative supplies of natural resources is normally likely to be lower than the long-term elasticity, simply because it
takes time to develop alternative sources. In certain cases, however, the limited size of
the alternative supply (such as obsolete scrap of certain metals) may mean that the
price elasticity of alternative supply will actually diminish over time.

These reservations and riders notwithstanding, the criteria mentioned offer some
guide to the possibility that any world commodity market can be successfully manipulated. The actions of the OPEC countries during 1973-74 provide an illuminating example of the criteria in operation. OPEC member countries in 1973 controlled 86 per
cent of world crude-oil exports, 52 per cent of world production and about 65 per cent
of proven world reserves. (The Arab states in the Organization of Arab Petroleum Exporting Countries [OAPEC] controlled 48 per cent of world exports, 31 per cent of
world production and about 50 per cent of world reserves.) There could be little doubt,
therefore, of OPEC's immediate ability to satisfy the first of the criteria mentioned.

As to the second of the criteria, both the short-term price elasticity of demand for
oil itself and the cross-elasticity in relation to substitute materials, at least on the evidence of 1973-74, were remarkably low. Between 1968 and 1973, the world demand
for oil grew at an average annual rate of 7.5 per cent. Between September, 1973, and
January, 1974, the f.o.b. price of the crude oil used as a benchmark (Arabian light
34° f.o.b. Ras Tanura) rose by 444 per cent, with other OPEC prices following suit.
World demand in 1974 was nevertheless almost one per cent *above* the 1973 level, although it falls by rather more than one per cent if the Soviet Union, Eastern Europe,
and China are excluded from the calculation. Obviously, therefore, the percentage fall
in demand was far less than the percentage rise in price, even when only market-economy countries are considered and allowance is made for the previously rising trend of
consumption. On the face of it, the demand in market-economy countries was about 8
per cent lower than the level previously projected, whereas the price had risen by well
over 400 per cent, implying an elasticity, even if all the demand constraints were at-
tributable to the higher price alone, of about (minus) 0.018! The figure itself may well
be relatively meaningless; a smaller increase in price might have provoked a comparable reduction of demand. It was nevertheless clear that the second of the criteria, at
least in the short term, was satisfied.

So, in that same short term, was the third. Indeed, there was never any serious
doubt in late 1973 that the short-term price elasticity of alternative, non-OPEC, oil

supplies was low. In contrast to the situation in 1956 and 1967, the balance between oil supply and demand was tight. The impact of the Arab export limitations was marginally reduced by the ability of some non-Arab producers in OPEC, such as Iran and Nigeria, to increase their production slightly (as well as by the refusal of Iraq to support the OAPEC policy on exports). There was very little scope, however, for a swift increase in production from non-OPEC sources, so that the short-term price elasticity of supply in relation to OPEC's action on prices was clearly low.

The fact that these first three criteria were amply satisfied helped, moreover, to ensure the satisfaction of the fourth. The circumstances of the market were so clearly favorable to collective action by OPEC members on prices that their cohesion as a group was hardly strained, despite the divergent characters of their governments and of their essential political objectives. At the same time, any tension or hesitation there might otherwise have been within OPEC was largely overwhelmed by the powerful sense of common political, as well as economic, purpose generated in October, 1973, among the group's Arab members. That is not to say that OPEC's cohesion will necessarily remain so high in all future circumstances. Some members of the group are more dependent than others upon the revenue from oil exports. Some are more vulnerable than others to countervailing economic pressure by consumers or by the forces of the market. Minor cracks already appeared in OPEC solidarity as the depression of demand, in late 1974 and early 1975, created pressure to constrain production and cut prices. The fact remains that solidarity was high in October, 1973, and that it remained so throughout the fifteen subsequent months.

On all this evidence, it is hardly surprising that OPEC was able to operate, in and after October, 1973, as a successful cartel. It does not follow, however, that the success can be sustained or repeated indefinitely. The short-term price elasticity of demand for oil may be low, but the medium-term and long-term elasticity is quite likely to be higher. Efforts in consuming countries to use oil more efficiently or to develop means of substituting other materials for oil will take some time to bear fruit, but they are calculated to reduce the relative demand for oil considerably when they do. The long-term price elasticity of alternative supply is also likely to be higher than its short-term counterpart. OPEC action in 1973–74 created an incentive to explore more actively for oil outside the OPEC world and to develop those non-OPEC reserves. The United States Congress was prompted, after apparently interminable delay, to approve the construction of a pipeline to bring oil from the North Slope in Alaska. A considerable stimulus was given to oil exploration and development in off-shore areas such as the North Sea. Interest was aroused or increased in the development of on-shore oil resources in Soviet Siberia and China. Again, the results of all this will not be apparent for some years, nor should their probable extent be exaggerated. Estimates vary, but my own rough calculations suggest that reactions to the 1973–74 crisis may turn out to add no more than 3 to 3.5 million barrels a day by 1980 to what would otherwise have been the level of production from "conventional" oil reservoirs outside the OPEC countries, although that additional element of production might rise in the longer term, say, by 1985, to something like 9 million barrels a day.[5] Nevertheless, it is clear that even by 1980 the ability of consumers, as a result of increases in OPEC prices, to draw oil from non-OPEC sources will be somewhat greater.

Any successful action to manipulate market prices is inherently likely to increase the longer-term price elasticity of supply for the commodity concerned. In the case of oil, the OPEC producers' actions in 1973–74 clearly did so. Ideally, a cartel seeks to raise the price to a level that gives its members a substantial monopoly profit but that remains below the level at which large-scale production by non-members becomes economical. The OPEC increases clearly went beyond this ideal limit, and made it economical for non-members to exploit even the high-cost oil reserves of an area such as the North Sea. They have also approached a level at which it may well be economical to produce some oil from "unconventional" sources as well, such as tar sands, shales, or even coal, though such sources can hardly contribute significantly to non-OPEC production capacity in the medium term.[6] In the long-term perspective of the nineteen-eighties and nineties, OPEC's insistence on so large a price increase in 1973–74 has opened the way to a substantial increase in non-OPEC supplies and, to that extent, may yet turn out to have undermined OPEC's own bargaining position.

By exploiting their position to an extent that has substantially increased the long-term price elasticity of both demand and supply, the OPEC countries may have purchased an immediate tactical victory at the expense of an eventual strategic defeat. The pattern is familiar. British cartelization of Malayan and Ceylonese rubber in the nineteen-twenties achieved an immediate price increase, but it promoted massive investment in rubber production in other countries, such as Indonesia, on a scale that provoked a subsequent price collapse. Brazilian efforts to boost the world coffee price in the late nineteen-twenties by restricting exports similarly produced an incentive to expand coffee planting that again led to an eventual slump in prices.

The OPEC countries might appear to be riding for a similar fall. However, the OPEC members in general and the Arab states in particular do have one factor of enormous importance on their side, namely, time. To establish a new underground coal mine, to develop a new off-shore oil or gas field, to build a new commercial plant for extracting oil from tar or shale, to design and construct a new nuclear reactor all take between six and twelve years. By that time, OPEC countries, whose underlying production costs represent only a very small part of the price now commanded by their product, may find themselves in a position to cut their prices and expand their production in order to maintain a dominant share of the market. In the interim, they will have had the benefit of several years of exports at a much higher price. At least for the near future, the primary factor that preserves OPEC from the fate that befell the over-ambitious cartels of the nineteen-twenties is that substitutes for OPEC's oil will be so long in coming.

IV

Any attempt to extrapolate from oil and its crisis to the cases of other resources must answer two questions. First, are there likely to be opportunities to manipulate other commodity markets to an economic end, comparable to the opportunity seized by OPEC in 1973–74? Second, are there opportunities and motives for producers of other commodities to withhold supplies in pursuit of a political end, as Arab oil producers withheld supplies in 1973–74?

The cases which have to be examined to answer these questions fall into two broad groups: 1) non-renewable resources—minerals, especially metals—and 2) renewable resources—agricultural products, especially foodstuffs. A number of useful surveys have been made for part or all of the field, prompted to some extent by the 1973–74 oil crisis itself.[7] It may be sufficient, therefore, to set out the situation in broad terms.

In the case of non-renewable mineral resources, one consideration that bears upon any attempt to extrapolate from the 1973–74 oil crisis is the unusual character of oil itself. There is probably no use for which crude oil is absolutely indispensable; internal combustion engines can be run on other fuels such as methanol, and even lubricants can be synthesized. But oil nevertheless enjoys extraordinary advantages in relation to other minerals because it performs so many functions better than any substitute can at a remotely comparable price. None of the metals can claim this high resistance to rational substitution over such a range of functions. Besides, metals can usually be recovered from scrap and recycled for a new purpose; oil once used is used forever. These qualitative peculiarities of oil are reflected in the quantitative position it has occupied in international trade. Even in 1968, when oil prices were at their lowest in modern history, no other mineral constituted so large a proportion of total world mineral production, by value, as did oil at 24 per cent (Table 1).

The character of the material aside, the relative bargaining power of the exporters of any resource depends on at least two further considerations: the extent to which the pattern of world production of the resource is geographically (and politically) matched to the pattern of its consumption, and the extent to which its injection into international trade is controlled by any particular group of exporting nations that might be inclined to act collectively. If the majority of consumers can achieve or approach self-sufficiency in production, those who export are unlikely to command great influence. If the major exporters are politically incapable of collective action, no cartel or other form of joint action to control the market is possible.

If we examine the cases of a few of the most important minerals exchanged in international trade and consider, in doing so, the extent to which demand in the industrialized market-economy countries can be satisfied from indigenous production, we find that oil is again an unusual case (Table 2). Although the United States provided 69 per cent of its own petroleum requirements in 1972 (a figure which had fallen by 1974 to 65 per cent), Western Europe and Japan were almost totally dependent on oil imports. In all other cases, including those of the three most important metals—iron, aluminum, and copper—some significant part of demand—at least 15 per cent—could be met from indigenous primary production or the recycling of scrap. At the same time, self-sufficiency was low enough in a number of cases to imply vulnerability to the restriction of imports by foreign suppliers. The United States produced only 15 per cent of its aluminum requirements. Western Europe produced only 41 per cent of the aluminum it needed and only 28 per cent of the copper. Japan, most vulnerable of all, could not satisfy its own requirements for more than 27 per cent of iron, 34 per cent of aluminum, 23 per cent of copper, and 35 per cent of coal.[8]

In certain cases, the contribution made by indigenous production could undoubtedly be increased if the pressure and the time available to do so were both sufficient. In

Table 1

Value of World Primary Mineral Consumption by Processing Sectors:
1968 Quantities in 1973 American Dollar Equivalents

	in billions of dollars	per cent
Energy		
Oil	38.7	23.9
Coal	14.8	9.1
Natural gas	6.0	3.7
Peat	3.1	1.9
Other	1.7	1.0
Total	64.3	39.6
Non-metallic		
Sand, gravel, stone	22.2	13.7
Clay	1.8	1.1
Chlorine	1.7	1.1
Other	12.7	7.8
Total	38.4	23.7
Non-ferrous metals		
Copper	7.7	4.7
Aluminum	6.1	3.8
Zinc	1.8	1.1
Lead	1.1	0.7
Other	7.5	4.6
Total	24.2	14.9
Ferrous metals		
Pig iron	32.2	19.9
Nickel	1.3	0.8
Other	1.8	1.1
Total	35.3	21.8
TOTAL (all sections)	162.2	100.0

Source: Adapted from P. Connelly and R. Perlman, *The Politics of Scarcity* (London and New York, 1975), pp. 10–11.

Table 2

Self-Sufficiency in Major Minerals (1972-73):
Area Production from Primary Sources and Scrap as Percentage of Area Consumption

Commodity	United States	Western Europe	Japan
Petroleum	69	3	0
Coal	100	83	35
Iron	75	86	27
Aluminum	15	41	34
Copper	91	28	23

Source: Adapted from P. Connelly and R. Perlman, *The Politics of Scarcity* (London and New York, 1975), p. 98.

particular, the recycling of scrap, which already provides 20 per cent of the world's aluminum production, 35 per cent of its steel, and 40 per cent of its copper, might be increased. The increase could not be instantaneous; processing plants would have to be installed or adapted and additional steps taken to promote the return flow of consumer-generated scrap (since a high proportion of industrial scrap is already recycled). Nor could it, in all cases, be substantial; some countries are already recycling much of the scrap they can obtain. Nevertheless, secondary production from scrap could do something toward cushioning the major OECD countries against the possible effects of collective action by metal exporters.

In some cases, major mineral importers might expand their own primary production in order to reduce the coercive power of outside suppliers. In the United States and some other industrialized countries, for instance, aluminum production from aluminiferous clays might be encouraged, even at higher costs. In general, however, an option of this sort is essentially one for the longer term. In the short term, the more important option, even more relevant to metals than to the case of oil, is that of building up substantial stockpiles of the commodity within importing countries. A stockpile of an imported metal sufficient to meet domestic demand for some months will be smaller in total volume and cost and will entail fewer problems of special storage than those associated with stockpiles of oil.

Whether the exporters of a commodity are in a position to act together as a cartel depends, as we have seen, not only upon their economic ability to do so effectively but also upon their political will. Since the establishment of OPEC in 1960, and especially since 1970, its members have sometimes been at loggerheads with each other, but their ability generally to hang together has nevertheless been impressive. A number of factors have operated in their favor: the geographical concentration of so many of the major sources of oil exports in the one region of the Middle East and North Africa; the preeminent role of a few international companies operating in almost all the OPEC countries and thus providing a common experience and, as it were, a common target for OPEC members; the fact that oil production and export, while contributing so much to government revenue and foreign-exchange earnings in exporting countries, requires relatively little labor and thus represents a relatively minor element in the pattern of local employment; the extent to which world prices for oil, unlike those for many metals, have been immune to sudden downward movements and have indeed, since the late nineteen-sixties, been rising steadily or dramatically. All these things have made it easier for oil-exporting countries to constitute and maintain an effective bloc. Two other factors may, however, be even more important to an explanation of OPEC's recent success. The first is the existence within OPEC of a subgroup of Arab countries, including some of the world's largest exporters, predisposed by considerations separate from oil to act together in pursuit of political ends. The politically motivated solidarity of OPEC's Arab members, especially with respect to the Arab-Israel conflict, has done much to provide a stimulus for collective action by the group as a whole, and has thus helped to preserve OPEC from some of the fissiparous pressures which, even in its own particularly favorable market circumstances, might otherwise have endangered its cohesion. The second factor is success. After a period of mixed fortunes in the nineteen-sixties, OPEC has experienced a history of notable achievement on behalf of its members. As with any group, nothing is more likely to bind the members of OPEC together than a recognition that collective action works.

No other major world mineral market is dominated by a group of exporters with so strong a predisposition to intervene collectively in the market as is the case with oil. In many cases, such as copper, the major exporters (Chile, Peru, Zaire, and Zambia) are geographically scattered. In others, such as manganese, the largest producers (South Africa and the Soviet Union) are politically antipathetic. Generally, in fact, the main producers and exporters of non-fuel minerals lack many of the attributes that have made OPEC's cohesion possible. The major exporters of most metals generally include developed and industrialized, as well as developing, countries. It would be dangerous to make too much of that point; some of the industrialized OECD countries have a longer and richer history of commodity-market manipulation than any in the Third World, and recent developments of policy in countries such as Australia and Canada point to the existence there of political pressures to control resource exports more closely. Nevertheless, it is likely to be considerably harder for countries at notably different stages of economic development to combine in an assertive cartel than it has been for the members of OPEC, despite their differences, to do so.

The manipulation of world markets by groups of exporters must also be considered for renewable resources and especially foodstuffs. The precedents are impressive: rubber, tea, coffee, cocoa, meat, sugar, and wheat were all involved, at some time between 1919 and 1939, in restrictive or coercive pricing arrangements. Some of them have since come to be the subjects of intergovernmental commodity agreements, involving consumer as well as producer governments. In the case of most renewable resources, however, the possibilities of rapid substitution are so considerable, and the general price elasticities of demand and alternative supply so large, that effective cartelization by producers, except as a limited response to falling prices, is implausible. Nor are the incentives to any cartel action involving the constraint of production so clear-cut in the case of renewable resources as in the case of minerals. Most of the tropical agricultural products, for example, represent an important component of employment in the countries concerned. Most agriculture also involves a relatively heavy investment that will be lost unless production is maintained. This, indeed, is an essential difference between the non-renewable and renewable resources: mineral reserves not extracted today remain, in principle, to be extracted in the future, but crops not harvested, or harvested and destroyed, are lost, with all their value, forever.

Perhaps the one renewable resource to which many of these remarks do not apply is grain. The international grain trade is dominated by so few exporters—the United States, Canada, and Australia provide over 85 per cent of the total—that cartelization might seem to be rather easy. In practice, there are fairly obvious political and economic reasons why an international grain cartel is unlikely. What is less implausible, but hardly less alarming, is that one producer alone might place restrictions on grain exports. The share of United States exports in the world grain trade is about two and a half times as large as the Saudi Arabian share of the world oil trade—and grain is at least as essential a commodity as oil. Moreover, on at least one occasion, the Arab use of the oil weapon in 1973–74 provoked what was taken to be a thinly veiled threat that the United States might react by restricting food exports to the countries responsible.[9] We may not seriously believe that the United States government would manipulate the international grain market for either economic or political ends—even if history has not always supported that belief—but it would be rash to ignore the sensitivity of grain importers to the theoretical possibility that it could.

With the conceivable exception of grain, it appears that few commodities are susceptible to the type of manipulation in which OPEC engaged in 1973–74. That would echo the conclusions reached in several recent and more detailed studies of individual resources.[10] Nevertheless, efforts to band together with the object of manipulating world markets may well be made by exporters of other resources. The major exporters of copper have been formally united in a producers' group, the Conseil Inter-gouvernemental des Pays Exportateurs de Cuivre (CIPEC), since 1967.[11] In March, 1974, the International Bauxite Association (IBA) was set up by the largest exporters of that material.[12] A meeting of the United Nations Conference on Trade and Development (UNCTAD) in October, 1973, recommended the establishment of a tungsten producers' association. Algeria, Italy, and Spain tried, in 1973–74, to operate jointly to raise the price of mercury.[13] Sri Lanka has sought to unite the major producers of natural rubber. Morocco, supported by Tunisia, took on the role of price leader in the world market for phosphate rock and has succeeded in matching OPEC's performance by more than quadrupling the price. Mexico tried, in 1971, to draw the other largest silver producers (Peru, Canada, Australia, and the United States) into a world pricing agreement. Costa Rica, El Salvador, and Mexico have taken the lead in setting up a new coffee producers' group, *Cafe Suaves Centrales* (CSC), even in the face of a cool reaction from larger producers such as Brazil and Colombia. The moral of the 1973–74 oil crisis is not, therefore, that analogous attempts will not be made to manipulate other commodity markets. It is rather that few such attempts are likely to succeed to anything like the extent that OPEC has succeeded.

The degree to which efforts to control commodity markets can succeed will depend upon the particular circumstance of the commodity in question. In the case of phosphates, for example, the success of the main exporters, led by Morocco, in multiplying the world price owed much to the fact that most of them, as developing African countries, felt some inclination to act together, to the fact that demand for phosphates for fertilizers and detergents was rising rapidly, and to the absence of any other material that could be substituted for phosphate rock. Nonetheless, in few cases will manipulation touch upon such a range of sensitivities as in the case of oil. More important, however, is the fact that very few cases indeed will find resource exporters with so much time on their side as there is on the side of the OPEC countries. Any commodity market can be controlled by a group of major producers—for a time, but few markets other than that for oil could offer the luxury to producers of anything like the same amount of time. A similar attack upon the price of, say, copper might be so ill-conceived—given the consequent incentive for consumers to turn more urgently to other sources or materials—as to be irrational. Still, it would be unwise to exclude the possibility of such irrational behavior on the part of all suppliers, especially in cases where economic apprehension or ambition may be reinforced by a sense of political grievance.

The question of political grievance is, in fact, a central one. The great increase in the price of oil has substantially and directly affected the relative costs of producing some of the other commodities. Sometimes that has been to the advantage of other resource producers, as in the case of the shift in the relative costs of natural and synthetic rubber, sometimes to their detriment, as in the case of agricultural products. However, the most important causal relationship between oil and other commodities lies in the political and emotional impact of OPEC's example. The 1973–74 crisis, perhaps by generating a combative atmosphere rather than by any more specific means, added impetus to the efforts of OPEC governments to wrest control of their national oil resources from foreign companies. The steps taken to expand state control of oil enterprises or to assume full ownership of them may be expected, in turn, to encourage parallel measures on the part of other resource-rich countries. A widespread movement to take national control of mineral production was underway well before October, 1973, in Chile, Guyana, Jamaica, Australia, and elsewhere, but its impetus is likely to be all the greater after the oil producers' success. At the same time, OPEC's success has stimulated the desire of other resource exporters in the developing world to increase or stabilize their income. A number of the OPEC states have not been reluctant to appear as the leaders, as well as the precursors, of that movement. Some have indirectly linked oil to other commodities by offering to use part of their surplus income, produced by higher oil prices, to support the prices of other raw materials such as copper or coffee. In economic terms, the potential ability of other resource producers to impose their will is still likely in most cases to be less than the ability of OPEC to control its own market. In political terms, the strength of the urge to follow OPEC's example is beyond doubt.

Nowhere is this more apparent than in the spate of international meetings that have been held since October, 1973. A springboard was provided by the special session of the United Nations General Assembly in early 1974, at which all aspects of the world trade in raw materials were touched upon, and by the proposal for a conference between oil producers and consumers, different versions of which were originally advanced by Saudi Arabia and France. Since then, activity has increased on a number of fronts. Algeria has taken the lead in arguing that the agenda for a conference of producers and consumers should be extended to cover all natural resources. The UNCTAD secretariat put forward plans for an eleven-billion-dollar buffer stock, financed by both exporters and importers, to support the prices of eighteen commodities that account for 60 per cent of world trade in non-fuel raw materials.[14] Those plans have since been taken up by the UNCTAD committee on commodities.

Meanwhile, the less formal group of non-aligned countries, again largely at the instigation of Algeria, has discussed proposals for a separate fund, ostensibly to be financed largely by OPEC states, that might be used to stabilize raw-material prices. Those proposals, reflected in the "Dakar Declaration" of February, 1975, by non-aligned states, contributed in turn to the debates in the Geneva meeting of the General Agreement on Tariffs and Trade (GATT), which also began in February, 1975, and in which the industrialized countries had pledged themselves to "secure additional benefits for the international trade of developing countries." The same proposals may also have contributed to the British decision, in the same month, to prepare a plan for commodity price stabilization to be discussed by Commonwealth governments, just as they added point to the final conclusion of the Lomé Convention in March, 1975. In the latter, the European Community committed itself to stabilize the foreign-exchange earnings of forty-six developing countries from twelve commodities. The non-aligned proposals, like those of the UNCTAD secretariat and the European Community, will provide an important part of the grist to be milled when the fourth UNCTAD conference meets in Nairobi in 1976.

The movement toward that critical meeting is, in fact, impressive in its scale. Much of what has so far been said about raising and stabilizing commodity prices or about the full control of natural resources by developing countries may turn out to be empty rhetoric. The overall strength of the thrust should not, however, be underestimated, even if the precision of its direction may be doubted. The actions of OPEC or of Arab oil producers in 1973-74 did not initiate the movement by developing countries and other natural-resource exporters to obtain fuller control of their production and better returns from it. Nor does the relative success of OPEC's efforts directly enhance the probability that other resource producers will succeed. What the oil crisis of 1973-74 has done, however, is to add enormously, by its example, to the inclination of other resource producers to attempt the imitation of OPEC's success. It has been said of nuclear explosives that the most important technical secret about them is that they work. OPEC's achievement is now seen by many other resource exporters in that light.

The probability of action by resource exporters suggests the possibility of reactions on the part of industrialized importing countries. The oil crisis itself provoked a movement on the part of many importers to arrange for preferential or assured access to future oil supplies through bilateral intergovernmental agreements. Similar efforts to guarantee supplies bilaterally in the case of other commodities may well proliferate. So may other reactions: allocation arrangements, stockpiling, and the promotion of alternative production, even from higher-cost indigenous sources. And one result, prompted in part by a desire to protect domestic development and production, is likely to be a growing tendency toward various types of import restriction by consuming countries, as a response to the export restrictions that so many resource producers now appear willing to envisage. Again, it would be wrong to ascribe the origin of such tendencies to the 1973-74 oil crisis, but its example has done a good deal to accelerate a movement toward what threatens to be a more confused and contentious pattern of international trading relationships during the coming years.

It has become habitual to focus the discussion of cartels and market manipulation primarily upon raw or partially processed materials. But it should not be forgotten that some of the most persistent and tightly organized measures in restraint of trade since 1945 have been those applied by the United States and its allies to the sale of high-technology equipment and certain capital goods and minerals to Communist countries. Computers, aerospace equipment, scarce metals, and even steel pipes have been embargoed in that way. Indeed, comparable restrictions have periodically been imposed on the sale of certain goods even within the same alliance: the United States, for example, has at different times withheld high-powered computers and special steels from France and advanced traveling-wave tubes from almost everyone. Meanwhile, a whole system of restrictions has been imposed by the United States, the Soviet Union, Britain, and, to some extent, France, individually or in concert, upon the export of special fissionable material or a range of nuclear equipment to non-nuclear-weapon states; Article I.2 of the Limited Test Ban Treaty and Article I of the Non-Proliferation Treaty embody clear undertakings to engage in the collective restraint of trade.

Insofar as the actions and successes of the OPEC countries in 1973-74 represent a general encouragement of the exploitation of commercial assets to political ends, it would be unwise to overlook the extent to which a very few advanced countries have an effective monopoly of the current supply of certain important manufactured goods and specialized materials, notably in high-technology categories, and are already accustomed to making use of it. As developing countries succeed in converting their raw-materials wealth into industrial development they will inevitably become more dependent upon that supply. Thus, any impression that the balance of effective bargaining power in international trade has shifted permanently, under the stimulus of the 1973-74 oil crisis, in favor of the Third World producers of natural resources is likely to prove erroneous in the long term. Meanwhile, however, the example of the oil crisis seems bound to encourage other raw-material exporters to reach for whatever short-term advantage their own customers may be forced to concede.

V

Far from being unique, by any general standard, the complex oil crisis of 1973 turns out to occupy a place in a much wider network of developments in international politics. It has neither created nor marked the creation of new forms, sources, or agencies of power. What it has done is to indicate and perhaps accelerate certain shifts in the patterns of power at the international level. By the same token, the crisis has not signaled the invention of new forms of assertive action by resource exporters. What it has done is to demonstrate the scale on which such action may be possible and the weight of the impact it may have, especially when economic and political pressures operate together. The exemplary influence of such a demonstration may be considerable.

To say that the oil crisis was an occasion of modulation rather than creation is not to diminish its significance for the understanding of international politics. A crisis is to the politician more or less what a cyclotron is to the physicist: a mechanism for raising components to a higher level of energy and bringing them into violent interaction. The result may be to provide evidence about the general nature of physical or political forces. Politicians and political scientists will long be occupied in analyzing the evidence thus produced by the 1973-74 oil crisis. For the moment, it may be enough to indicate one or two of the main areas on which their analysis must concentrate.

If we treat the crisis as a window through which to view the dynamics of politics, what can be learned about the relative status and relationships of the political groups involved? We have spoken of countries, for example, as either "producers" or "consumers," "exporters" or "importers," "developed" or "developing." One thing the crisis and the subsequent events have demonstrated is that such convenient dichotomies are even more misleading than was apparent in the past. In particular, the crisis has demonstrated the inaccuracy of the dichotomy between "developed" and "developing," at least to the extent that those terms were taken to be synonymous with "rich" and "poor." The multiplication of oil prices by OPEC states, almost all of which have been considered "developing," has borne heavily upon many other "developing," as well as most "developed," nations, forcing them into a common predicament. "Developed" countries have themselves been affected in very different degrees, depending on their relative dependence on OPEC oil and their widely varying ability to pay its higher price from reserves or additional export earnings. "Developing" countries have been divided in a similar manner. Meanwhile, the OPEC states themselves, although still, in the main, "developing," are on the way toward ownership of an unprecedented proportion of international financial reserves. On the one hand, that points both to the imprecision of any concept of a homogeneous Third World and to some obvious strains within the Western world of "developed" nations; on the other hand, it undermines the conventional equation of "developed" with rich (or "have") and "developing" with poor (or "have-not").

The old distinction becomes even more difficult when we notice the urgency with which some who have been considered "developing" but are certainly rich now seek to imitate others who have been thought "developed" but, in many respects, are in fact poor. With hardly any exception, the OPEC countries are seeking to use as much as possible of their new wealth to industrialize their societies at the maximum possible rate. In the process, from being treated essentially as producers and exporters of other than finished manufactures, they must become relatively more significant, and also more dependent, as consumers and importers—of capital goods and even of resources such as alumina and iron ore for their new industries. The immediate impression created by the 1973-74 crisis may be that the major exporters of oil have gained enormously in bargaining power vis-à-vis the developed industrial nations. So they have, but in a limited context and for only a limited time. In the longer term, they can hardly avoid using that bargaining power to increase their own dependency upon others: upon the established industrial nations for technology and its products, and upon other primary-resource producers for industrial raw materials and the wider range of foodstuffs that a wealthier population will demand.

This, perhaps, is the paradox of development that events since the beginning of the 1973-74 crisis have illustrated—development in the form of industrialization and diversification entails a multiplication of dependencies, and the bargaining power that the dependency of others bestows can only be exerted for development purposes at the expense of assuming a reciprocal dependency. The corollary is equally striking—the industrial nations, in order to survive their dependency upon others, have no alternative but to allow those others not only to move toward them on the scale of "development" but also, in the process, to become increasingly dependent upon them.

The policies of the OPEC countries are, of course, the products of their governments. Here again, the course of events since October, 1973, points to some of the ways in which the character of political actors is changing. Some of the governments concerned are long established; others have only recently emerged from a colonial past. Almost all, however, now find their field of activity and responsibility to be expanding. In the first place, most of them are face to face for the first time with a controlling interest in the production and sale of their own oil. Ineluctably, they are thus drawn into the expansion of their own management role—an expansion that, however conducted, must influence the distribution of power within the governmental structure. In the second place, they are confronted by an unprecedented responsibility for the management and disposition of financial resources. At present, they have nothing but rudimentary banking institutions of their own: a few men with some experience in the conservative management of very limited funds. They must depend, therefore, upon banking structures within the jurisdiction of the oil-importing countries and on international financial agencies over which the oil importers exercise decisive influence. To safeguard their own interests in those circumstances, the OPEC governments must again expand their roles and capabilities in an unfamiliar direction. In the third place, the governments of the OPEC countries are forced to assume new responsibilities and to assemble new capabilities as an adjunct to the process of industrial development. For the moment at least, they may have no choice but to depend upon governmental agencies or commercial firms from the developed world to manage that process on their behalf—as they may have to depend on foreign companies to manage their oil or foreign banks to manage their money. Once more, however, they cannot avoid expanding their own management function, whether to supplant foreign contractors or merely to control what they are doing.

OPEC governments, therefore, are fated to undergo a remarkable degree of change. They must henceforth take on more complex political and administrative tasks, not only at the national but also at the international level. They will have to give more hostages to fortune, internationally as well as domestically. Individuals and groups within them will find more routes to power and must adapt to more fluid patterns of power. In some ways, the governments of these countries may become more prudent as a result; in others, the effect may be the reverse. In any event, they will not be the same kinds of international actors as those to which other governments were accustomed before 1973.

The events since October, 1973, have illuminated and partly impelled an analogous modification in the role of governments in oil-importing countries as well, and especially in the governments of the most developed countries. With hardly any exceptions, those governments were previously content to delegate to other groups, and especially commercial firms, almost all the responsibility for managing the procurement and distribution of oil, together with a great deal of the responsibility for making and executing energy policy in general. They were equally willing to leave management of international financial transactions largely in the hands of a commercial banking industry, lightly harnessed by the control of the central banks. All the assumptions upon which those elaborate exercises in delegation were founded have now been rudely shaken. As a consequence, we have witnessed a helter-skelter race to set up new government departments and agencies and new inter-governmental organizations to deal with oil and energy. We are now seeing a more leisurely—perhaps too leisurely—movement to augment governmental and inter-governmental instruments for the international management of finance.

It is clear that these trends are bound to modify the overall character of the governments concerned and that, in doing so, they will also modify their character as actors at the international level. The major Western governments now feel more immediately responsible for certain functions—the procurement and distribution of oil, the formulation and execution of energy policies, and the management of international financial transactions—than at any time in the past. It may be overly optimistic to expect that they will therefore always act more wisely at the international level than in the past, but they will certainly act on the basis of altered assumptions.

Not only governments are in the course of rapid change; changes of comparable magnitude are even more obviously affecting non-governmental international actors, especially oil companies and banks. The ways in which the role and personality of central banks will change are, for the present, difficult to fathom. As to the oil companies, the impact of the 1973-74 crisis and its aftermath is already dramatic. They are in the process of losing the greater part of two important functions: as proprietors in the production and sale of crude oil, and as the *de facto* agents of most importing governments in managing petroleum procurement and distribution. The outcome is, to some extent, uncertain, but many of the oil companies, having been deprived of the vertical integration for which they strove in the past, may pursue horizontal extension—that is, expansion into general energy enterprises and processing conglomerates—all the more actively. In any case, the oil companies will also not be the same kinds of actors on the international scene as in the past.

Accordingly, the same political and economic forces that converged in the 1973-74 oil crisis are now operating to modify the composition, character, and psychology of some of the most important actors in international politics. That much is revealed by the evidence of the crisis and of subsequent events. What, however, can we say about the mechanisms and processes, as distinct from the actors, of international politics? The question is obviously too large to address in any comprehensive fashion. There are aspects of it, however, that stand out too far to be ignored: the role of force in international affairs and the nature of international bargaining.

It would be overly facile to assert that, because force was not used in an effort to resolve the oil crisis, the role of force in international politics has been diminished. The crisis itself arose, to an important extent, from the international use of force between the Arab states and Israel. Besides, force has rarely been used, domestically or internationally, as a means of regulating prices. It is nevertheless a fact that a small number of countries whose military strength, separately or together, is relatively trivial were able to impose a politically motivated embargo on nations much stronger militarily without even having to consider seriously the possibility of a military reaction. The Western countries, against which Arab economic strength was primarily turned, did not seek to transform their own superior military strength into countervailing power. Instead, they entered into a process of bargaining with Arab governments in which they confined themselves to drawing on components of economic and diplomatic strength even though that could only partially offset the economic strength of the oil exporters. At the same time, they sought to achieve a longer-term equilibrium of bargaining strength through diplomatic efforts to mediate in the Arab-Israel conflict and by striving to reduce their own dependence on Arab oil.

There were many reasons for the industrialized oil-importing countries to refrain from using military force in 1973-74: constraints of a nuclear world, fears of creating serious tensions with the Soviet Union and within the West as well, thoughts of long-term damage to relations with all Arab governments, doubts about the effectiveness of military action, and, above all, the intuitive perception that certain new norms of international behavior had come to prevail. Their restraint cannot be taken as necessarily permanent or unqualified—nor even as an entirely dependable guide to their actions in any future crisis. It does, however, indicate that it is no longer useful to rank the nations of the world only or primarily in terms of their military strengths and weaknesses. In extreme ambition or desperation, it is still to be expected that a nation's resources will be focused upon the creation of military strength and its transformation into power. Short of such extremes, it appears less likely now than in the past that the great majority of nations will resort to the use of force. Instead, they will seek to transform into effective bargaining power other dimensions of national strength, offsetting relative bargaining weakness in one area by exploiting bargaining strength in another. Diplomatic or political weakness may be offset by technical, scientific, or financial strength. A weakness in natural resources may be offset by a relative strength in industrial capacity, scientific and technical skill, or financial sophistication.

One implication of this diversification of approaches to international bargaining, which was seen in the oil crisis, is that bargaining processes are likely to become more intricate. In the past, the perception that force might be invoked operated as a great simplifier. The *ultimate* relevance of force has not been destroyed; it may not even have been diminished. But, perhaps because the potential scale and effect of international violence have increased so greatly during the last quarter of a century, the threshold of force has also been raised. At least between the larger and stronger countries of the world, force is seen to be an option only at the highest levels of tension and risk. One effect is to accord a much larger field of action to non-forcible bargaining. Within that field, the need to exploit relative strength in one area in order to offset relative weakness in another without resorting to force means that it becomes increasingly unusual for any international bargaining context to be isolated. The task of foreign policy becomes more intricate because the reluctance to use force encourages the interconnection of different bargaining processes.

Another implication of the glimpses that the 1973-74 oil crisis has given of international bargaining in action refers back to what was said at an earlier stage about the paradox of development. Just as development multiplies reciprocal dependencies among nations, so the multiplication of such dependencies leads to a pattern of offsetting strengths and weaknesses in international bargaining. As the course of development proceeds, more nations come to be involved together not only in a larger number of markets but also in a wider network of reciprocal dependencies. As that happens, their bargaining power vis-à-vis each other comes to be the sum of their relative strengths and weaknesses in all those contexts, yielding a growing complexity of political cooperation and conflict, which force, except at the limit, can no longer simplify.

In these complex circumstances, the role of any national foreign policy can only be to seek—and accept—some composite equilibrium of gains and losses over a growing range of different bargaining contexts. The implication is that the current tendency of international politics is, in fact, egalitarian. The outcome of an international confrontation can no longer be predicted by referring to some simple hierarchy of omnicompetent national power. One lesson of the oil crisis is similar to that which might have been drawn from the Vietnam war: that, more because of changes in the implicit ethics of international relations than because of shifts in the balance of material strength, relatively small and weak countries can now achieve bargaining equality with larger and stronger states *in particular contexts*. One reason is that they are sometimes prompted to exploit a particular bargaining strength with special fervor or at high risk just because they aspire to an equality of status. Although the OPEC governments in 1973-74 sought to exploit their bargaining strength over oil primarily to their economic advantage and, in the Arab case, to a particular political end, it would be rash to dismiss the insistent undercurrent of status aspirations in their policies. In word and in action, both during and since the oil crisis, they have persistently sought to be recognized by the traditional great powers as equal partners in the operation of an international system. For that reason, no ingenious device to pay for their oil, develop their countries, or manage their money will satisfy the OPEC governments if it is a device over which they themselves have little if any control.

The OPEC countries have found their own particular lever with which to raise their international status. The price they pay will eventually be measured by their assumption of more numerous and onerous dependencies on others. The example they provide will remain. Other developing countries rich in natural resources will be tempted, perhaps unwisely, to imitate them. Few who do so will have the same success, at least in the longer term. By one means or another, however, those other countries will seek to establish and exploit their own bargaining levers and thus to reach for equality of status.

In the face of this general movement, governments in the industrial states of the Western world must decide whether they are determined to act in their traditional guise as great powers, by trying to diminish their particular dependencies on resource exporters while assuming the responsibility for restoring a "proper" order to international trade, international finance, and international politics. Their only alternative is to concede some real and growing role in the management of the international environment to those who, like the members of OPEC, will not be satisfied or even placated without it. They cannot avoid choosing some elements of the great power role; certain needs are pressing, and certain habits die hard. In the short term, indeed, the tactics of the great power may bring success. With the longer term in mind, however, a failure to recognize the secular pressure by other countries for status and participation may mean either turning back the clock to long before October, 1973, or accepting a future of unprofitable and unavailing conflict.

REFERENCES

[1]A valuable general survey of alternative approaches to the study of crisis in international politics is Charles F. Hermann, "International Crisis as a Situational Variable" in J. N. Rosenau, ed., *International Politics and Foreign Policy* (New York, 1969), pp. 409-421. The ideas of system change and system over-

load are discussed more fully in Charles A. McClelland, "The Acute International Crisis" in K. Knorr and S. Verba, eds., *The International System: Theoretical Essays* (Princeton, N.J., 1961), pp. 182-204. One of the relatively rare attempts to apply a definition of crisis systematically in the study of international economic affairs is Edward L. Morse, "Crisis Diplomacy, Interdependence, and the Politics of International Economic Relations" in R. Tanter and R. H. Ullman, eds., *Theory and Policy in International Relations* (Princeton, N.J., 1972), pp. 123-150, especially pp. 126-132.

[2]Consider, for instance, the frequency of allusions to a "dollar crisis" or a "sterling crisis" whenever pressure bears upon the parities of the currencies concerned.

[3]Estimates vary, but the accumulation of OPEC surpluses in the first eleven months of 1974 was reportedly estimated by U.S. Treasury officials as $53 billion, of which some 70 per cent appears to have been in short-term funds (*The Banker*, CXXV.587 [January, 1975], p. 5).

[4]*Trade in Primary Commodities: Conflict or Co-operation?* A Tripartite Report by Fifteen Economists from Japan, the European Community, and North America. (Washington D.C., The Brookings Institution, 1974), p. 28. See also Bension Varon and Kenji Takeuchi, "Developing Countries and Non-Fuel Minerals," *Foreign Affairs*, 52:3 (April, 1974), 497-510.

[5]These calculations are based upon the following components of *additional* production, beyond that anticipated before October, 1973, for the dates in question:

	1980	1985
	(in millions of barrels per day)	
Northwest Europe (off-shore)	0.5-0.6	1.2-2.4
United States (including Alaska)	1.4-1.6	2.5-3.0
China and Southeast Asia	0.9-1.0	1.6-2.0
Other	0.2-0.3	1.0-1.6
Total	3.0-3.5	6.3-9.0

[6]An optimistic estimate might include the following components of "unconventional" oil production in the non-Communist world in 1980:

	(in millions of barrels per day)
Tar sands	0.15-0.20
Shales	0.40-0.50
Coal liquefaction	0.05
Total	0.60-0.75

[7]In addition to the works cited in note 5, a valuable summary is that presented by Hans H. Landsberg in testimony before the Subcommittee on Foreign Economic Policy of the Committee on Foreign Affairs of the U.S. House of Representatives on May 15, 1974, an abbreviated version of which is in *Resources*, 47 (September, 1974), 1-3. The mineral cases are usefully discussed in P. Connelly and R. Perlman, *The Politics of Scarcity: Resource Conflicts in International Relations* (London and New York, 1975), especially Chapters, 2, 4, and 5.

[8]In all these estimates, self-sufficiency is measured by comparing domestic consumption with domestic production from both indigenous primary sources and scrap.

[9]Speech by Vice-President (subsequently President) Ford to the Manufacturing Chemists Association, January 8, 1974.

[10]Including the studies cited in notes 5 and 8.

[11]The members of CIPEC are Chile, Peru, Zaire, and Zambia.

[12]The members of IBA are Australia, Guinea, Guyana, Jamaica, Sierra Leone, Surinam, and Yugoslavia.

[13]The effort to raise and maintain the mercury price was notably unsuccessful. Initially, the price rose from a 1973 low of $260 per 76-pound flask to $350 in early 1974. But by November, 1974, it was $240, and by January, 1975, only $175.

[14]The commodities concerned are corn, wheat, rice, sugar, coffee, cocoa, tea, cotton, jute, wool, rubber, bauxite, alumina, copper, iron ore, lead, tin, and zinc.

Selective Chronology of the Oil Crisis

Date	Location	Event
1973		
September 1	Tripoli	Libya nationalizes 51 per cent of the interests of Esso Libya/Sirte, Mobil, Shell, Gelsenberg, Texaco, Standard Oil of California, Libyan-American (Atlantic Richfield), and Grace.
September 5-9	Algiers	Conference of less-developed countries approves forming "producers' associations," calls for withdrawal of Israeli forces from occupied Arab lands.
September 15-16	Vienna	35th OPEC conference designates Abu Dhabi, Iran, Iraq, Kuwait, Qatar, and Saudi Arabia (the "Gulf Six") to negotiate collectively with the companies over prices. Other OPEC members to negotiate individually.
October 6	Sinai and Golan	Beginning of the Arab-Israeli war.
October 7	Baghdad	Iraq nationalizes the holdings of Exxon and Mobil in the Basrah Petroleum Company, corresponding to 23.75 per cent equity in the company.
October 8-10	Vienna	OPEC ministerial committee of the Gulf Six meets with oil companies' representatives to discuss revision of the 1971 Teheran agreement and oil prices. Negotiations fail.
October 16	Kuwait	The Gulf Six unilaterally raise the posted price of Saudi marker crude from $3.011 to $5.119 per barrel.
October 17	Kuwait	Arab oil ministers agree on the use of oil weapon in the Arab-Israeli conflict, a mandatory cut in exports, and a recommended embargo against unfriendly states.
October 19-20	Riyadh *et al.*	Saudi Arabia and other Arab states proclaim an embargo on oil exports to the United States.
October 23-28		Oil embargo extended to the Netherlands by Arab states.
November 4-5	Kuwait	Arab oil ministers agree to cut production by 25 per cent of the September level, the cutback to include volumes embargoed to the unfriendly states; Arab oil policy to be explained in consumer countries by special emissaries.

November 18	Vienna	Arab oil ministers cancel the scheduled 5 per cent cut in production for EEC.
November 19-20	Vienna	OPEC's 36th conference endorses decisions for the Gulf Six.
November 23	Algiers	Arab summit conference adopts open and secret resolutions on the use of the oil weapon. Embargo extended to Portugal, Rhodesia, and South Africa.
December 9	Kuwait	Arab oil ministers announce a further production cut of 5 per cent for January; friendly countries exempted.
December 22-24	Teheran	OPEC ministerial committee of the Gulf Six decides to raise posted price of marker crude to $11.651 per barrel effective January 1, 1974.
December 24-25	Kuwait	Arab oil ministers cancel the 5 per cent production cut scheduled for January and agree on a change in the production cut from 25 to 15 per cent of the September level.

1974

February 12-14	Algiers	Heads of state of Algeria, Egypt, Syria, and Saudi Arabia discuss oil strategy in view of the progress in Arab-Israeli disengagement.
March 13	Tripoli	Arab oil ministers agree to end the embargo against the United States and to restore production to pre-October levels. Formal announcement to be postponed.
March 17	Vienna	Arab oil ministers announce end of the embargo against the United States.
June 1-3	Cairo	Arab oil ministers decide to end most restrictions on exports of oil, but continue embargo against the Netherlands, Portugal, South Africa, and Rhodesia.
July 10-11	Cairo	Arab oil ministers lift the embargo against the Netherlands.

APPENDIX B

Statistical Tables

GENERAL NOTE: Additional statistical materials of general interest are included as part of the individual articles contained in this issue. There may in some cases be differences between figures in these Tables and figures for ostensibly similar indicators in the volume, arising from occasional differences in definitions, sources, or units employed.

Table A-1

TOTAL ENERGY CONSUMPTION, OIL CONSUMPTIONS, AND NET IMPORTS OF OIL, 1962 AND 1972
(IN MILLION METRIC TONS OIL EQUIVALENT)

	Total Energy Consumption		Oil Consumption		Net Imports or Exports (−) of oil	
	1962	1972	1962	1972	1962	1972
United States	1,187	1,864	498	776	99	230
Canada	76	154	44	80	10	(−9)
Western Europe	704	1,111	264	704	265	680
Japan	120	310	49	237	48	235
Oceania	34	60	15	30	14	15
Other non-Communist Countries	203	394	172	367	(−410)	(−1,138)
Communist Countries	948	1,585	175	396	(−42)	(−46)
TOTAL WORLD	3,272	5,478	1,217	2,590	—	—

Source: Derived from, and partially estimated on the basis of, data in various issues of UN, *World Energy Supplies*, Series J; and British Petroleum Co., *Statistical Review of the World Oil Industry*.

Table A-2

OIL CONSUMPTION, PRODUCTION, AND NET TRADE, BY MAJOR WORLD REGIONS, 1972

	Consumption 1,000 b/d	Production 1,000 b/d	Net Imports 1,000 b/d	% of cons.	Net Exports 1,000 b/d	% of prod.
United States	15,980	11,180	4,515	28.3%		
Canada	1,665	1,835			170	9.3%
Caribbean	1,195	3,650			2,450	67.1
Other Western Hemisphere	2,105	1,325	770	36.6		
Western Europe	14,205	435	13,735	96.7		
Middle East	1,145	17,975			16,830	93.6
North Africa	370	3,745			3,375	90.1
West Africa	200	2,085			1,885	90.4
East and South Africa, South Asia	975	86	880	90.3		
South East Asia	1,430	1,295	125	8.7		
Japan	4,800	143	4,765	99.3		
Oceania	635	306	300	47.2		
U.S.S.R., Eastern Europe, China	7,990	8,865			880	9.9
TOTAL	52,695	52,925				

Note: Includes natural gas liquids. Regional net imports or net exports may not precisely equal the difference between consumption and production, and world-wide consumption and production are not precisely equal, because different sources employ somewhat different definitions. Indonesia is included within "S.E. Asia" and Mexico within "Other Western Hemisphere."

Sources: Derived from: British Petroleum Co., Statistical Review of the World Oil Industry 1972 (1973); The Petroleum Economist, March, 1974; and UN, World Energy Supplies 1969–1972, Series J, No. 17 (1974).

Table A-3

PATTERNS OF DEMAND FOR MAJOR PETROLEUM PRODUCTS, SELECTED REGIONS, 1972

	United States 1,000 b/d	per cent	Canada 1,000 b/d	per cent	Western Europe 1,000 b/d	per cent	Japan 1,000 b/d	per cent
Gasoline	6,960	43.6	575	34.5	2,880	20.3	915	19.1
Middle distillates*	3,935	24.6	520	31.2	4,750	33.4	835	17.4
Residual fuel oil	2,300	14.4	315	18.9	4,740	33.4	2,080	43.3
Other products**	2,785	17.4	255	15.3	1,835	12.9	970	20.2
TOTAL OIL CONSUMPTION	15,980	100.0	1,665	100.0	14,205	100.0	4,800	100.0

*Comprising principally "inland" consumption of heating oils, diesel, and kerosene

**Comprising a variety of energy (e.g., liquified petroleum gases) and non-energy (e.g., lubricants) products. Also includes bunker fuels, such as diesel and residual, whose "inland" consumption is included in other lines of the table.

Source: British Petroleum Co., Statistical Review of the World Oil Industry 1972 (1973).

Table A-4

Retail Prices (Including Taxes) of Gasoline (Standard Grade) and Heating Oil
(U.S. Cents per Gallon)

		1963	1965	1970	1972	1973
United Kingdom	-gasoline	na	na	62	62	72
	-heating oil	18	18	16	21	23
United States	-gasoline	30	31	35	35	39
	-heating oil	16	16	19	20	23
Germany	-gasoline	54	54	57	69	87
	-heating oil	12	10	12	13	26
Netherlands	-gasoline	48	53	63	82	98
	-heating oil	10	7	12	14	23
France	-gasoline	75	72	72	83	95
	-heating oil	14	13	15	20	24
Belgium	-gasoline	57	57	66	79	99
	-heating oil	16	15	16	19	28
Italy	-gasoline	58	67	79	98	99
	-heating oil	11	10	10	12	15

Note: Qualities and specifications of heating oil differ among countries. The Italian quality in particular is lower than that of other countries.

na = not available

Source: 1. United States: prices of both gasoline and heating oil taken from American Petroleum Institute, *Petroleum Facts and Figures 1971*, updated where necessary by Consumer Price Index
for gasoline and #2 fuel oil.

2. Other countries: European Economic Community-Eurostate, *Energy Statistics Yearbook
1969–72* (converted at market rates of exchange).

Table A-5

ESTIMATED OIL EXPORTS AND REVENUES OF MAJOR OPEC COUNTRIES

A. GOVERNMENT OIL REVENUES	millions of U.S. dollars				
	1970	1971	1972	1973	1974
Saudi Arabia	1,200	2,149	3,107	5,100	20,000
Kuwait	895	1,400	1,657	1,900	7,000
Iran	1,136	1,944	2,380	4,100	17,400
Iraq	521	840	575	1,500	6,800
United Arab Emirates	233	431	551	900	4,100
Qatar	122	198	255	400	1,600
Libya	1,295	1,766	1,598	2,300	7,600
Algeria	325	350	700	900	3,700
Nigeria	411	915	1,174	2,000	7,000
Venezuela	1,406	1,702	1,948	2,800	10,600
Indonesia	185	284	429	900	3,000
	7,729	11,979	14,374	22,800	88,800

B. OIL EXPORTS	millions of barrels				
Saudi Arabia	1,359	1,707	2,163	2,850	3,060
Kuwait	1,082	1,169	1,176	1,078	905
Iran	1,318	1,562	1,752	2,037	2,072
Iraq	545	594	382	705	670
United Arab Emirates	253	339	384	555	614
Qatar	133	156	176	208	189
Libya	1,188	989	813	794	547
Algeria	358	276	373	382	350
Nigeria	376	531	628	715	785
Venezuela	1,288	1,206	1,133	1,170	998
Indonesia	265	273	345	426	460
	8,165	8,802	9,325	10,920	10,650

GOVERNMENT OIL REVENUES PER BARREL OF EXPORTS	U.S. dollars				
Saudi Arabia	.883	1.259	1.437	1.789	6.535
Kuwait	.828	1.197	1.409	1.763	7.734
Iran	.862	1.246	1.358	2.012	8.397
Iraq	.957	1.415	1.507	2.127	10.149
United Arab Emirates	.920	1.272	1.434	1.621	6.677
Qatar	.915	1.264	1.445	1.923	8.465
Libya	1.090	1.786	1.966	2.896	13.893
Algeria	.907	1.268	1.877	2.356	10.571
Nigeria	1.093	1.722	1.870	2.797	8.917
Venezuela	1.092	1.411	1.719	2.393	10.621
Indonesia	.698	1.040	1.243	2.112	6.521
AVERAGE PER-BARREL REVENUES ALL COUNTRIES	.946	1.360	1.541	2.087	8.338

Source: The Petroleum Economist, March, 1975

Table A-6

1972 CRUDE-OIL PRODUCTION IN NON-COMMUNIST REGIONS BY EIGHT MAJORS
(THOUSAND BARRELS PER DAY)

	Exxon	BP	Shell	Texaco	Socal	Gulf	Mobil	CFP	Total by Majors	Total Production	Majors' Share of Regional Product
Western Europe	59	26	39	31	1	4	22	—	182	332	54
Far East/ Oceania	190	1	292	429	427	—	36	—	1,375	1,882	73
Africa	307	603	760	122	122	452	272	151	2,789	5,526	50
Middle East	2,283	3,995	1,649	2,100	2,078	1,881	1,135	895	16,016	18,337	87
Latin America/ Caribbean	1,519	6	945	275	63	250	140	—	3,198	4,913	65
North America	1,376	18	819	1,063	592	688	589	5	5,150	13,046	39
TOTAL	5,734	4,649	4,504	4,020	3,283	3,275	2,194	1,051	28,710	44,036	65
Per cent of Production	13.0	10.6	10.2	9.1	7.5	7.4	5.0	2.4	65.2	100.0	

Source: The Petroleum Economist, March, 1974

Table A-7

CALCULATION OF PERSIAN GULF CRUDE-OIL PRICES

	(U.S. dollars per barrel)				Per cent Increase Jan. 1, 1974, over
	Jan. 1, 1973	Oct. 1, 1973	Nov. 1, 1973	Jan. 1, 1974	Jan. 1, 1973
Saudi Arabian 34° gravity oil.[1]					
1. Posted price	2.591	3.011	5.176	11.651	350
2. Royalty (12½ per cent of 1)	.324	.376	.647	1.456	—
3. Production cost	.100	.100	.100	.100	—
4. Profit for tax purposes (1) minus (2 plus 3)	2.167	2.535	4.429	10.095	—
5. Tax (55 per cent of 4)	1.192	1.394	2.436	5.552	—
6. Government revenue (2 plus 5)	1.516	1.770	3.083	7.008	362
7. Oil company cost (3 plus 6)	1.616	1.870	3.183	7.108	340
8. Estimated oil company profit	.500	.500	.500	.500	—
9. Estimated sales price (f.o.b.) (7 plus 8)	2.116	2.370	3.683	7.608	260
10. Estimated transportation cost[2] (to U.S. Gulf coast)	1.480	1.480	1.480	1.480	—
11. Estimated sales price (c.i.f.) (to U.S. Gulf coast)	3.596	3.850	5.163	9.088	153

[1]Price increases shown are for Saudi Arabian light crude oil 34° API gravity. Saudi light is used as the benchmark for Persian Gulf crude because it is the largest single type of crude oil produced there and represents a good average between higher priced low-sulfur crude and lower priced heavier oil.

[2]Using tanker rates of Worldscale 100.

Source: Annual Report of the Council on International Economic Policy, 1974; as reproduced in U.S. Senate, *Implications of Recent Organization of Petroleum Exporting Countries (OPEC) Oil Price Increases*, Committee on Interior and Insular Affairs, 93rd Congress, second session, 1974

Table A-8

WORLD "PUBLISHED PROVED" OIL RESERVES AT END 1973

Country/Area	Share of Total (in per cent)	Billion Barrels
United States	6.3	41.8
Canada	1.4	9.3
Caribbean	2.9	17.6
Other Western Hemisphere	2.2	14.0
Total Western Hemisphere	12.8	82.7
Western Europe	2.6	16.4
Middle East	55.4	349.7
Africa	10.4	67.3
U.S.S.R., E. Europe & China	16.3	103.0
Other Eastern Hemisphere	2.5	15.6
Total Eastern Hemisphere	87.2	552.0
World (excl. U.S.S.R., E. Europe & China)	83.7	531.7
WORLD	100.0	634.7

Source: Shown in British Petroleum Co., *Statistical Review of the World Oil Industry 1973* (1974) which in turn cites as its sources: U.S.A., American Petroleum Institute; Canada, Canadian Petroleum Association; all other areas, estimates published by the *Oil & Gas Journal* (Worldwide Oil issue, December 31, 1973).

Notes (as shown in the BP document):

1. Proved reserves are generally taken to be the volume of oil remaining in the ground which geological and engineering information indicates with reasonable certainty to be recoverable in the future from known reservoirs under existing economic and operating conditions.
2. The recovery factor, i.e., the relationship between proved reserves and total oil in place, varies according to local conditions and can vary in time with economic and technological changes.
3. For the United States and Canada, the data include oil which it is estimated can be recovered from proved natural-gas reserves.
4. The data exclude the oil content of shales and tar sands.

Notes on Contributors

ALBERTO CLÔ, born in 1947 in Bologna, Italy, is an energy economist with an Italian petrochemical firm, and is associated with the Centro di Economia e Politica Industriale at the University of Bologna. He is the author of several articles on the politics of oil.

JOEL DARMSTADTER, born in 1928 in Germany, is senior research associate, Resources for the Future, Washington, D. C. Among his recent publications are *Conserving Energy: Prospects and Opportunities in the New York Region* (1975), and *Energy in the World Economy* (1971).

NORMAN GIRVAN, born in 1941 in Jamaica, is director of the Caribbean Center for Corporate Research. His publications include *Foreign Capital and Economic Underdevelopment in Jamaica* (1972).

MARSHALL I. GOLDMAN, born in 1930 in Elgin, Illinois, is professor of economics at Wellesley College and an associate of the Russian Research Center of Harvard University. His publications include *Doing Business with the Soviets* (1975), and *The Spoils of Progress: Environmental Pollution in the Soviet Union* (1972).

KLAUS KNORR, born in 1911 in Germany, is William Todd Professor of Public and International Affairs, Woodrow Wilson School, Princeton University. His most recent publications are *The Power of Nations* (1975), *Power and Wealth* (1973), and *Military Power and Potential* (1970).

HANS H. LANDSBERG, born in 1913 in Germany, is director of the Division of Energy and Resource Commodities, Resources for the Future, Washington, D. C. He has contributed to a number of publications dealing with energy and natural resources.

ULF LANTZKE, born in 1927 in Reierort, Germany, is executive director of the International Energy Agency and Special Counsellor to the Secretary General of the OECD on energy questions.

GEORGE LENCZOWSKI, born in 1915 in Poland, is professor of political science at the University of California, Berkeley. Among his publications are *Soviet Advances in the Middle East* (1972), *Oil and State in the Middle East* (1960), and *The Middle East in World Affairs* (1952).

JAMES W. MCKIE, born in 1922 in Los Angeles, California, is dean of the College of Social and Behavioral Sciences, University of Texas at Austin, and was a member of the Committee on Emergency Energy Capacity of the National Academy of Sciences in 1973.

ZUHAYR MIKDASHI, born in 1933 in Beirut, Lebanon, is professor of business administration at the American University of Beirut. He is the author of numerous articles on the oil industry and on other aspects of business and economics.

EDITH PENROSE, born in 1914 in California, is professor of economics at the School of Oriental and African Studies, University of London. Her publications include *The Growth of Firms, Middle East Oil and Other Essays* (1971), and *The Large International Firm in Developing Countries: The International Petroleum Industry* (1968).

ROMANO PRODI, born in 1939 in Reggio Emilia, Italy, is professor of industrial organization and director of the Centro di Economia e Politica Industriale at the University of Bologna. His publications include *La diffusione delle innovazioni nell'industria italiana* (1971).

IAN SMART, born in 1935 in Caterham (Surrey), England, is deputy director and director of studies at the Royal Institute of International Affairs, London. He is the author of numerous publications dealing with strategic studies, arms control, Middle Eastern affairs, and the political role of force.

ROBERT B. STOBAUGH, born in 1927 in McGehee, Arkansas, is professor of business administration at Harvard University. His publications include *Nine Investments Abroad and Their Impact at Home* (1975), and *Petrochemical Manufacturing and Marketing Guides* (2 vols., 1966-68).

YOSHI TSURUMI, born in 1935 in Fukuoka, Japan, is visiting associate professor at the Harvard Business School and the author of several articles on Japanese business activities abroad.

RAYMOND VERNON, born in 1913 in New York City, is director of the Center for International Affairs and Herbert F. Johnson Professor of International Business Management at Harvard University. His publications include *Manager in the International Economy* (1968), and *Sovereignty at Bay: The Multinational Spread of U.S. Enterprise* (1971).

MIRA WILKINS, born in 1931 in New York City, is professor of economics at Florida International University. Her publications include *Maturing of Multinational Enterprise: American Business Abroad from 1914 to 1970* (1974), and *The Emergence of Multinational Enterprise: American Business Abroad from the Colonial Era to 1914* (1970).

Index